Handbook of Cannabis Production in Controlled Environments

Handbook of Cannabis Production in Controlled Environments

Edited by Youbin Zheng

CRC Press
Taylor & Francis Group
Boca Raton   London

CRC Press is an imprint of the
Taylor & Francis Group, an **informa** business

First edition published 2022
by CRC Press
6000 Broken Sound Parkway NW, Suite 300, Boca Raton, FL 33487–2742

and by CRC Press
4 Park Square, Milton Park, Abingdon, Oxon, OX14 4RN

CRC Press is an imprint of Taylor & Francis Group, LLC

© 2022 Taylor & Francis Group, LLC

Library of Congress Cataloging-in-Publication Data
Names: Zheng, Youbin, editor.
Title: Handbook of cannabis production in controlled environments / edited
    by Youbin Zheng.
Description: First edition. | Boca Raton : CRC Press, 2022. | Includes
    bibliographical references and index.
Identifiers: LCCN 2021061214 (print) | LCCN 2021061215 (ebook) |
    ISBN 9780367713546 (hardback) | ISBN 9780367712570 (paperback) |
    ISBN 9781003150442 (ebook)
Subjects: LCSH: Marijuana industry. | Cannabis.
Classification: LCC HD9019.M382 H36 2022 (print) | LCC HD9019.M382
    (ebook) | DDC 338.1/7379–dc23/eng/20220131
LC record available at https://lccn.loc.gov/2021061214
LC ebook record available at https://lccn.loc.gov/2021061215

ISBN: 978-0-367-71354-6 (hbk)
ISBN: 978-0-367-71257-0 (pbk)
ISBN: 978-1-003-15044-2 (ebk)

DOI: 10.1201/9781003150442

Typeset in Times LT Std
by Apex CoVantage, LLC

Contents

Contents

Preface

Cannabis sativa, commonly called cannabis or marijuana, is a plant that we have used for thousands of years for many different purposes, including medical, recreational and religious activities, as well as a source of food and fibers. In modern history, cannabis has transitioned from a legal and widely used plant to an illegal crop, but many countries and regions are now reverting toward legalization again, such as Canada and some U.S. states. During these decades of prohibition, there have been many advancements in horticultural science and technology, especially in controlled environment plant cultivation, but most were not been applied or tested on cannabis. Recently, there has been increasing demand for cannabis products, such as dried inflorescence and a variety of extracts and processed products. When used as medicine or for recreational purposes, cannabis is most often produced in controlled environments (e.g., greenhouse and indoor growing rooms) to ensure consistent high quality.

As a result, there are burgeoning operations in controlled environment cannabis cultivation in regions where cannabis is legalized. These operations—and individual growers who cultivate cannabis for personnel consumption—need scientific information on how to cultivate cannabis more effectively and efficiently. People have accumulated enormous amounts of knowledge and experience on how to cultivate this plant in small scale during the period when cannabis was illegal, but there has been very little formal research to verify many of these practices. Many legal operations utilized talent from the legacy market at the beginning of their operations, and these individuals brought valuable knowledge and experience with them, making significant contributions to the initiation and establishment of legal operations. However, since legal operations tend to be larger and are heavily regulated, there are issues of scaling up. Also, due to the high value and history of this plant, there are many myths created by the industry to sell products without any scientific evidence to support them.

In the meantime, people have been using controlled environments to cultivate various plant species for many years, ranging from low-tech greenhouses to more advanced vertical farms. Supported by governments from all over the world and the industry itself, billions of dollars have been spent conducting research on how to cultivate plants in controlled environments in recent decades, and enormous amounts of scientific knowledge and technology have been developed for large-scale production. However, when developing and then started teaching the cannabis production course for senior university students at the University of Guelph (Guelph, Canada) in the beginning of 2019, I realized that there was very limited scientific reading materials for controlled environment cannabis cultivation. After its recent legalization, many scientists—especially in Canada—have started scientific studies on how to efficiently and effectively cultivate cannabis in controlled environments. This book was created to summarize recent developments in this area. It can serve as a handbook for people involved in commercial cannabis cultivation, as a textbook for students, and as an information source for individuals who grow their own cannabis for personal consumption.

The authors, especially the senior authors of each chapter, are either world-known scientists or leading practitioners in the cannabis industry with strong scientific background in their fields. This book is based on the most recent scientific research on cannabis production, as well as knowledge and experience gained through the authors' many years of hands-on practices. Based on the fact that new scientific research results are rapidly becoming available, we intend to update this book when sufficient information is available and deemed necessary in the future.

On behalf of all the authors, I would like to thank our industry partners and government organizations, such as the Natural Sciences and Engineering Research Council of Canada and Ontario Centres of Excellence, for providing financial support, research space, cannabis plant material and knowledge during our scientific explorations, which enabled us to gain hands-on knowledge and to make the creation of this book possible. We would also like to thank our families, colleagues and friends for their endless support during the writing of this book.

Youbin Zheng

Editor

Dr. Youbin Zheng is a professor in the Ontario Agricultural College at the University of Guelph, and the president of the Canadian Society for Horticultural Science. He joined the University of Guelph in 2000, and has more than 25 years research experience in controlled environment plant production. His research has resulted in about 150 peer-reviewed scientific papers, more than ten book chapters and numerous extension articles and scientific reports. His recent research focus includes indoor cannabis and other medicinal and nutraceutical plant production, greenhouse ornamental and vegetable production, and nursery and landscape plant production and maintenance, especially of rootzone management and lighting. Dr. Zheng has been actively involved in indoor cannabis production research since early 2010. His group has published North America's few peer-reviewed scientific papers on indoor cannabis production. He developed the world first university-level cannabis production course in 2019 and has been teaching this course since. Dr. Zheng has been frequently invited to speak about his research results, both nationally and internationally.

Contributors

Dr. Deron Caplan is a senior consultant at Sostanza Global, specializing in controlled environment plant production. He earned North America's first Ph.D. with research focused on indoor cannabis production, then led operations at The Kelowna Research Station—the world's first facility dedicated to advancing cannabis cultivation techniques and systems. He was the Director of Research and Development at The Flowr Corporation, managing a team of ten scientists and operators, and working closely with partners at The Hawthorne Gardening Company, a subsidiary of The Scotts Miracle-Gro Company. Deron has provided expert commentary on cannabis production to many media outlets and the government of Canada.

Dr. Mike Dixon is a professor in the School of Environmental Sciences and Director of the Controlled Environment Systems Research Facility (CESRF), University of Guelph (Guelph, Canada). He served as Chair of the Department of Environmental Biology from 2003–2008. Dr. Dixon earned his Ph.D. from Edinburgh University in Scotland. As project leader for the Canadian research team investigating the contributions of plants to life support in space, Dr. Dixon formed the Space and Advanced Life Support Agriculture (SALSA) program at the University of Guelph. This program currently represents Canada's main contribution to the international space science objectives in biological life support and collaborates with NASA and the Canadian and European space agencies. The CESRF is among the world's leading research venues for technology developments and research dedicated to studying plant and microbial interactions in advanced life support systems. The technical "pull" of space exploration has aided the development of a wide range of technologies that have spun off into applications in terrestrial agri-food sectors and most notably the phyto-pharmaceutical (medicine from plants) sector in recent years.

Juan David Gutierrez is a co-founder and senior consultant at Sostanza Global, providing design, technical, operational and strategic consulting to the cannabis industry. He was the Director of Cultivation at MedReleaf for almost four years, where he led efforts to develop industry-leading and award-winning high-quality and high-yield cannabis cultivation practices. Once MedReleaf was acquired by Aurora Cannabis, Juan led assessment and improvement programs for all its cultivation facilities in Canada and supported similar projects in South America and Europe. Apart from cannabis experience, Juan has a deep knowledge and expertise in commercial plant production. Juan holds a B.Sc. in horticulture from the University of Guelph.

Dr. Max Jones is an associate professor in the Gosling Research Centre for Plant Preservation, Department of Plant Agriculture, University of Guelph. He teaches a variety of courses in the areas of plant propagation and tissue culture, and maintains an active research program developing micro-propagation and biotechnologies for a variety of plant species. In 2018, he received a cannabis research license, making him among the first researchers to legally cultivate the species on a university campus. Since then, he has trained several graduate students and published more than 15 peer-reviewed articles focused specifically on cannabis propagation and biotechnology, making him a global leader in this field.

Jason Lemay completed his B.Sc. and M.Sc. degrees at the University of Guelph. He studied the biology and management of carrot rust fly (*Psila rosae*) during his M.Sc. Jason published three peer-reviewed manuscripts and presented at 11 events ranging from international scientific conferences to local grower meetings. He then went to work as the director of research and development, and pest management consultant for Eco Habitat Agri Services, a consulting firm located in southern Ontario. Focused on integrated pest management (IPM) in greenhouse ornamentals and vegetables, he began to work with cannabis growers, leading up to its legalization. His work focused on developing and optimizing IPM programs for his clients. Through weekly scouting visits, Jason developed a deep understanding of applied biological control. Jason is currently completing his Ph.D. at the University of Guelph, focusing on developing biological control strategies for cannabis in controlled environments. He is interested in the tritrophic interactions that affect the performance of biological control agents used to manage arthropod pests in cannabis.

David Llewellyn is a researcher in the Controlled Environment Systems Research Facility research group at the University of Guelph, where he specializes on the optimization of production inputs for greenhouse and indoor-grown crops, including cannabis. Crop lighting is a major research focus, especially with the advent of modern LED technologies which have broadened the scope for manipulating the crop lighting environment as a production tool. He has conducted lighting research in a broad range of horticultural commodities, including several recent publications on lighting in cannabis production. His aim is to serve the horticulture industry by performing grower-relevant research that will improve production efficiency while reducing the industry's environmental footprint.

Dr. Philipp Matzneller is a co-founder and senior consultant at Sostanza Global, providing design, technical, operational and strategic consulting to the cannabis industry. He was the senior scientist of cultivation at MedReleaf and director of applied cultivation after Aurora Cannabis acquired the company. Under his guidance, a team of a dozen scientists and assistants across multiple facilities collected data and conducted experiments that resulted in a substantial increase in yield and quality of the production. Previously, Philipp worked in the ornamental greenhouse industry and as a research associate conducting climate impact studies. He holds B.Sc. and M.Sc. degrees in agriculture from the University of Bologna (Italy) and a Ph.D. in agroclimatology from the Humboldt University of Berlin.

Adrian S. Monthony is undertaking his doctorate in computational plant biology at the Université Laval in Quebec City, Canada. He currently studies the genetic, epigenetic and phytohormone regulation of sex determination in *Cannabis sativa*. He holds an M.Sc. in Plant Agriculture from the University of Guelph, and a B.Sc. in biochemistry and molecular biology from the University of British Columbia (Canada). Adrian is passionate about advancing our understanding of plants for society's benefit. He has been at the forefront of cannabis research since recreational legalization in Canada; authoring five well-received peer-reviewed papers on cannabis biology during his M.Sc., including the development of a floral reversion-based cannabis micropropagation method. He brings an interdisciplinary approach to his research, drawing from his past experiences in biochemistry, biotechnology, tissue culture, analytical chemistry, conservation, pharmaceutical production and computational biology. Outside the lab, Adrian likes to advocate for improved equity, diversity and inclusivity in sciences.

Dr. Zamir Punja: After completing a B.Sc. degree in plant sciences at the University of British Columbia, followed by M.Sc. and Ph.D. degrees in plant pathology from the University of California, Davis. Zamir joined the Campbell Soup Company and worked jointly with North Carolina State University, Raleigh on carrot diseases. He was appointed manager of plant biotechnology research for Campbell's up to the time he joined Simon Fraser University (Canada) in 1989. His research interests include the etiology and management of plant diseases on vegetable and horticultural crops, and the applications of plant biotechnology for disease management. His lab was the first to create genetically modified carrot, hemp and ginseng plants. He has worked closely with a number of industries, including greenhouse vegetables, ginseng, blueberry and wasabi, and more recently cannabis. His latest research on cannabis pathogens has revealed a number of previously unreported pathogens. He is a Fellow of the Canadian Phytopathological Society and has received numerous research and teaching awards, including the Sterling Prize for Controversy for his work on GMO foods. He was editor-in-chief of the *Canadian Journal of Plant Pathology* for 18 years. Zamir's research group currently focuses on cannabis pathology and methods to improve quality of greenhouse-grown cannabis.

Cameron Scott is a recent graduate of Simon Fraser University, where he completed his M.Sc. in plant pathology in Dr. Zamir Punja's lab. He has spent the last four years working with licensed producers of cannabis across Canada. This work has focused largely on the diagnosis and management of established and emerging fungal diseases, both pre- and post-harvest, in cannabis production. While working in this industry, Cameron has been an author on several peer-reviewed publications, including those focusing on pathogen characterization, management of powdery mildew, and root and crown rot pathogens of cannabis, as well as general pathogens and molds that affect the production and quality of cannabis. He has presented this research at conferences in Canada, as well as the United States. Cameron previously worked for Bayer Crop Science for four growing seasons, where he was involved in the production of canola parent seed on various scales. He holds a B.Sc. in biology from the University of the Fraser Valley (Canada).

Dr. Cynthia Scott-Dupree is a professor and was the Bayer Chair in Sustainable Pest Management (2014–2019) in the School of Environmental Sciences at the University of Guelph, and has been a faculty member there since 1986. She received her Master of Pest Management (1983) and Ph.D. (1986) from Simon Fraser University. Over the years, she has supervised 61 graduate students and four postdoctoral fellows, edited three books and five book chapters, and published 89 refereed scientific papers, 37 refereed proceedings papers, 80 technical reports and 30 extension publications. Her current research interests include sustainable management (IPM) of insect crop pests using environmentally compatible control methods in horticultural, field and greenhouse cropping systems including cannabis, management of invasive alien insect species and impact of agro-ecosystems on non-target organisms, including beneficial insects such as honey bees, bumble bees, native bees and natural enemies of insect pests (i.e., biological control agents primarily for greenhouse IPM). She is investigating the potential of a number environmentally compatible IPM control tactics including conservation and classical biological control, novel technologies such as RNAi and sterile insect technique (SIT); management of invasive alien insect species; impact of agro-ecosystems on beneficial insects such as honey bees, non-*Apis* bees (i.e., bumble bees and leafcutter bees) and natural enemies; and development of standardized pesticide risk assessment methods for non-*Apis* bees. She has been involved with risk assessment method development for studying the impact of pesticides in agro-ecosystems on bumble bees and leafcutter bees in lab and semi-field situations; survey and development of IPM strategies for carrot weevil, pepper weevil, carrot rust fly, brown marmorated stink bug and spotted wing *Drosphila*—all economically important insect pests in Ontario; and in-package fumigation of fruits and vegetable to control insect pests post-harvest using innovative controlled release technologies.

Dr. Ernest Small is a principal scientist with Agriculture Canada. He specializes in the evolution and classification of economically important plants, dealing particularly with food, forage, biodiversity and medicinal species. He has authored about 400 journal publications and 15 books, most recently the 600-page volume *Cannabis: A Complete Guide*. Dr. Small selected the standard strain of marijuana that was officially employed in Canada for 13 years following the legalization of medical marijuana in 2001, and used to treat over 100,000 patients. His classification of cannabis on the basis of THC content has been adopted by numerous countries as the foundation of legislation governing hemp and marijuana cultivation. He has received a dozen scientific society and book awards, and is a member of the Order of Canada, the country's highest honor.

Dr. Michael Stasiak is a research scientist at the Controlled Environment Systems Research Facility in the School of Environmental Sciences at the University of Guelph. He is a plant physiologist with varied academic interests that include astrobiology, advanced life support, CEA automation and control, sensor technology, medicinal plant production and plant growth chamber design, construction and operation.

1 Introduction

Youbin Zheng and Ernest Small

CONTENTS

1.1 CANNABIS SATIVA

Cannabis has been a dominant concern of society for several decades, and so most people have acquired appreciable knowledge (and some misunderstandings) of the subject. This book deals with technical aspects of commercial horticultural production in controlled environment facilities. Many unfamiliar with the phrase "controlled environment" may think of hospitals or airplanes, where certain environmental factors are controlled for the comfort and health of people. In the context of this book, the phrase denotes control of environmental factors that influence quality, yield and efficiency of plant production. "Controlled environment" here can include growing facilities such as indoor growth rooms, greenhouses and high tunnels. The individual chapters, prepared by specialists, often assume that the reader is acquainted with some of the following background information.

1.1.1 HISTORY

The species *Cannabis sativa* has been grown as a field crop for thousands of years. Until the 20th century, Western countries cultivated it for fiber (called "hemp") from the stem and edible seeds (called hempseed), while many Asian and African countries predominantly grew the plant for drug ("marijuana," "cannabis" hereafter) usage. In the 20th century, concern about recreational usage led to prohibition in most of the Western world. This was followed by the development of a huge illicit drug trade and an associated black market which to this day supplies most cannabis consumption. In the 1990s, the medical use of cannabis was accepted in some nations, and subsequently also its recreational use, but governments insisted that the plants be cultivated in very secure indoor facilities, not outdoors. During the prohibition period, amateur breeders had selected high-yielding, early-maturing cannabis varieties ideally suited to indoor growth in basements and garages. These provided foundational material for the establishment of controlled environment cannabis facilities, which now represent multi-billion–dollar businesses in several countries. Today, cannabis production advice is available from innumerable "gray market" (unauthorized, but not significantly prosecuted) sources, from advisory booklets, online instructions, mail-order seed companies and horticultural equipment supply sources. However, while the roots of the cannabis industry trace to knowledge and strains developed clandestinely by amateurs, the scale and quality requirements of modern operations necessitate professional scientific and technological inputs. Indeed, in recent years, there have been many significant advances. This book reviews the science and technology that govern efficient controlled environment cultivation of cannabis.

DOI: 10.1201/9781003150442-1

1.1.2 Appearance and Sexual Reproduction

Cannabis sativa is an annual, although some female plants can be maintained for more than a year under controlled environment conditions. Male and female plants cannot be reliably distinguished by appearance until they mature sufficiently to produce flowers. At the flowering stage, males are usually taller (typically 10–15%) but tend to be more delicate than the relatively robust females, and die after they have shed their pollen. The species is usually described as an herb, although the stems can become quite woody. The distinctive leaves with several palmately arranged leaflets (Figure 1.1) are so well known that *C. sativa* is probably the most widely recognized plant in the entire world. In both sexes, most of the flowers are in clusters. Typical of wind-pollinated plants, the flowers are quite small (just a few millimeters long) but very numerous, and lack big colorful petals or perfume that would attract insect pollinators. The male flower clusters are quite feathery, and take on a yellowish hue from the pollen within the five stamens of the male flowers (Figure 1.2 right). The female flowers are scarcely recognizable as flowers, consisting of a tiny ovary, enveloped by a tiny bract (elementary leaf) with protruding whitish stigmas (the pollen-receptive parts of flowers), giving the cluster of female flowers the appearance of being covered by whitish hairs (Figure 1.2 left). A female flower will mature into a single seed if pollinated (Figure 1.3)

The species is remarkably polymorphic (i.e., has many appearances), the result of evolution (natural selection) in nature, domestication (artificial selection of useful varieties), and considerable plasticity (i.e., different responses to growth conditions). Plants grown for fiber (always outdoors) are tall (typically 2 m or more) and have limited branching. Plants grown for oilseed (also always outdoors) may resemble fiber plants, but recent varieties have been bred to be quite short (sometimes about 1 m or less) and relatively unbranched. Wild plants (mostly "weeds") in hostile conditions (outdoors, of course) may be much shorter than 1 m and unbranched, but in good conditions some develop into highly branched giants sometimes exceeding 3 m. Wild plants are virtually always dioecious (with separate male and female plants), and until the latter half of the 20th century, so were fiber and oilseed varieties, but in recent times, many monoecious varieties (with male and female flowers on the same plant) have been bred.

FIGURE 1.1 Leaf of *Cannabis sativa*.

Source: Photo (public domain) by Christopher Thomas

FIGURE 1.2 Cannabis plants with female flowers (left) and male flowers (right).

Source: Photos by Youbin Zheng

FIGURE 1.3 Dried cannabis inflorescence and achenes ("seeds"). Two seeds are particularly apparent at the bottom, close to the thumb.

Source: Photo by Youbin Zheng

In Asia, the homeland of cannabis drug production, the plants are traditionally—and to this day—cultivated outdoors. The varieties grown are "landraces" (a phrase denoting domesticated varieties selected locally by farmers, but more variable than modern cultivars created by modern breeders). In Chapter 3, the two principal kinds of cannabis landraces ("sativa-type" and "indica-type") are discussed. The Asian cannabis plants are reproduced by seeds, and since they are dioecious, about half are male and half are female. Asian cannabis drug farmers long ago discovered that when pollen from the males fertilize the females so that they produce seeds, the quality of the females is much reduced. Moreover, the males are much inferior to the females as sources of drugs. The solution adopted was to rogue away (kill) the male plants as soon as they can be recognized (i.e., when they first produce male flowers). Black market growers who produce cannabis outdoors frequently follow this old practice. Well-cared for outdoor cannabis plants (provided with good soil, irrigation, a location open to the sun, and spaced at last 1 m apart) typically are well-branched and 2 m or more in height (Figure 1.4). They are impressive, but as explained in this book, their genetics and appearance are usually inappropriate for controlled environment production.

As noted previously, with prohibition of cannabis, much production shifted from field to indoors, and compact varieties (conventionally called "strains") were created. Although sexual reproduction played a large role in the hybridizations that led to the selection of numerous strains, propagation was found to be most easily accomplished by using stem cuttings (a form of vegetative reproduction)

FIGURE 1.4   A *Cannabis sativa* plant about 3 m tall, grown in the author's garden in Ontario, Canada.

Source: Photo by Dr. Weiduo Si

of female clones. Clones are simply genetically identical organisms, like identical twins, but with numerous copies. Clonal reproduction is well known in horticulture and agriculture (potato and apple varieties, for example, are clones). Commercial cannabis production in controlled environments is based mostly on female clones replicated by cuttings (with tissue culture increasingly supplementing cuttings). Clones are advantageous particularly because they produce very uniform plants, especially with regard to the feature or features that are valued. Nevertheless, growth conditions can significantly influence efficiency and quality of production, as documented in this book.

All crops are basically systems for conversion of light into useful harvested materials, but in the case of cannabis, light also plays a special role because the maturation of most strains into females (also males, although they are usually not needed) is controlled by daily duration of uninterrupted dark period. *Cannabis sativa* is a "short-day" plant, and in nature, shortening daylight (strictly, lengthening nights of uninterrupted darkness) initiates flowering. This is the key requirement in controlled environments for cannabis to flower (and produce cannabinoid-bearing buds) several times annually, whereas normally outdoor-grown plants can produce only one seasonal harvest. "Day-neutral" or "autoflowering" plants, as discussed in this book, are races that are indifferent to day length, which arose near the northern limits of survival of *C. sativa*, and are programmed to come into flower quickly to mature seeds before being killed by early winter. Autoflowering plants normally flower within 30–50 days after seeds germination, regardless of the day length. In theory, such plants (or their hybrids) could be advantageous, as they will mature rapidly under continuous light, but their usefulness remains to be demonstrated.

Commercial controlled environment cannabis production cycles are divided into three stages: propagation, vegetative growth and flowering. Each stage can have different lengths, depending on the genetics of the plants, the cultivation system and horticultural strategies. Many commercial growers employ soilless cultivation (with roots in nutrient solution or in solid soilless medium), but culture in natural soil is also practiced (See Chapter 5). Use of rooting stem cuttings is currently the most common propagation method used (see Chapter 4). Short stem cuttings are inserted in growing media and stimulated to produce roots. This is usually done under relatively low light, and takes about two weeks. The subsequent vegetative growth stage in short-day strains depends on whether natural light cycles only are employed, but in well controlled facilities, experience and market conditions dictate when short days (i.e., long nights) should be employed. The length of the uninterrupted dark period is genetic-dependent and can range from 9–12 hours, with the majority of strains requiring about 12-hour uninterrupted dark periods (see Chapter 6). Normally less than two weeks of short-day treatment suffice for flowering. With continued growth of the flowers, they produce congested clusters known as "buds" which are the chief commercial product (Figure 1.5) of controlled growth operations. The buds are ready to be harvested 5–10 weeks after the initial appearance of flowers.

1.1.3 Phytochemical Constituents

The recreational (and consequently, also the commercial) value of drug forms of *C. sativa* is determined by two classes of chemicals: cannabinoids and terpenes. The medicinal value of the plants for humans and other animals is also determined by these, and perhaps also by other compounds, such as flavonoids. However, hundreds of other chemical constituents have been identified, and studies are underway to determine their potential medicinal, therapeutic and industrial values.

"Cannabinoids" is a comprehensive term that includes different classes of chemicals, including some that occur naturally in humans and play vital physiological roles. "Phytocannabinoids" are cannabinoids that occur in plants, especially a class of chemicals mostly with a C_{21} terpenophenolic skeleton that occurs primarily in *Cannabis sativa*. Cannabinoids can be classified into 11 types: (-)-delta-9-*trans*-tetrahydrocannabinol (Δ^9-THC), (-)-delta-8-*trans*-tetrahydrocannabinol (Δ^8-THC), cannabigerol (CBG), cannabichromene (CBC), cannabidiol (CBD), cannabinodiol (CBND), cannabielsoin (CBE), cannabicyclol (CBL), cannabinol (CBN), cannabitriol (CBT) and miscellaneous-type cannabinoids.

FIGURE 1.5 Cannabis buds (dried inflorescences, i.e., congested clusters of unfertilized flowers), the principal commercial product of controlled environment cannabis facilities, on sale in a dispensary.

Source: Photo by My 420 Tours (CC BY SA 4.0)

More than a dozen of these phytocannabinoids are currently the subjects of intensive medical research, but the commercial cannabis industry (which is mainly based on the euphoric qualities of the plant) is especially concerned with the following cannabinoids: Δ^9-THC, and to a very minor extent also the isomer Δ^8-THC, are the only natural phytocannabinoids that are euphoric. CBD significantly modifies or moderates the effects of THC in ways that are appreciated recreationally and medicinally. The most popular commercial cannabis strains have a cannabinoid profile dominated by Δ^9-THC, but many strains additionally contain CBD, and the percentages of these two cannabinoids are commonly advertised. In living plants, cannabinoids exist predominantly in the form of carboxylic acids (i.e., a -COOH radical is attached to the molecule). These decarboxylate into their neutral counterparts (the molecules lose the acidic -COOH radical, leaving an H atom), under the influence of light, time (such as prolonged storage), alkaline conditions or when heated as occurs when cannabis is smoked or cooked (e.g., in brownies). THC slowly degenerates (oxidizes) to cannabinol (CBN), which has much less euphoric ability. After harvest, the shelf life of cannabis drugs may be extended by cold storage and protection from light and oxygen.

Authorized cannabis sellers generally advertise the percentage content of the major cannabinoids in their offerings, at least with respect to Δ^9-THC and CBD. The best practice is to report the total of the neutral and acidic forms of the cannabinoid, since in most forms of consumption the effective dose is the combination (for example, for THC this would include both the neutral form + the acidic form). Should the information provided give separate analyses for the neutral and acidic forms, the effective (total) dosage can be determined after converting the acidic percentage to the equivalent neutral amount with the equation: Neutral content = 0.8772 × Acidic content. This is because major cannabinoids (THC, CBD, CBG, CBC, not CBN) have a molecular formula of $C_{21}H_{30}O_2$. Decaboxylation causes cannabinoid acids to loss a CO_2.

It should be appreciated that there are complex patterns of biochemical conversions of canna-binoids. Phytocannabinoids are biosynthesized from olivetolic acid and the mevalonate-pathway intermediate geranyl pyrophosphate (GPP) to first form cannabigerolic acid (CBGA, which is the precursor to Δ^9-THCA), cannabidiolic acid (CBDA) and cannabichromenic acid (CBCA). From THCA, CBDA and CBCA other cannabinoids are generated. For example, through photoxidation or pyrolysis, CBD can be converted to CBE; Δ^9-THC can be degraded to Δ^8-THC under heat, or to CBT and CBN when exposed to oxygen; CBC can be converted to CBL-type phytocannabinoids when exposed to light. That is why Δ^9-THC(A), CBD(A) and CBC(A) are sometimes called primary phytocannabinoids.

"Terpenes" and "terpenoids" are hydrocarbons. Terpenoids are oxygen-containing terpenes, and they are distinguished in different ways, or the terms are employed as synonyms, the practice in this book. Terpenes are volatile compounds that are responsible for the different aromas or odors of cannabis strains. Terpenes are very widespread in the plant kingdom, and indeed, all of the terpenes identified to date in cannabis also occur in other plant species. Cannabis strains differ considerably in odor, and because consumers have distinctive preferences for these odors, terpene profile is an important commercial consideration. Additionally, many consumers and some scientists believe that terpenes contribute to the medicinal properties of cannabis. The odor from cannabis terpenes may require suppression in commercial cannabis production facilities to avoid annoying neighbors.

In cannabis, there are monoterpenes ($C_{10}H_{16}$, with a molecular weight [mw] of 136), sesquiter-penes ($C_{15}H_{24}$, mw 204), diterpenes (20C) and triterpenes (30C). Some commonly existing terpenes in cannabis include α-pinene, β-caryophyllene, β-myrcene, limonene, terpinolene, linalool, selina-3,7(11)-diene, γ-selinene, 10-epi-γ-eudesmol, β-eudesmol, α-eudesmol, bulnesol and α-bisabolol. There are two pathways for terpene biosynthesis: the plastidial methylerythritol phosphate (MEP) pathway and the cytosolic mevalonate (MEV) pathway. Both pathways start from the 5C isoprenoid diphosphate precursors through different enzymatic modifications and eventually end up with the different terpenes.

Both the cannabinoids and terpenes of cannabis plants occur exclusively in tiny specialized hairs called glandular trichomes, specifically in a swollen glandular area at the tip of each trichome. This co-occurrence of cannabinoids and terpenes reflect common biosynthetic pathways for the two classes of chemicals. There are also many non-glandular prickly and unpalatable trichomes which protect the plant from being consumed, so it is important to specify that the glandular trichomes are under discussion. The biggest glandular heads burst on contact, and can make the flower clusters quite sticky. The largest glandular glands and the greatest concentration of these occur in the buds (Figure 1.6). Cannabis drug strains have much larger clusters of female flowers than hemp or wild

FIGURE 1.6 Cannabis bud covered with glandular trichomes.

Source: Photo by Thomas Elliott (public domain)

plants, which is why they produce higher quantities of cannabinoids. The buds are inconspicuous when young—but when mature, they constitute the basic product of commercial environmentally controlled facilities. They may contain 20% or more cannabinoids (on a dry weight basis). Smaller and much less frequent secretory hairs occur on the foliage, which can contain up to 5% cannabinoids. Very few secretory hairs are on the stems, twigs, and roots, which are regarded as waste because they contain few or no cannabinoids. Generally, for drug-type cannabis, THC content can be 2.5–> 25% in the flowers and 0.2–6% in the leaves, but can be 0–0.3% in the stems, 0.002–0.02% in the seeds and 0.0–0.003% in the roots.

Flavonoids in cannabis are mainly flavones and flavonols. They are present highest in cannabis leaves and then in inflorescence, but not detectable in roots, stems and seeds. Cannabis uses the phenylpropanoid and flavonoid biosynthetic pathways to build its core flavonoid skeletons. Research has demonstrated that cannflavins can have anti-inflammatory activity of about 30 times that of aspirin.

Cannabis flowers, leaves, stems and roots can contain different phytochemicals. The phytochemicals can be other than the aforementioned, such as sterols which are rich in roots. All parts of cannabis plants may have different medicinal and therapeutical values.

The contents of cannabinoids, terpenes and flavonoids in cannabis are dependent on genetics and the growing environments. Therefore, cultivation methods can play a major role in these second metabolites production when the genetics are chosen.

1.1.4 Origin and Classification

The classification and geography of different kinds of plants are discussed in Chapter 3. An international code of nomenclature governs many aspects of scientific names of all plants. Regardless, scientists are free to recognize groups as members of different ranks, such as species and subspecies. Most authorities accept only one species of the genus, *C. sativa*, but a few segregate plants with considerable THC as *C. indica*, and some segregate wild European plants as *C. ruderalis*. A second international code dictates aspects of the names of cultivated (domesticated) varieties ("cultivars"), and this code is employed for kinds of plants grown for fiber and oilseed (i.e., "hemp"). However, cultivated euphoriant plants (with high THC) are mostly recognized in the trade as "strains," a highly ambiguous, non-scientific category that is specifically rejected under the cultivated plant code. There are thousands of strain names, some of which legitimately designate distinctive kinds comparable to cultivars, but most of which are indistinguishable. Relative amounts of cannabinoids are often employed for practical purposes in practical classifications of *C. sativa*. In numerous countries, a level of 0.3% THC by dry weight in the female flowering parts is accepted as a legal distinction between "hemp" (< 0.3%) and "cannabis" (> 0.3%); this was lowered to 0.2% in much of Europe. Alternatively, many authors accept a division of *C. sativa* into "chemotypes," the first three based on relative THC (including THCA) and CBD (including CBDA), the fourth based on CBG (including CBGA) and the last representing mutants with very little cannabinoids:

Chemotype I: THC > 0.3% and CBD < 0.5%
Chemotype II: THC > 0.3% and CBD > 0.5%
Chemotype III: THC < 0.3% and CBD > 0.5%
Chemotype IV: CBGA/THC > 1; CBG > 0.3%
Chemotype V: Total cannabinoid content < 0.02%

Chemotypes I and II are mainly used for recreational and medicinal cannabis; hemp used for CBD production belong to Chemotype III; Chemotype IV is rare; and Chemotype V is a very rare mutant form maintained in cultivation.

For more details on cannabis origin and scientific nomenclature, please read Chapter 3 and some of the related references listed in the Bibliography of this chapter.

1.2 WHY CONTROLLED ENVIRONMENT PRODUCTION

"Controlled environment" in this book refers to a plant growing environment with key or at least some of the influential environmental factors controlled. The environmental factors—which can affect cannabis growth and development, and ultimately determine the yield and quality—can include radiation (lighting), temperature, gases (e.g., oxygen and carbon dioxide [CO_2]), water (e.g., water in the air and at rootzone), and nutrients. The majority of large commercial controlled environment cannabis facilities are using soilless culture systems (Chapter 5).

In Canada, the world's first developed country to legalize cannabis for recreational use, the vast majority of government-licensed cultivation operations are currently using controlled environments for producing cannabis (Chemotypes I & II are chiefly grown). Even for field cultivation (either for Chemotype I & II or III), seedlings or transplants are best established in controlled environments before transfer to the outdoors. This is especially true for short-season regions such as in Canada, where there is danger of spring frost. Also, transplant production in controlled environments can ensure plants start in the field with homogeneity in size.

Numerous ornamentals and several food crops are grown commercially in controlled environments, primarily to supply out-of-season demand in nearby cities. There are five special advantages to producing cannabis indoors that do not apply to most other controlled environment crops.

1. Only female plants are grown because pollen from male plants causes the females to develop seeds, lowering cannabis quality. The wind carries cannabis pollen long distances, and it is easier to keep the females protected from males indoors.
2. In most cannabis varieties, flowering is induced by providing uninterrupted long dark periods, as occurs outdoors in the autumn. Because light can be artificially controlled indoors, plants can be scheduled to flower when desired. Up to six crops are possible annually, whereas outdoors, usually only one crop is possible in a year.
3. The very high security required by governments is much easier to provide in buildings compared to fields.
4. A few cannabis "autoflowering" varieties grow well under continuous light, which can be provided indoors, so plants grow more rapidly.
5. Cannabis inflorescences and buds can be sticky, and outdoors they can be contaminated by insects and wind-blown soil. Extremely clean material can be produced in controlled environments.

Table 1.1 presents some advantages and disadvantages of cannabis cultivation in controlled vs. field environments. For more in-depth discussion, please read Chapter 2.

1.3 THE ORGANIZATION OF THIS BOOK

This book reviews the essential aspects of cannabis cultivation, including genetics, breeding, transplant propagation, vegetative- and flowering-stage plant cultivation, harvest and post-harvest. The topics of the ten chapters are summarized in the following.

Chapter 1 introduces the basic biology, phytochemical constituents, and controlled environment production of cannabis. For more detailed and in-depth discussions on these topics, readers are referred to other chapters of this book and the listed literature in the Bibliography.

Chapter 2 provides an overview of the science and technology behind the control of temperature, humidity, carbon dioxide, air distribution and lighting, and discusses how these key environment factors affect cannabis growth and development.

Chapter 3 outlines cannabis genetics and breeding. It discusses cannabis evolution, scientific nomenclature, sources of genetics for cannabis breeding, cannabis breeding techniques and related issues.

TABLE 1.1

Advantages and Disadvantages of Cultivating Cannabis in Controlled vs. Field Environments

Controlled Environments	Outdoors
Advantages	
1. Some or all growing conditions can be controlled.	1. No need for artificial light.
2. Growing environment (e.g., light intensity and spectrum) manipulation can be used to improve cannabis quality (e.g., improve secondary metabolite contents).	2. Low cost of startup.
	3. Low cost per unit of cannabis product.
	4. More forgiving of mismanagement.
3. Can cultivate year-round and harvest five or more crops per year.	5. Can have large cultivation area.
4. Product quality can be more homogeneous and consistent.	
5. Easier to apply disease and insect pest control technologies (e.g., biocontrol); therefore, less plant loss and contamination.	
6. Both yield and quality are more predictable.	
7. Does not need fertile soil, so cultivation facility can be built on any land.	
8. Can have multi-level cultivation; therefore, far more efficient in land use.	
9. Can be built close to cities where the human resources are readily available and with less transportation needed.	
10. Can have higher water and fertilizer use efficiency; therefore less environmental impact.	
11. Can be used to propagate homogenous transplants in early spring for field production.	
12. Better working environments for workers.	
Disadvantages	
1. High initial costs for infrastructure construction.	1. Weather dependent.
2. Need electrical lights.	2. Potential contamination from pollutants (e.g., heavy metals, pesticides) in the soil and air.
3. High energy consumption for growing environment control.	3. Insects and diseases control can be difficult, especially when few agents are allowed in cannabis production.
4. Need highly trained people in managing the plant cultivation.	4. Mold contamination when the weather is too wet before harvesting.
	5. Yield and quality are not controllable or predictable.
	6. Certain genotypes need physical support to avoid branches broken by wind or heavy rain.
	7. Potential to be accidentally pollinated by male plants.
	8. Can be difficult to black out for inducing flowering if short-day genotypes are used.
	9. Most of the autoflowering genotype plants are small and their inflorescences have lower cannabinoids content than most of the indoor genotypes.
	10. Only one crop per year in most regions.

Source: Adopted from Zheng (2021)

Chapter 4 is about propagation by seeds, stem cuttings and in vitro technologies.

Chapter 5 discusses water sources, water quality, water and nutrient solution treatments, nutrition and fertilization, growing media and their selection, different rootzones used in cannabis cultivation and integrated rootzone management for different cultivation systems.

Chapter 6 discusses how CO_2 affects cannabis production and should be supplied during cultivation. It also examines the basics of radiation, lighting technologies and how to use lighting (e.g., light intensity, quality and photoperiod) in cannabis cultivation.

Chapter 7 reviews plant canopy management. First, from the view point of plant physiology, it discusses why canopy management is important. Next, it describes in detail how to manage canopies for stock (i.e., mother) plants for production of cuttings, and to manage vegetative and flowering plants. The techniques discussed include pruning, de-leafing, training, topping, bending, trellising, staking, etc.

Chapter 8 discusses general plant pathology, disease management principles and specific disease control practices for cannabis production. It also reviews the most prominent cannabis diseases, include crown rot, root rot, damping off, powdery mildew, bud rot and post-harvest decay.

Chapter 9 is about insect pest management. It first introduces the basic concepts of integrated pest management for controlled environment cannabis cultivation, then discusses in detail the biology and management strategies for major insect pests of cannabis including aphids, mites, thrips, whiteflies and fungus gnats.

Chapter 10 covers when and how to harvest, post-harvest trimming, drying, curing and storage.

Each chapter has a bibliography or references list at the end for readers to pursue more details and related information.

BIBLIOGRAPHY

Bautista, J. L., S. Yu, and L. Tian. 2021. Flavonoids in *Cannabis sativa*: Biosynthesis, bioactivities, and biotechnology. *ACS Omega* 6: 5119–5123.

Elsohly, M., and W. Gul. 2016. Constituents of *cannabis sativa*. In *Handbook of cannabis*, ed. R. G. Pertwee. Cambridge: Oxford University Press, 3–22.

Gülck, T., and B. L. Møller. 2020. Phytocannabinoids: Origins and biosynthesis. *Trends in Plant Science* 25: 985–1004.

Hanuš, L. O., and Y. Hod. 2020. Terpenes/terpenoids in cannabis: Are they important? *Medical Cannabis and Cannabinoids* 3: 25–60.

Jin, D., K. Dai, Z. Xie, and J. Chen. 2020. Secondary metabolites profiled in cannabis inflorescences, leaves, stem barks, and roots for medicinal purposes. *Scientific Reports* 10: 3309. https://doi.org/10.1038/s41598-020-60172-6.

Luo, X., M. A. Reiter, L. d'Espaux, et al. 2019. Complete biosynthesis of cannabinoids and their unnatural analogues in yeast. *Nature* 567: 123–126.

McPartland, J. M. 2018. *Cannabis* systematics at the levels of family, genus, and species. *Cannabis and Cannabinoid Research* 3: 203–212.

McPartland, J. M., and G. W. Guy. 2017. Models of *Cannabis* taxonomy, cultural bias, and conflicts between scientific and vernacular names. *Botanical Review* 83: 327–338.

McPartland, J. M., and E. Small. 2020. A classification of endangered high-THC cannabis (*Cannabis sativa* subsp. *indica*) domesticates and their wild relatives. *PhytoKeys* 177: 81–112.

Pacifico, D., F. Miselli, M. Micheler, et al. 2006. Genetics and marker-assisted selection of the chemotype in *Cannabis sativa* L. *Molecular Breeding* 17: 257–268.

Small, E. 2015. Evolution and classification of *Cannabis sativa* (marijuana, hemp) in relation to human utilization. *Botanical Review* 81: 189–294.

Small, E. 2017. *Cannabis: A complete guide*. Boca Raton: CRC Press, Taylor & Francis.

Small, E. 2018. *Botanical determinants of medical properties of marijuana*. Proceedings 2018 Canadian Consortium for the Investigation of Cannabinoids annual conference, Toronto. https://cciceducation.com/posts/botanical-determinants-of-medical-properties-of-marijuana.

Zheng, Y. 2021. Soilless production of drug-type *Cannabis sativa*. *Acta Horticulturae* 1305: 376–382. https://doi.org/10.17660/ActaHortic.2021.1305.49.

Chapter 4 is about propagation by seeds, stem cuttings and by mini tissue lapses.

Chapter 5 discusses water sources, water quality, water and nutrient solution treatments, nutrition and fertilization, growing media and their selection, different rootzones used in common cultivation and integrated rootzone management for different cultivation systems.

Chapter 6 discusses how CO_2 affects cannabis production and should be supplied during cultivation. It also examines the basics of radiation, lighting technologies and how to use lighting (e.g. light intensity, quality and photoperiod) in cannabis cultivation.

Chapter 7 reviews plant canopy management. First, from the view point of plant physiology, it discusses why canopy management is important. Next, it describes in detail how to manage cannabis plants for stock (i.e. mother) plants for production of cuttings, and to manage vegetative and flowering plants. The techniques discussed include pruning, de-leafing, training, topping, bending, trellising, staking, etc.

Chapter 8 discusses general plant pathology, disease management principles and specific disease control practices for cannabis production. It also reviews the most prominent cannabis diseases which include crown rot, root rot, damping off, powdery mildew, bud rot and post harvest decay.

Chapter 9 is about insect pest management. It first introduces the basic concepts of integrated pest management for controlled environment cannabis cultivation, then discusses in detail the biology and management strategies for major insect pests of cannabis including aphids, mites, thrips, whiteflies and fungus gnats.

Chapter 10 covers when and how to harvest, post-harvest trimming, drying, curing and storage.

Each chapter has a bibliography or references list at the end for readers to pursue more details and related information.

BIBLIOGRAPHY

Hamilton, J. P., Yu, Paul L. Task. 2021. Flavonoids in Cannabis Sativa: Biosynthesis, bioactivities, and biotechnology. ACS Omega 6: 5119–5123.

ElSohly, M., and W. Gul. 2016. Constituents of cannabis Sativa. In Handbook of cannabis, ed. R. G. Pertwee. Cambridge: Oxford University Press, 3–22.

Gülck, T., and B. L. Møller. 2020. Phytocannabinoids: Origins and biosynthesis. Trends in Plant Science 25: 985–1004.

Hanus, L. O., and Y. Hod. 2020. Terpenes/terpenoids in cannabis: Are they important? Medical Cannabis and Cannabinoids 3: 25–60.

Ho, H., K. O. L. Z. Sin, and J. Chan. 2020. Secondary metabolites purify. A cannula, tuberculosis, newer compounds and uses for traditional plays and medicine. Review. Int. Pest Improvement 1-5: 1-13.

Izzo, S., N. A. Raso, L. M Pesma, et al. 2019. Cannabinoids: Inhibition of cannabicides and their anti-cancer therapeutic power. Master Rev. 11: 1–30.

McPartland, J. M. 2018. Cannabis systematics at the levels of family, genus, and species. Cannabis and cannabis Res. New. 1(1): 203–212.

McPartland, J. M., and R. W. Guy. 2017. Models of cannabis taxonomy, cultural bias, and conflicts between scientific and vernacular names. Botanical Review 83: 327–338.

McPartland, J. M., and E. Small. 2020. A classification of endangered high-THC cannabis (Cannabis sativa subsp. indica) domesticates and their wild relatives. PhytoKeys 177: 81–112.

Pollastro, F., T. Minassi, M. Manzera et al. 2018. Cannabis phenolics and isolated selection of the chemotype in cannabis sativa L. Phytochemistry 135: 157–158.

Small, E. 2015. Evolution and classification of Cannabis sativa (marijuana, hemp) in relation to human utilization. Botanique Review 81: 189–294.

Small, E. 2017. Cannabis: A complete guide. Boca Raton: CRC Press. Taylor & Francis.

Small, E. 2016. Results of the taxonomy of modern resources of marijuana. Proceedings, 2016 Canadian Consortium for the Investigation of Cannabinoids annual conference. Toronto. Impact of endocannabinoids wholesale rehabilitants of medical properties on marijuana.

Zager, J. 2021. Surface production of marijuana Cannabis sativa Area floral glands 1707, 1706–302. Impact Biotechnol 2000/Acid Oxide 2021, 1303–09.

2 Growing Facilities and Environmental Control

Michael Stasiak and Mike Dixon

CONTENTS

There is nothing more critical to consistent plant production—both in terms of yield and expression of secondary metabolites—than optimization and control of the growing environment. A growing

DOI: 10.1201/9781003150442-2

environment that changes from crop to crop will potentially result in very different chemistries in the final product. For medicinal and recreational products that demand repeatable results, the need for cannabinoid and terpene profiles to be the same from batch to batch and year to year is highly desirable, and this can only be achieved through tight, homogeneous control over the environment variables of light, temperature, carbon dioxide, air movement, vapor pressure deficit (VPD), water and nutrients.

BOX 2.1

CEA: Controlled Environment Agriculture uses advanced technological methods in enclosed spaces such as buildings or greenhouses to improve plant production in terms of both yield and quality.

The legalization of cannabis in many parts of the world has brought production out of basements and closets into the realm of controlled environment agriculture (CEA). Cannabis is a unique plant; however, the basic techniques of CEA plant production still apply, albeit with some tweaks to coax certain properties out of the plant and speed up production. This chapter will provide an overview of the variables that can be controlled in an indoor plant production environment, their effect on plant productivity and common methods used to achieve precise control over them. In addition, the two primary methods of CEA production using either natural light (greenhouse) or completely artificial light (indoor) will be reviewed.

2.1 GREENHOUSE PRODUCTION

One of the most common methods of plant production on a year-round basis in challenging climates is found in the glass- or plastic-covered box known as a greenhouse. Although the greenhouse concept has been around since Roman times (~30 AD), the modern greenhouse concept used for large-scale plant production did not get its start until World War II, when vegetables were grown hydroponically to supplement the diet of troops stationed around the world.

There are many different greenhouse design options in both freestanding and gutter-connected builds, and each has its own advantages and disadvantages; however, information on greenhouse structure is beyond the scope of this book. For readers looking for more information on greenhouse design, the *Ball Redbook* (Beytes 2021) is a valuable resource.

FIGURE 2.1 Greenhouse production of cannabis takes advantage of sunlight, but energy costs for cooling and heating can offset this otherwise free environmental input.

Source: Image credit Michael Stasiak

Much like a field-grown crop, greenhouse production is limited to a single horizontal production plane. Because natural sunlight is used, artificial lighting is required during the shorter days of the winter, and blackout curtains are needed to eliminate light during the summer months to provide the needed 12:12-hour photoperiod to induce flowering.

Although the measured and controlled environment variables of a greenhouse are generally identical to that of indoor cultivation, the methods to achieve control can differ considerably, as can the reliability of maintaining control over environment variables within specific limits. These differences will be outlined as each parameter is discussed in the subsections that follow. As with any plant production method, there are both advantages and disadvantages to greenhouse cannabis production.

2.1.1 Advantages

Year-round production	Greenhouse cover allows the grower to maintain close to ideal climate conditions for crop production throughout the year.
Control over temperature and humidity	The greenhouse hardware and control system are able to maintain climate conditions that are within a range favorable for the growing crop.
Carbon dioxide	Because the environment is closed, the ability to add supplemental CO_2 to the growing crop to improve production is possible.
Protection from extreme weather events	By growing inside a covered greenhouse, the effect of extreme weather events can be eliminated or minimized.
Greater control over insect pests and disease	Closed structures like greenhouses, with the proper mitigation controls in place, can reduce pest and disease outbreaks through the process of exclusion and/or chemical (with regulatory approval) and/or biological control.
More efficient use of resources	The use of more water-efficient growing systems (NFT, drip irrigation, ebb and flood, etc.) is easier to set up and maintain in a greenhouse.
Improved safety and comfort	A greenhouse provides protection from the external environment for staff who are looking after cannabis cultivation.

2.1.2 Disadvantages

High construction cost	Compared to growing outdoors, the cost of the greenhouse structure requires a large initial investment. Costs can be minimized if using plastic film structures (hoop houses, air houses), but these have shorter lifespans. More expensive glass and composite plastic (e.g., polycarbonate, acrylic) structures will last longer and allow for greater control of the environment. Additionally, cannabis greenhouse production requires additional expenses for installation of blackout curtains to manage photoperiod during the flowering stage of cannabis plants, and lighting systems to extend the photoperiod during the winter months.
High operating expenses	Glass and plastic structures have low insulation values, so cooling/ventilation in the summer and heating in the winter are ongoing operational costs that must be maintained.
Pests/pathogens	A protected environment, while helping to reduce insect and disease entry, unfortunately also offers ideal conditions for pests and pathogens to thrive if care is not taken to prevent their entrance and movement throughout the greenhouse.
Inconsistent lighting	Because the major light input for production is from the sun, differences in day-to-day irradiation levels (cloud cover, seasonal changes) do not allow for consistent batch-to-batch production quality. This problem can be mitigated, however, using variable-spectrum, variable-intensity LED systems that can reduce the intensity changes caused by clouds and seasonal photoperiods.

Light pollution	Greenhouse operations, when located close to communities, are often cited for light pollution—excess lighting that, mostly during the winter months when supplemental lighting is needed, is an annoyance to the local population.

2.2 INDOOR PRODUCTION

Indoor production, as the name implies, is the production of cannabis in boxes (grow rooms) with all lighting provided by electrical lighting sources. This method of production can provide the highest degree of control over all of the environment variables of light (intensity and spectrum), carbon dioxide, temperature, humidity (expressed as vapor pressure deficit: VPD), air distribution, water and nutrients. Indoor production with a high level of control, while being the most expensive option for cannabis production, offers the highest potential for quality and batch-to-batch consistency compared to field and greenhouse-grown production.

As with greenhouse production, there are many ways to build a box and equip it with lighting and environment control hardware. Systems range from repurposed shipping containers and warehouses to custom plant growth chambers and systems with clean-room pedigrees. Any of these systems can perform well provided the environment (temperature, VPD, air movement, light, CO_2) is *homogeneous*: the same from top to bottom, side to side and throughout the plant canopy. A high degree of control within an indoor space can easily be achieved by following the principles outlined in the following subsections. With proper delivery of conditioned air flow to each plant—that is at the right temperature, VPD and concentration of CO_2, coupled with lighting systems optimized for the style of growth and cultivar—consistent batch-to-batch crop production can be achieved. The remainder of this chapter will focus on systems needed for indoor cannabis production. Greenhouse systems for plant production are well described elsewhere (Beytes 2021). As with greenhouse production, indoor cannabis production has a number of advantages and disadvantages.

FIGURE 2.2 Indoor cannabis production allows for a very controlled environment; however, energy costs are considerably higher than greenhouse production.

Source: Image credit Michael Stasiak

2.2.1 ADVANTAGES

Year-round production	As with greenhouse production, indoor facilities allow the grower to maintain ideal climate conditions for crop production throughout the year.
Control over temperature and humidity	Without the effect of the external climate on the indoor environment, the control system in a properly designed growing system is able to precisely maintain climate conditions that are ideal for the growing crop 365 days of the year.
Protection from extreme weather events	By growing inside an indoor facility, the effect of extreme weather events can be eliminated.
Greater control over pests and disease	As with greenhouses, an indoor controlled environment production system provides excellent conditions for many pests and pathogens. However, pest management procedures can be isolated and focused on specific challenges with far greater reliability. The addition of good manufacturing practices and "clean-room" standards are used in numerous cannabis production facilities.
More efficient use of resources	By including techniques such as fertigation and hydroponics, it is possible to reduce the consumption of water and fertilizers as crops only consume what they need for their development, minimizing production costs. Water condensed during humidity control can be directly returned to the fertigation system or stored for later use.
Greater safety and comfort	Indoor working conditions are far more favorable for staff who are performing crop management tasks.
No light pollution	Because the growth is entirely contained indoors, there is no light escaping from the facility.
Odor control	The air handling system is easier to manage in terms of odor abatement in an enclosed building than it is in a greenhouse that relies on ventilation for temperature and humidity management.

2.2.2 DISADVANTAGES

High construction cost	A fully contained plant production environment requires a well constructed enclosure equipped with appropriate mechanical and electrical systems capable of establishing and sustaining the desired growing environment. Capital cost for this class of infrastructure is the highest of all options for growing cannabis. Nevertheless, a broad range of examples of this type of system has been exploited in the cannabis industry, with varying degrees of success. A detailed cost/benefit assessment is highly recommended before selecting an option. Additionally, the degree of control demanded by indoor systems requires higher-quality sensors and fully integrated control systems which cost more than those traditionally used in greenhouse systems.
High operating expenses	Since all the environment variables require some level of feedback control to maintain the desired environment, energy costs—especially for lighting—are high.
Pests/pathogens	While being the best option for reduced pest and pathogen problems, a protected environment offers ideal conditions for biological organisms if care is not taken to mitigate their entrance and movement throughout the growing facility. As a product designated for human consumption, the use of chemical pesticides is severely limited, making pathogen control difficult should it be needed.

Overall, both greenhouse and indoor production provide a much higher level of control over the plant environment than in the field, but indoor production provides the highest level of control that can ensure batch to batch and year to year product consistency that is required for medically approved products (Table 2.1).

TABLE 2.1

Summary of Outdoor, Greenhouse and Indoor Cannabis Production Parameters

	Outdoor	Greenhouse	Indoor
Degree of environment control	None	Moderate	High
Startup cost	Low	Moderate	High
Cost of production	Low	Moderate	High
Quality (subjective)	Moderate	Good	High
Consistency from batch to batch	Low	Moderate	High
Number of crops per year	1	5–7	5–7
Control over secondary metabolite expression between batches	None	Moderate	High

2.3 THE PLANT ENVIRONMENT

2.3.1 TEMPERATURE

2.3.1.1 Theory

Temperature is a measurement of how hot or cold something is using a definite scale (Fahrenheit, Celsius, Kelvin). The temperature that most biological life process can occur normally, both in plants and other organisms, ranges from 0–40°C. Organisms that survive outside of this range generally do so through adaptive processes such as dormancy, hibernation or other evolutionary modifications that allow for growth and survival in extreme environments. Different stages of plant growth can have different optimal temperatures; however, within the normal range of growth room temperatures, these differences are generally minimal. Cannabis, an annual dioecious plant adapted to temperate climates, has an optimal temperature range between 20°C and 30°C, depending on the stage of growth as well as the species and cultivar. At the cloning or seedling stage, temperature is best between 20°C and 24°C, while the vegetative stage should be warmer to encourage faster growth, with the range extending to 25°C. To encourage mobilization of photosynthate (sugars) to the developing flowers, temperatures during bloom should be raised above that of the vegetative stage, with a range from 23–28°C. Exact temperature parameters are species- and cultivar-specific, and will be better characterized as research in this field progresses. If temperatures are too high, lighter terpenes will begin to evaporate, resulting in a decrease in terpene yield and quality.

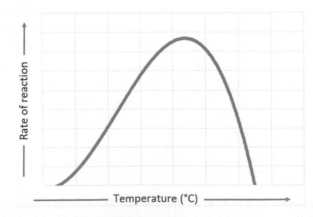

FIGURE 2.3   The effect of temperature on plant productivity.

Source: Image credit Michael Stasiak

The effect of temperature on plant productivity directly relates to the effect of temperature on the nature and speed of the chemical processes within the plant. For the chemical reactions that take place during plant growth, the upper limit for temperature occurs when surfaces begin to dry out and proteins begin to denature, eventually leading to the death of the plant. Conversely, as temperatures decrease, the chemical processes become slower, and the rate of growth is decreased. In CEA, temperature is generally not the limiting factor to growth and productivity, as long as it falls within an optimal range for growth and development.

2.3.1.2 Measurement

Accurate measurement of temperature within the growing environment is critical for consistent growth and productivity in cannabis. There are five types of sensors that are applicable to measurement in the range of temperatures suitable for plant production. All but one can be integrated into control devices or systems, but each type has uses in plant production CEA. All require regular attention to calibration protocols to maintain reliability and accuracy.

2.3.1.2.1 Thermocouples

Thermocouples are devices that indicate temperature measurement with a change in voltage. As temperature goes up, the output voltage of the thermocouple rises. These sensors take advantage of a thermoelectric effect often observed between two dissimilar metals (Seebeck Effect). The most common type (J) uses iron and constantan as the two metals, and has a measuring range between −40°C and 750°C. While the range is well beyond anything needed in plant growth systems, their low cost and long life makes them popular in electronic temperature sensing devices. Other thermocouples commonly used include the K (chromel/alumel, −200–1350°C) and T (copper/constantan, −200–350 °C) types. Their reliability and accuracy rely on robust electronic reference circuitry and regular calibration.

2.3.1.2.2 Resistive Sensors

Resistive temperature sensors are also electrically operated; however, rather than sensing a change in voltage, they show a change in resistance with a change in temperature. The two main types of resistive sensors are RTDs (resistive temperature devices) and thermistors. RTDs increase in a positive direction, with resistance going up as temperature rises, while thermistors will decrease in resistance as temperature rises. A platinum RTD sensing range is between −200°C and +850°C, whereas a thermistor senses in the −55°C to +150°C range. These sensors are the most common type found in electronic temperature devices used in heating, ventilation and air conditioning (HVAC) hardware due to their low cost and reliability.

2.3.1.2.3 Infrared Sensors

Infrared (IR) sensors are non-contact sensors so can tell you the temperature of an object from a distance. These are useful in grow systems for assessing the operation of hardware (heat exchangers, lighting) and that of the plant canopy. In a well watered crop with adequate light, the leaf canopy will be cooler than the surrounding air because of the evaporative cooling effect of transpiration. A large canopy-to-air difference is indicative of a healthy canopy, whereas a low canopy-to-air difference can indicate a canopy under water stress. Common IR sensors will usually read in the range of 0–100°C.

2.3.1.2.4 Bimetallic Devices

As with thermocouples, bimetallic temperature devices take advantage of dissimilar properties of metals, but in this case with respect to expansion and contraction of metals as they are heated and cooled. In these devices, two metals are bonded together (hence "bimetallic") and connected mechanically to a pointer or electronic switch. When heated, one side of the bimetallic strip will expand more than the other, initiating movement in the pointer or switch. These are most common in simple non-electronic thermostats and can be useful as fail-safe sensors to prevent over- or under-heating or cooling, as they don't require a power source. Their range can be from −50–300°C.

2.3.1.2.5 Thermometers
Thermometers are fluid-filled tubes that take advantage of the thermal expansion characteristics of mercury or alcohol. Like bimetallic devices, thermometers do not require power to indicate temperature. They are useful in grow rooms as a visual indicator that will corroborate electronic sensor operation. Their measuring range is generally between −70°C to 300°C, although they can be manufactured with smaller ranges that show high precision and accuracy. Because of toxicity, mercury thermometers should never be used in facilities that produce products for human consumption.

2.3.1.3 Control
Control of temperature is obtained by adding or removing heat—thermal energy—from the system you want to control. That transfer of heat from one place to another can occur in three ways: thermal radiation, heat conduction and convection (Figure 2.4). By definition, *thermal radiation* is the electromagnetic energy given off by all matter that is above the temperature of absolute zero and is caused by the thermal motion of atomic particles. It's the heat you feel from the sun, HPS lamps or infrared heaters and is a property of light from the ultraviolet (UV) spectrum through visible and into far and infrared spectra. *Heat conduction* is the movement of heat within and between solid objects. An example of this is the movement of heat from your hot steering wheel in the summer to your hands, or from the hot water in a greenhouse heating system which transfers heat to the metal heat exchanger fins during the winter. The final type of heat transfer is through *convection*. This is the concept exhibited with hot air rising and cool air falling. In the greenhouse radiator example, the heat can be felt from a distance rising off the metal radiator fins. There are a number of methods available to add and remove heat from an environment; those most common to greenhouses and indoor growing systems are shown in Table 2.2. Heat can be added to a system with radiation, conduction or convection; however, cooling is only possible with conduction, as heat must be

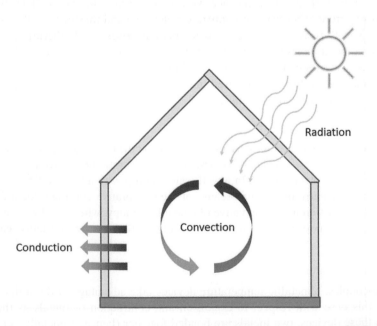

FIGURE 2.4 The processes of convection, conduction and radiation in relation to a simple greenhouse.

Source: Image credit Michael Stasiak

TABLE 2.2
Types of Heating and Cooling Systems Used in Controlled Environment Plant Production

System	Heating or Cooling?	Details
Unit heaters	Heating	More common in greenhouses, unit heaters provide local fan-driven heat to an environment. They can be fueled by natural gas, propane or oil, as well as hot water, electricity or steam. These only provide bulk heat to an environment. They do not provide homogeneously distributed heating or canopy penetration; for this, additional air movement systems are required.
Hot water	Heating	Hot water is used in both active (forced air through heat exchangers) and passive radiant systems using perimeter or under bench finned piping. Passive systems require additional air movement with ducting or stationary fans to ensure proper air movement through the plant canopy. Heat sources can utilize natural gas or oil-fired boilers, as well as electric heating and ground-source heat pumps.
Chilled water	Cooling/ condensing	Chilled water systems are often used for cooling and condensation for VPD control. When combined with a distributed air ducted system and hot water heating, highly efficient systems can be built. Chilled water is generally produced using conventional AC refrigeration-based systems that can include ground-source heat pumps and/or cooling towers to improve efficiency.
Steam	Heating	Steam-based heating systems function similarly to hot water systems; however, they are less efficient, so their use is declining. The relatively high temperature of steam distribution systems precludes deployment close to plants and homogeneity in the aerial environment is more difficult to achieve. Steam is pumped through pipes to provide heat, and the water that condenses during the cooling process must be returned to the boiler.
Split systems	Both	Split AC systems where the condenser is outside the building and the evaporator is within the growing space can provide localized cooling to smaller growing spaces. Advanced heat pump units can provide heating, as well, but these units operate in only one mode at a time.
Resistance	Heating	Electric heaters convert electricity to heat with 100% efficiency and can be used in smaller systems due to their simple operation. Cost of operation depends on local electricity rates and are generally too expensive to operate where rates are high.
Infrared	Heating	Infrared heaters are common in greenhouses and are gas-fired with propane or natural gas. Infrared heat is a radiant heat source, so it will heat objects directly rather than the air.
Ground-source heat pump	Both	Ground-source heat pumps are gaining momentum in heating and cooling systems due to their energy efficiency. These systems take heat from the ground and release it to the indoor growing environment. When cooling is needed, heat is removed from the indoor environment and dumped into the ground. These systems use conventional refrigeration technology.
Air source heat pump	Both	Air source heat pumps extract heat from the air using the same technology as refrigeration systems. When heating, heat is extracted from the outdoor air and released indoors. Cooling is the reverse of this process, whereby heat from the inside is extracted and then dumped to the building exterior in the same way heat is extracted from your refrigerator to keep food cold.
Solar	Heating	Depending on the location of the growing facility, solar powered hot water heaters can provide considerable amounts of heat that can be fed into a recirculating hot water system. Solar-based systems are highly efficient, requiring only sunlight and a small amount of power to circulate water to storage tanks. Although counter-intuitive, the hot water supplied by a solar system can be used to provide chilled water using absorption chilling technology.
Absorption chillers	cooling	Absorption chillers, while complex, are a highly energy efficient method of using low-grade hot water to produce chilled water. This method is still relatively new, but should be considered in any new large-scale production facility. Absorption chillers are a fascinating technology, and further information can be found in the book *Absorption Chillers and Heat Pumps* (Herold et al. 2016).

transferred to a different medium and removed from the CEA environment—for example with a chilled water heat exchanger or AC condenser unit.

The most energy efficient and simplest design for heating and cooling in CEA is through the use of hot and chilled water systems with forced air distribution. Production of the hot and chilled water can be achieved using a variety of methods, and can be customized based on the facility location, utility costs and potential availability of low-cost sustainable inputs.

Using a chilled water heat exchanger and a hot water heat exchanger (Figure 2.5), the temperature of the air flowing through the coils can be modulated using feedback control. The advantage of this system is that the chilled water coil can also be used as the condenser for VPD control, eliminating the wasteful use of separate cooling and dehumidification systems. In this case, the chilled water coil removes excess water while cooling the air (Figure 2.6), and the hot water coil provides re-heat to bring the temperature to the correct setpoint. The biggest advantages of using a chilled water–based

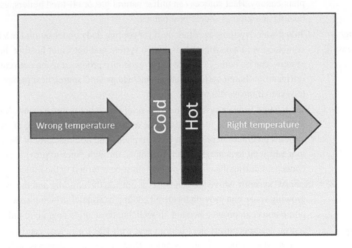

FIGURE 2.5   Basic method for temperature control using hot and chilled water heat exchangers.

Source: Image credit Michael Stasiak

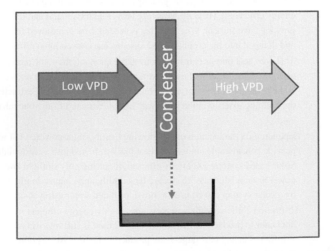

FIGURE 2.6   Collecting condensed water using a chilled water heat exchanger. If the condenser temperature is below the dewpoint, water will collect and drip into the collection reservoir.

Source: Image credit Michael Stasiak

condensing system for VPD control are that water removal can be measured to provide an estimate of evapotranspiration, and the collected water can be returned to the irrigation system, provided the water is properly conditioned prior to reuse (removal of pathogens and possible dissolved metal ions).

2.3.2 VPD (HUMIDITY)

2.3.2.1 Theory

Vapor pressure deficit (VPD) defines the difference between the amount of water that air at a given temperature can hold (called the saturation vapor pressure, or SVP), and the actual amount of water vapor in the air. This value is more important in plant production than relative humidity alone, as it describes the driving force for water movement out of the leaf stomata into the environment and is independent of temperature. As the VPD value includes air temperature in its calculation, it is a better indicator of the plant environment and the driving force for transpiration than relative humidity (RH). Although the SI unit for pressure is the Pascal (Pa), it is generally shown as either kilopascals (kPa) or millibars (mb). One kPa is equivalent to 10 mb or 1,000 Pa.

Why do we use units of pressure? The components of the air—nitrogen, oxygen, water vapor, carbon dioxide and a number of minor gases—all contribute of the total air pressure in proportion to their concentrations (i.e., partial pressures). At sea level, the total air pressure averages around 101.3 kPa (1,013.0 mb); of that, around 78 kPa is nitrogen, 20.9 kPa is oxygen, 0.01–4 kPa is water vapor (large range because it is highly dependent on the temperature) and around 0.04 kPa is carbon dioxide. Trace gases including argon, methane, nitrous oxide, hydrogen, carbon monoxide and ozone make up the remainder.

Movement of water and nutrients through the plant are driven by the process of transpiration, and faster transpiration draws nutrients to the growing points of the plant at a faster rate (Figure 2.7). Transpiration is driven by the difference between the vapor pressure of water in the stomata (considered to be at saturation vapor pressure in a well-watered plant) and the vapor pressure of water in the atmosphere. The

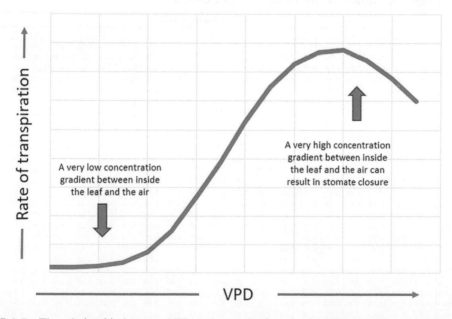

FIGURE 2.7 The relationship between VPD and transpiration. As the VPD increases, transpiration will increase as the driving force for evaporation from the leaf stomata increases. If the air becomes too dry, transpiration becomes excessive, and most plant species will respond with stomatal closure to conserve water. The closure of the stomates also shuts off the source of CO_2 for photosynthesis.

Source: Image credit Michael Stasiak

combination of transpiration from the plant and evaporation from other surfaces (soil, growing media) and open water (e.g., nutrient solution in cultivation system) is called evapotranspiration (ET).

The VPD is a direct indication of this driving force, regardless of the temperature. The higher the VPD, the higher the driving force to move water from the stomata to the atmosphere. Similarly, the higher the VPD value, the lower the relative humidity. At a given temperature, a high VPD (low relative humidity) has less water in the air than a low VPD (high relative humidity). A lower VPD has a much lower driving force for evaporation, meaning that water is less likely to evaporate from the stomatal crypt into the surrounding air.

For proper vegetative plant growth, a VPD in the range of 0.8–1.2 kPa will allow transpiration to take place at an optimal rate (Table 2.3). When the VPD is too low, the difference between the inside of the stomate and the outside is small, water is less likely to evaporate and conditions become optimal for bacterial and fungal growth. When the VPD is too high, the plant senses this and stomata close in response to conserve water and avoid potential drought stress. During flowering, cannabis plants should have a higher VPD to reduce bacterial and fungal growth within the developing inflorescence (Table 2.4). The higher VPD in this case increases the driving force for water vapor to move out of the tight developing buds.

2.3.2.2 Measurement

VPD is not directly measured as it is a calculated value that integrates both temperature and humidity, so to obtain a VPD value, both temperature and RH must be measured. There have been numerous methods developed for measuring the amount of moisture in the air, including the use of chilled

TABLE 2.3
Optimal VPD Values for a Range of Temperature and Relative Humidity in Vegetative Growth Stage Are Shown by the Values in Dark Green

			Relative Humidity																				
	C	**F**	**100**	**95**	**90**	**85**	**80**	**75**	**70**	**65**	**60**	**55**	**50**	**45**	**40**	**35**	**30**	**25**	**20**	**15**	**10**	**5**	**0**
Temperature	**10.0**	**50.0**	0.00	0.06	0.12	0.18	0.24	0.31	0.37	0.43	0.49	0.55	0.61	0.67	0.73	0.80	0.86	0.92	0.98	1.04	1.10	1.16	1.22
	11.0	**51.8**	0.00	0.07	0.13	0.20	0.26	0.33	0.39	0.46	0.52	0.59	0.65	0.72	0.79	0.85	0.92	0.98	1.05	1.11	1.18	1.24	1.31
	12.0	**53.6**	0.00	0.07	0.14	0.21	0.28	0.35	0.42	0.49	0.56	0.63	0.70	0.77	0.84	0.91	0.98	1.05	1.12	1.19	1.26	1.33	1.40
	13.0	**55.4**	0.00	0.07	0.15	0.22	0.30	0.37	0.45	0.52	0.60	0.67	0.75	0.82	0.90	0.97	1.04	1.12	1.19	1.27	1.34	1.42	1.49
	14.0	**57.2**	0.00	0.08	0.16	0.24	0.32	0.40	0.48	0.56	0.64	0.72	0.80	0.88	0.96	1.04	1.11	1.19	1.27	1.35	1.43	1.51	1.59
	15.0	**59.0**	0.00	0.08	0.17	0.25	0.34	0.42	0.51	0.59	0.68	0.76	0.85	0.93	1.02	1.10	1.19	1.27	1.36	1.44	1.53	1.61	1.70
	16.0	**60.8**	0.00	0.09	0.18	0.27	0.36	0.45	0.54	0.63	0.72	0.81	0.91	1.00	1.09	1.18	1.27	1.36	1.45	1.54	1.63	1.72	1.81
	17.0	**62.6**	0.00	0.10	0.19	0.29	0.39	0.48	0.58	0.68	0.77	0.87	0.96	1.06	1.16	1.25	1.35	1.45	1.54	1.64	1.74	1.83	1.93
	18.0	**64.4**	0.00	0.10	0.21	0.31	0.41	0.51	0.62	0.72	0.82	0.92	1.03	1.13	1.23	1.34	1.44	1.54	1.64	1.75	1.85	1.95	2.05
	19.0	**66.2**	0.00	0.11	0.22	0.33	0.44	0.55	0.66	0.77	0.87	0.98	1.09	1.20	1.31	1.42	1.53	1.64	1.75	1.86	1.97	2.08	2.19
	20.0	**68.0**	0.00	0.12	0.23	0.35	0.47	0.58	0.70	0.81	0.93	1.05	1.16	1.28	1.40	1.51	1.63	1.74	1.86	1.98	2.09	2.21	2.33
	21.0	**69.8**	0.00	0.12	0.25	0.37	0.49	0.62	0.74	0.87	0.99	1.11	1.24	1.36	1.48	1.61	1.73	1.86	1.98	2.10	2.23	2.35	2.47
	22.0	**71.6**	0.00	0.13	0.26	0.39	0.53	0.66	0.79	0.92	1.05	1.18	1.31	1.45	1.58	1.71	1.84	1.97	2.10	2.23	2.37	2.50	2.63
	23.0	**73.4**	0.00	0.14	0.28	0.42	0.56	0.70	0.84	0.98	1.12	1.26	1.40	1.54	1.68	1.82	1.95	2.09	2.23	2.37	2.51	2.65	2.79
	24.0	**75.2**	0.00	0.15	0.30	0.44	0.59	0.74	0.89	1.04	1.19	1.33	1.48	1.63	1.78	1.93	2.08	2.22	2.37	2.52	2.67	2.82	2.97
	25.0	**77.0**	0.00	0.16	0.31	0.47	0.63	0.79	0.94	1.10	1.26	1.42	1.57	1.73	1.89	2.05	2.20	2.36	2.52	2.68	2.83	2.99	3.15
	26.0	**78.8**	0.00	0.17	0.33	0.50	0.67	0.83	1.00	1.17	1.34	1.50	1.67	1.84	2.00	2.17	2.34	2.50	2.67	2.84	3.01	3.17	3.34
	27.0	**80.6**	0.00	0.18	0.35	0.53	0.71	0.89	1.06	1.24	1.42	1.59	1.77	1.95	2.12	2.30	2.48	2.66	2.83	3.01	3.19	3.36	3.54
	28.0	**82.4**	0.00	0.19	0.38	0.56	0.75	0.94	1.13	1.31	1.50	1.69	1.88	2.06	2.25	2.44	2.63	2.82	3.00	3.19	3.38	3.57	3.75
	29.0	**84.2**	0.00	0.20	0.40	0.60	0.80	0.99	1.19	1.39	1.59	1.79	1.99	2.19	2.39	2.59	2.78	2.98	3.18	3.38	3.58	3.78	3.98
	30.0	**86.0**	0.00	0.21	0.42	0.63	0.84	1.05	1.26	1.47	1.68	1.90	2.11	2.32	2.53	2.74	2.95	3.16	3.37	3.58	3.79	4.00	4.21
	31.0	**87.8**	0.00	0.22	0.45	0.67	0.89	1.11	1.34	1.56	1.78	2.01	2.23	2.45	2.68	2.90	3.12	3.34	3.57	3.79	4.01	4.24	4.46
	32.0	**89.6**	0.00	0.24	0.47	0.71	0.94	1.18	1.42	1.65	1.89	2.12	2.36	2.60	2.83	3.07	3.30	3.54	3.77	4.01	4.25	4.48	4.72
	33.0	**91.4**	0.00	0.25	0.50	0.75	1.00	1.25	1.50	1.75	2.00	2.25	2.50	2.74	2.99	3.24	3.49	3.74	3.99	4.24	4.49	4.74	4.99
	34.0	**93.2**	0.00	0.26	0.53	0.79	1.06	1.32	1.58	1.85	2.11	2.37	2.64	2.90	3.17	3.43	3.69	3.96	4.22	4.49	4.75	5.01	5.28
	35.0	**95.0**	0.00	0.28	0.56	0.84	1.12	1.39	1.67	1.95	2.23	2.51	2.79	3.07	3.35	3.62	3.90	4.18	4.46	4.74	5.02	5.30	5.58
	36.0	**96.8**	0.00	0.29	0.59	0.88	1.18	1.47	1.77	2.06	2.36	2.65	2.95	3.24	3.53	3.83	4.12	4.42	4.71	5.01	5.30	5.60	5.89
	37.0	**98.6**	0.00	0.31	0.62	0.93	1.24	1.56	1.87	2.18	2.49	2.80	3.11	3.42	3.73	4.04	4.35	4.67	4.98	5.29	5.60	5.91	6.22
	38.0	**100.4**	0.00	0.33	0.66	0.99	1.31	1.64	1.97	2.30	2.63	2.96	3.28	3.61	3.94	4.27	4.60	4.93	5.25	5.58	5.91	6.24	6.57
	39.0	**102.2**	0.00	0.35	0.69	1.04	1.39	1.73	2.08	2.43	2.77	3.12	3.46	3.81	4.16	4.50	4.85	5.20	5.54	5.89	6.24	6.58	6.93
	40.0	**104.0**	0.00	0.37	0.73	1.10	1.46	1.83	2.19	2.56	2.92	3.29	3.65	4.02	4.39	4.75	5.12	5.48	5.85	6.21	6.58	6.94	7.31

TABLE 2.4

Optimal VPD Values for a Range of Temperature and Relative Humidity during Flowering Stage Are Shown by the Values in Dark Green

		Relative Humidity																			
C	F	100	95	90	85	80	75	70	65	60	55	50	45	40	35	30	25	20	15	10	5	0
10.0	50.0	0.00	0.06	0.12	0.18	0.24	0.31	0.37	0.43	0.49	0.55	0.61	0.67	0.73	0.80	0.86	0.92	0.98	1.04	1.10	1.16	1.22
11.0	51.8	0.00	0.07	0.13	0.20	0.26	0.33	0.39	0.46	0.52	0.59	0.65	0.72	0.79	0.85	0.92	0.98	1.05	1.11	1.18	1.24	1.31
12.0	53.6	0.00	0.07	0.14	0.21	0.28	0.35	0.42	0.49	0.56	0.63	0.70	0.77	0.84	0.91	0.98	1.05	1.12	1.19	1.26	1.33	1.40
13.0	55.4	0.00	0.07	0.15	0.22	0.30	0.37	0.45	0.52	0.60	0.67	0.75	0.82	0.90	0.97	1.04	1.12	1.19	1.27	1.34	1.42	1.49
14.0	57.2	0.00	0.08	0.16	0.24	0.32	0.40	0.48	0.56	0.64	0.72	0.80	0.88	0.96	1.04	1.11	1.19	1.27	1.35	1.43	1.51	1.59
15.0	59.0	0.00	0.08	0.17	0.25	0.34	0.42	0.51	0.59	0.68	0.76	0.85	0.93	1.02	1.10	1.19	1.27	1.36	1.44	1.53	1.61	1.70
16.0	60.8	0.00	0.09	0.18	0.27	0.36	0.45	0.54	0.63	0.72	0.81	0.91	1.00	1.09	1.18	1.27	1.36	1.45	1.54	1.63	1.72	1.81
17.0	62.6	0.00	0.10	0.19	0.29	0.39	0.48	0.58	0.68	0.77	0.87	0.96	1.06	1.16	1.25	1.35	1.45	1.54	1.64	1.74	1.83	1.93
18.0	64.4	0.00	0.10	0.21	0.31	0.41	0.51	0.62	0.72	0.82	0.92	1.03	1.13	1.23	1.34	1.44	1.54	1.64	1.75	1.85	1.95	2.05
19.0	66.2	0.00	0.11	0.22	0.33	0.44	0.55	0.66	0.77	0.87	0.98	1.09	1.20	1.31	1.42	1.53	1.64	1.75	1.86	1.97	2.08	2.19
20.0	68.0	0.00	0.12	0.23	0.35	0.47	0.58	0.70	0.81	0.93	1.05	1.16	1.28	1.40	1.51	1.63	1.74	1.86	1.98	2.09	2.21	2.33
21.0	69.8	0.00	0.12	0.25	0.37	0.49	0.62	0.74	0.87	0.99	1.11	1.24	1.36	1.48	1.61	1.73	1.86	1.98	2.10	2.23	2.35	2.47
22.0	71.6	0.00	0.13	0.26	0.39	0.53	0.66	0.79	0.92	1.05	1.18	1.31	1.45	1.58	1.71	1.84	1.97	2.10	2.23	2.37	2.50	2.63
23.0	73.4	0.00	0.14	0.28	0.42	0.56	0.70	0.84	0.98	1.12	1.26	1.40	1.54	1.68	1.82	1.95	2.09	2.23	2.37	2.51	2.65	2.79
24.0	75.2	0.00	0.15	0.30	0.44	0.59	0.74	0.89	1.04	1.19	1.33	1.48	1.63	1.78	1.93	2.08	2.22	2.37	2.52	2.67	2.82	2.97
25.0	77.0	0.00	0.16	0.31	0.47	0.63	0.79	0.94	1.10	1.26	1.42	1.57	1.73	1.89	2.05	2.20	2.36	2.52	2.68	2.83	2.99	3.15
26.0	78.8	0.00	0.17	0.33	0.50	0.67	0.83	1.00	1.17	1.34	1.50	1.67	1.84	2.00	2.17	2.34	2.50	2.67	2.84	3.01	3.17	3.34
27.0	80.6	0.00	0.18	0.35	0.53	0.71	0.89	1.06	1.24	1.42	1.59	1.77	1.95	2.12	2.30	2.48	2.66	2.83	3.01	3.19	3.36	3.54
28.0	82.4	0.00	0.19	0.38	0.56	0.75	0.94	1.13	1.31	1.50	1.69	1.88	2.06	2.25	2.44	2.63	2.82	3.00	3.19	3.38	3.57	3.75
29.0	84.2	0.00	0.20	0.40	0.60	0.80	0.99	1.19	1.39	1.59	1.79	1.99	2.19	2.39	2.59	2.78	2.98	3.18	3.38	3.58	3.78	3.98
30.0	86.0	0.00	0.21	0.42	0.63	0.84	1.05	1.26	1.47	1.68	1.90	2.11	2.32	2.53	2.74	2.95	3.16	3.37	3.58	3.79	4.00	4.21
31.0	87.8	0.00	0.22	0.45	0.67	0.89	1.11	1.34	1.56	1.78	2.01	2.23	2.45	2.68	2.90	3.12	3.34	3.57	3.79	4.01	4.24	4.46
32.0	89.6	0.00	0.24	0.47	0.71	0.94	1.18	1.42	1.65	1.89	2.12	2.36	2.60	2.83	3.07	3.30	3.54	3.77	4.01	4.25	4.48	4.72
33.0	91.4	0.00	0.25	0.50	0.75	1.00	1.25	1.50	1.75	2.00	2.25	2.50	2.74	2.99	3.24	3.49	3.74	3.99	4.24	4.49	4.74	4.99
34.0	93.2	0.00	0.26	0.53	0.79	1.06	1.32	1.58	1.85	2.11	2.37	2.64	2.90	3.17	3.43	3.69	3.96	4.22	4.49	4.75	5.01	5.28
35.0	95.0	0.00	0.28	0.56	0.84	1.12	1.39	1.67	1.95	2.23	2.51	2.79	3.07	3.35	3.62	3.90	4.18	4.46	4.74	5.02	5.30	5.58
36.0	96.8	0.00	0.29	0.59	0.88	1.18	1.47	1.77	2.06	2.36	2.65	2.95	3.24	3.53	3.83	4.12	4.42	4.71	5.01	5.30	5.60	5.89
37.0	98.6	0.00	0.31	0.62	0.93	1.24	1.56	1.87	2.18	2.49	2.80	3.11	3.42	3.73	4.04	4.35	4.67	4.98	5.29	5.60	5.91	6.22
38.0	100.4	0.00	0.33	0.66	0.99	1.31	1.64	1.97	2.30	2.63	2.96	3.28	3.61	3.94	4.27	4.60	4.93	5.25	5.58	5.91	6.24	6.57
39.0	102.2	0.00	0.35	0.69	1.04	1.39	1.73	2.08	2.43	2.77	3.12	3.46	3.81	4.16	4.50	4.85	5.20	5.54	5.89	6.24	6.58	6.93
40.0	104.0	0.00	0.37	0.73	1.10	1.46	1.83	2.19	2.56	2.92	3.29	3.65	4.02	4.39	4.75	5.12	5.48	5.85	6.21	6.58	6.94	7.31

mirrors and human hair. Modern instruments for the measurement of relative humidity (hygrometers) use solid-state capacitive sensors which measure the effect of humidity on the dielectric constant of a polymer or metal oxide material. Their accuracy, when properly calibrated, can be +/−2% in the range of 5–95% RH.

The calculation of VPD from air temperature and RH is done using the formula:

$$VPD = SVP - AVP$$

Where: VPD = vapor pressure deficit (kPa)
SVP = saturated vapor pressure (kPa)
AVP = actual vapor pressure (kPa)

The first step is to calculate the amount of water that the air can hold at a given temperature. In this case, we are using Teten's formula to calculate the SVP:

$$SVP = 0.61078 \exp\left(\frac{17.27T}{T + 237.3}\right)$$

Where: T = the air temperature (°C)
We can now use the calculated SVP and the air RH to calculate the AVP:

$$AVP = \frac{RH * SVP}{100}$$

Where: RH = the relative humidity of the air

Finally, with values of the SVP and AVP known, the VPD can be calculated simply by subtracting the AVP from the SVP as shown in the first equation. The resulting number will be in kPa. For a VPD value in millibars, multiply the result by 10.

2.3.2.3   Control

BOX 2.2

BTU: British Thermal Unit, where one BTU refers to the amount of energy needed to increase the temperature of one pound of water by 1°F. Kilojoules are the metric equivalent unit, but are less commonly used in the industry.

When a cannabis crop is grown indoors with an optimized spacing, the need to remove water from the atmosphere will begin soon after transplantation to the vegetative stage after cloning. The demand for water removal will increase through veg and into bloom for the remainder of the crop

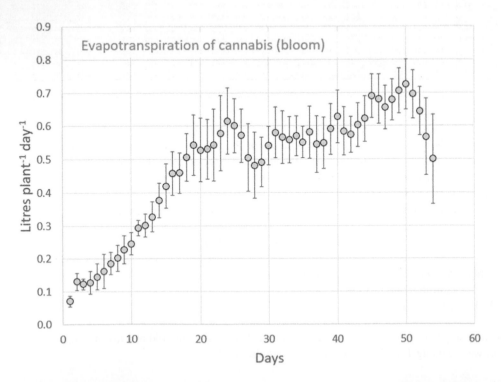

FIGURE 2.8   Evapotranspiration (ET) during bloom in a typical cannabis crop. ET on a per-plant basis showed declines at 24 days (trimming) and 50 days (lowered light levels) after the switch to a 12:12-hour photoperiod.

Source: Image credit Michael Stasiak

cycle, with peak ET of 0.5–0.8 liters per plant per day or 5–8 liters per square meter per day in a closed canopy. The removal of this amount of water from the atmosphere is energy intensive. It takes 2,343 BTUs to remove 1 liter of water from the air. In an indoor grow with, for example, 1,000 plants, that can equate to 800 liters per day (33 liters per hour on average) with an energy demand of 78,000 BTU just to remove water vapor. In terms of tons of chilling (1 ton = 12,000 BTU), a chiller sized to remove 6.5 tons of heat to maintain the chilled water temperature during the condensation process would be needed if using a condensing type system for water removal. There are three main methods used to remove excess water from the air of a closed environment.

2.3.2.3.1 Condensing Systems

In order to remove water from the air, the air must pass over a condensing surface that is chilled either through the use of compressors for AC refrigerant or chilled water. Chilled water systems are capable of a finer degree of control, as the chilled water stream is easier to modulate (flow and temperature control) than an AC unit. Refrigerant-based condensers cause direct cooling of the condenser, are more difficult to modulate and are prone to icing.

In these systems (Figure 2.9), air from the grow room is passed through the heat exchanger that is cooled to below the dewpoint temperature, where water is condensed and removed from the air. This process inherently cools the air to a temperature unsuitable for direct return to the grow room. A second heat exchanger reheats the air to the proper temperature before return. A suitably sized recirculating air handling system moves the air from the room, through the heat exchangers and back at a sufficient rate to maintain air temperature and VPD near optimal levels.

2.3.2.3.2 Desiccants

Another method of removing water from the air is through the use of desiccants, materials that can reversibly adsorb water onto their surface. In commercial systems, desiccant dehumidifiers work by using a large revolving wheel (rotating drum) filled with a chemical adsorbent that removes moisture from the air as it passes through the wheel. As the wheel turns, a second air stream of

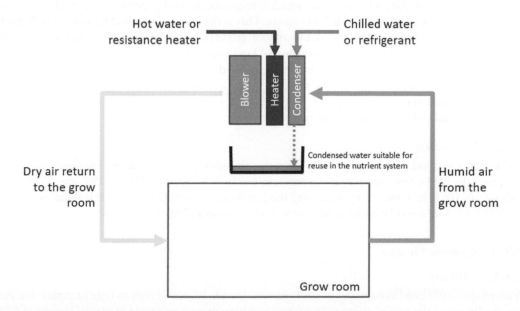

FIGURE 2.9 Basic operation of a condenser-based dehumidification system.

Source: Image credit Michael Stasiak

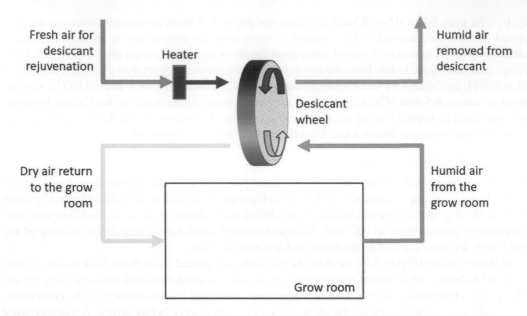

FIGURE 2.10   Basic operation of a desiccant-based dehumidification system.

Source: Image credit Michael Stasiak

heated air strips the moisture out of the desiccant so the process can begin again, as shown in Figure 2.10.

2.3.2.3.3   Ventilation

Ventilation involves replacing and mixing indoor air with fresh outdoor air. As long as the external environment is at a higher VPD than that which is required, the exchange of air can dilute the water vapor in the indoor air to acceptable VPD limits. This is the common method for control of humidity in greenhouses, but could be utilized by indoor growers to improve overall system energy efficiency when coupled with artificial intelligence (AI) and outdoor temperature and humidity sensors. At many times of the year, outdoor air is capable of modulating indoor temperature and VPD, and reducing use of energy-intensive HVAC systems. Although temperature and VPD can be addressed under some conditions with this approach, it represents either a "leak" or dilution of aerial CO_2 in systems that add CO_2 to enhance photosynthesis.

2.3.2.3.4   Heating

While not a method of removing water from the air, adding heat to the environment effectively increases the VPD. For example, if the temperature is 20°C with an RH of 75%, the VPD is an undesirable 0.6 kPa; however, by increasing the temperature to 26°C while the RH remains at 75%, the VPD is 0.8 kPa and far more conducive to transpiration and photosynthesis.

2.3.3   Carbon Dioxide

2.3.3.1   Theory

With temperature conditions under control, carbon dioxide is second only to light in inputs that can drastically affect the rate of photosynthesis and resulting growth and yield of plants. Uptake of CO_2 in plants is driven by the concentration gradient between the stomatal crypt and the environment in the same way (but opposite direction) as transpiration is limited by the concentration of water vapor in the atmosphere. The higher the external concentration, the larger the driving force moving CO_2 into the leaf through the stomata.

The rate of photosynthesis generally increases with increasing CO_2 concentration until limited by another factor, such as stomatal closure or availability of light (Figure 2.11A) (See Chapter 6 for more detailed discussion). For many plants, transpiration decreases as the CO_2 concentration increases because stomata do not need to open as much to maintain an optimal photosynthetic rate

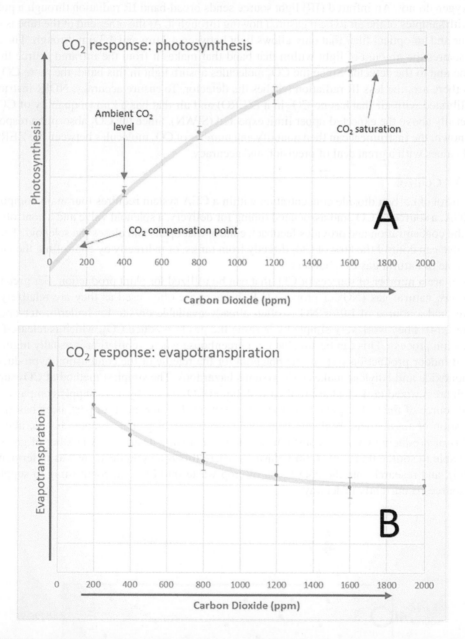

FIGURE 2.11 Effect of increasing carbon dioxide concentration on the rate of photosynthesis (A) and evapotranspiration (B). Evapotranspiration is the combined measurement of evaporation from the plant growing medium and plant transpiration.

Source: Image credit Michael Stasiak

(Figure 2.11B). In addition to increased productivity, higher atmospheric CO_2 increases water use efficiency (plant growth per unit of water uptake).

2.3.3.2 Measurement

The most accurate and cost-effective method of measuring the concentration of carbon dioxide in the air is with a nondispersive infrared (NDIR) sensor. The NDIR takes advantage of the fact that carbon dioxide absorbs radiation between 4.2 µm and 4.3 µm (4,200–4,300 nm), whereas nitrogen and oxygen do not. An infrared (IR) light source sends broad-band IR radiation through a polished tube with samples of the air to be measured flowing through it. At the other end of the tube is a light detector and an optical filter that only allows light between 4.2 µm and 4.3 µm through. The detector measures the amount of light within that band that makes it from the infrared source through the tube and to the detector. Since the CO_2 molecules absorb light in this band, the more CO_2 molecules there are, the less IR radiation reaches the detector. To ensure accuracy, NDIR instruments are calibrated with air that has no CO_2 in it (ZERO) and air that has a known quantity of CO_2 that is generally above the expected upper limit expected (SPAN). Since the CO_2 absorption response is well known, the instrument can then quantify any number of CO_2 molecules between the ZERO and SPAN values with a great deal of precision and accuracy.

2.3.3.3 Control

The control of carbon dioxide concentration within a CEA system requires four main components: the IRGA, a source of CO_2 and associated tubing for delivery, a solenoid valve and a controller that reads the concentration and provides feedback control that opens and closes the solenoid. The solenoid valve can control the flow of CO_2 directly from tanks, or indirectly by controlling the gas flow to propane or natural gas CO_2 burners.

There are a number of sources of CO_2 that can be utilized for plant production. For greenhouse operations, natural gas (NG) or propane (LP) burners are often used as they are relatively inexpensive, and a source of LP or NG is often already available on-site for boiler/heater operation (Figure 2.14). They work very simply by burning the gas to produce CO_2, which is released in the combustion process. This can be suitable for a greenhouse where ventilation is usually high, but in a closed indoor production system, the potential for the release of the combustion byproducts carbon monoxide and ethylene makes such systems hazardous. The simplest method of CO_2 supply is through pressurized tanks that are refilled with liquid CO_2 on-site by a gas supply company.

The source of the CO_2 supplied to growing operations has, for the most part, been based on the combustion of fossil fuels. With the heightened awareness of global climate change, alternative methods of producing CO_2 from sustainable sources is highly desirable. A number of gas suppliers are able to source their CO_2 from industries where this is a byproduct, such as in fermentation (alcohol), and research into the conversion of crop waste into CO_2 via composting to supplement other sources is currently underway.

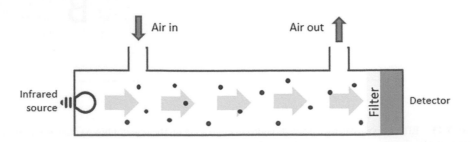

FIGURE 2.12   Basic operation of a nondispersive infrared gas analyzer (IRGA).

Source: Image credit Michael Stasiak

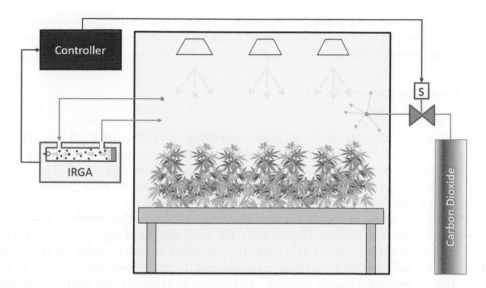

FIGURE 2.13 Basic configuration of a feedback control system for atmospheric CO_2 control. The basic components are a CO_2 source, a solenoid valve (S) and a CO_2 analyzer (IRGA). Feedback control would be performed using a standalone CO_2 valve controller or a fully automated control system.

Source: Image credit Michael Stasiak

FIGURE 2.14 Examples of gas fired (left) and tank supplied (right) CO_2 supply systems.

Source: Image credit Johnson CO_2 (left) Linde plc (right)

Delivery of CO_2 within the growing environment is effectively done through direct injection into the air handling system in an indoor growing environment. Greenhouse gas–fired CO_2 burners are equipped with blowers or fans to effectively distribute the CO_2 throughout the production zones.

2.3.4 Radiation (Lighting)

2.3.4.1 Theory

Cannabis plants, like all green plants, are photosynthesizing organisms that depend on light energy to grow and reproduce. Photosynthesis is the process whereby the energy from light, coupled with CO_2 from the atmosphere, water and minerals, is used to form sugars and oxygen. The sugars are used by the plant to drive chemical processes that produce all the amino acids, proteins, carbohydrates and fats needed to build a plant.

At low light intensities, the rate of the light-dependent reaction of photosynthesis increases linearly with increasing light intensity (Figure 2.15). The more photons of light that fall on a leaf, the faster the rate of photosynthesis. As light intensity is increased further, however, the rate of photosynthesis is eventually limited by some other factor—usually the chemical reactions can no longer keep up with the amount of energy received—and the rate plateaus. At very high light intensities, chlorophyll may be damaged, and the rate of photosynthesis drops (not shown in the graph). As indicated previously, increasing the concentration of carbon dioxide in the atmosphere will increase the rate of photosynthesis (Figure 2.15: 1,200 ppm vs. 400 ppm) but the limitations of excess light still prevail at high light intensities.

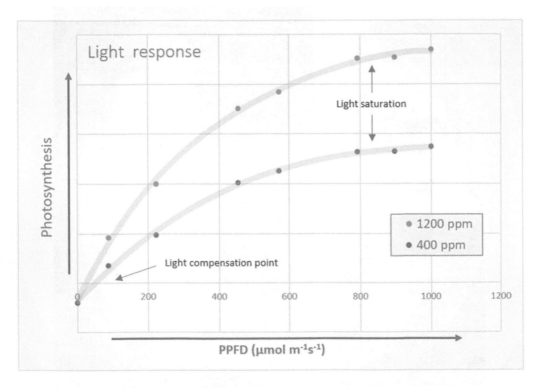

FIGURE 2.15 Light response curve showing the change in the rate of photosynthesis in response to increasing light levels. The addition of a higher concentration of CO_2 results in a shift of the curve upwards.

Source: Image credit Michael Stasiak

Much more detail and in-depth discussions on the effects of radiation on cannabis are covered in Chapter 6.

2.3.4.2 Measurement

See Chapter 6.

2.3.4.3 Control

Before the development of horticultural LEDs, lighting control for controlled environments was simply either on or off. Control of these lighting systems utilized high power relays (also called contactors) connected to an automated control system that was programmed to turn the lights on and off as needed by the grower. Even simpler systems utilized mechanical timers for the same task. For the control of basic high-pressure sodium, metal halide or fluorescent lighting, both of these methods worked perfectly well.

Step forward to the early 2010s and the availability of light-emitting diode (LED) technology brought a finer degree of control to plant lighting. Not only could intensity be varied, but spectral manipulation was brought to plant production with multi-spectrum control from ultraviolet (UV) through to far-red (FR) by early adopters such as Heliospectra and Intravision Lighting. With intensity and wavelength (quantity and quality of the light) now available for control, on-and-off systems with relays and mechanical timers are no longer sufficient to gain full lighting system control. The ability to modify light spectrum is a powerful tool in plant production, as simple changes to the light recipe can have a dramatic impact on the production of plant secondary metabolite production (Figure 2.16). It is these types of plant responses to light spectrum that could be manipulated to provide improved production of the medicinal compounds in cannabis and contribute to standardizing the profiles of these medicinal compounds.

Currently available methods for the control of LED lighting are shown in Table 2.5. These methods are known to interface with modern greenhouse/growth room control systems; however, there are many proprietary lighting control systems on the market which may not interface directly with the many variations of available control hardware.

FIGURE 2.16 The effect of red, white and blue light with equal photosynthetic photon flux densities on the color of red leaf lettuce. All plants in this photo are the same cultivar.

Source: Image credit Michael Stasiak

TABLE 2.5

Methods for the Control of LED Lighting

Control Type	Description
0–10 VDC	Control of lighting via a simple change in the voltage of a two-wire control line was the standard for lighting control of dimmable light fixtures in industrial settings for decades. If used in multispectral systems, each channel requires a separate 0–10 VDC output.
DALI	The Digital Addressable Lighting Interface (DALI) is an open standard that any lighting manufacturer can adopt and use in their hardware. Control through a two-wire cable can daisy chain between lights and sensors, and are fully addressable to provide a high degree of control. DALI allows full control of intensity and spectrum in lighting systems equipped with this protocol. Control response is delayed, with changes taking seconds to implement.
DMX-512	This lighting protocol is common to theatrical systems; however, it is suitable for use in horticultural applications due to the high degree of control available. Unlike DALI, response is nearly instantaneous.
MODBUS	This is another open standard communication protocol widely used for the control of electronic equipment. MODBUS uses basic serial two-way communication over ethernet and is capable of a high degree of customization and control. It is common to many control systems, including programmable logic controllers (PLC). Because of its open standard and simplicity, new sensors, actuators and lighting systems are becoming available with MODBUS control.
Proprietary	Many horticultural lighting suppliers use their own proprietary control methods for lighting and can be either wired or wireless. Most suppliers will be able to provide interfaces that will bridge their systems to more convention control methods.

2.3.5 AIR DISTRIBUTION

2.3.5.1 Theory

BOX 2.3

The boundary layer is a thin zone of unstirred (calm) air that surrounds each leaf. The thickness of the boundary layer is directly related to the air speed. The higher the air velocity, the smaller the boundary layer and the easier it is for gases and energy to exchange between the leaf and the surrounding air.

Proper air distribution is critical. Ideally, it provides air of the right temperature, the correct VPD and the perfect carbon dioxide concentration to every leaf on every plant. It also carries away the water vapor from transpiration and heat generated during photosynthesis and from thermal lighting loads. Insufficient air distribution may result in poor growth and can allow disease to take hold and thrive under conditions of high humidity. The movement of air (flow, velocity, speed) is measured in meters per second (m/s). A great deal of research was performed between 1950 and 1990 to characterize the optimal air speed for plant growth and productivity. In general, air speeds between 0.3 m/s and 0.7 m/s through the canopy are considered optimal for plants grown in controlled environments. Air speeds that are too slow result in poor environment homogeneity and poor growth, as the boundary layer resistance is too high to allow efficient movement of CO_2 into and water and oxygen out of the leaves through the stomates. If the air speed is too high, mechanical damage to leaves and stems can occur, along with excessive transpiration, leading to water stress (Figure 2.17).

Air distribution through a plant canopy can occur in three ways: bottom-up, top-down and from the side (Figure 2.18). The least desirable method of moving air through a canopy is from the side. In this method, air is either pushed or pulled through a perforated plenum (wall) using a ducted

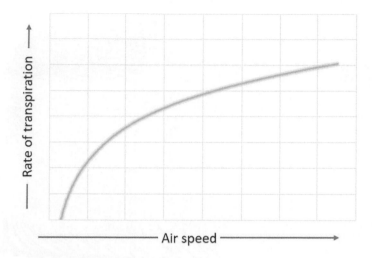

FIGURE 2.17   The rate transpiration in response to air speed.

Source: Image credit Michael Stasiak

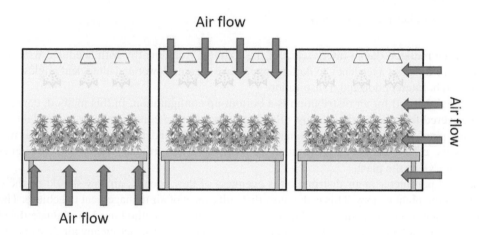

FIGURE 2.18   Air distribution in plant canopies showing bottom-up (left), top-down (center) and side (right) air flow configurations.

Source: Image credit Michael Stasiak

centralized blower or pushed using stationary fans, allowing the air to flow across and through the plant canopy to the other side of the growing space, where it is then returned to the HVAC for reconditioning and return in a continuous loop. The primary problem with this method is that conditioned air leaving the plenum quickly picks up additional heat and moisture from the plants as it passes through and over them. The desired homogeneity is lost as plants on one side are exposed to very different conditions compared to plants on the other side.

A more desirable method is top-down distribution. In this case, a perforated plenum in the ceiling delivers air from above the canopy and the air is pushed through the growing plants and returned to the HVAC system from below. This is an improvement over side distribution, as each plant is receiving the same conditioned air throughout the canopy field. The drawback to this method is that heat from the lighting systems is picked up and must be compensated for. Depending on the lighting configuration (high-pressure sodium, ceramic metal halide, LED), significant alteration to the flow

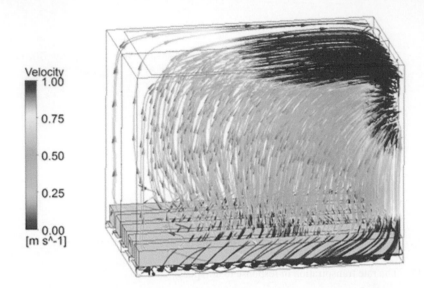

FIGURE 2.19 An example of computational fluid dynamics modeling in a small growth chamber showing the direction and velocity of air flow.

Source: Image credit European Space Agency

of air can be realized due to interference of the fixtures. Additionally, with adequate wind speed, top-down air flow has a tendency to flatten upper leaves against lower and alters leaf angles, thereby reducing light interception and transpiration.

The ideal method for air distribution is a bottom-up configuration. In this method, conditioned air is delivered from below the canopy, allowing proper flow of conditioned air to all plants in the horizontal plane. As the majority of stomata are located on the undersides of the leaves, this method improves the effectiveness of transpiration by removing moisture and heat, while enhancing the delivery of CO_2 to the plant.

The final parameter in air distribution is evenness of flow. Ideally, airflow should be the same throughout the plant canopy. This is the most difficult aspect of air management to achieve. The use of computational fluid dynamics (CFD), a computer simulation method used to evaluate the effect of various parameters on the environment, is suggested before constructing any air delivery system to ensure adequate air flow (Figure 2.19).

2.3.5.2 Measurement

Within the range of airflows found in a controlled environment, few wind speed sensors (anemometers) have the ability to measure the relatively low air velocities through and around the plant canopy. There are three main types of anemometers that are applicable to—and cost effective for—measurement of air speed with respect to plant growth (Figure 2.20). Other types exist—for example, laser Doppler, acoustic and ultrasonic—but these are expensive and rarely used in commercial applications.

The cup anemometer uses cups that rotate on a central bearing shaft. The cups rotate at a speed proportional to the speed of the air. Vane anemometers use fan blades that rotate in response proportionately to the speed of the air flowing through them. Finally, the hot wire anemometer uses a heated wire on the end of a probe that is maintained at a constant temperature. As wind speed increases, the amount of power needed to maintain the temperature is increased as heat is driven off the wire. This is the most suitable device for measuring air velocity indoors, as it is the most sensitive to low air speeds. Models are available that can measure as low as 0.1 m/s. As air speeds below 0.3 m/s are insufficient for plant production, this type of tool is ideal.

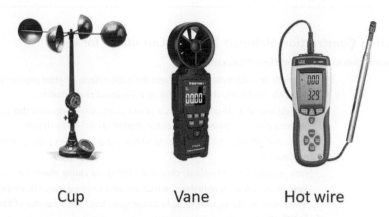

Cup   Vane   Hot wire

FIGURE 2.20   Three main types of anemometers used to measure air velocity.

Source: Image credit Sean Linehan (left), Dongguan Habotest Instrument Technology Co., Ltd (center), and CEM Instruments (right)

Blower   Fan

FIGURE 2.21   Examples of a blower and a fan.

Source: Image credit Dayton Electric Mfg. (left) and GGS Structures Inc. (right)

2.3.5.3   Control

There are two methods of moving air that are common to indoor and greenhouse growing, blowers and fans (Figure 2.21). Blowers move large volumes of air with moderate increase in pressure and are best at delivering air through ducting. Fans, on the other hand, move large volumes of air with very little increase in pressure and are ideal for moving air around in a room. Both come in a variety of sizes, shapes and configurations.

As blowers are generally already a part of an indoor air handling system designed to heat and condition the air, the inclusion of a properly designed ducting system can provide homogeneous air flow throughout a plant canopy.

2.3.6   SOURCES OF CONTAMINATION

2.3.6.1   Building Materials

The choice of building materials used in the construction of indoor plant growth systems is extremely important. Numerous products are known to off-gas or leach chemicals that are toxic to plants, or in the case of medical cannabis production, toxic to humans. A list of suitable and unsuitable materials are shown in Table 2.6. Also it is important to note that there are galvanic incompatibilities between metals, especially when used in nutrient systems (Table 2.7). Using incompatible metals can enhance their degradation and potentially release plant-toxic metals into the nutrient solution.

TABLE 2.6
Plant Production Construction Materials That Are Unsuitable for Use

Unsuitable Construction Materials for Plant Production

Plywood	Although formaldehyde may not have definitive effects on plant productivity, it is a probable carcinogen that can build up in enclosed environments.
Polyethylene	Polyethylene will off-gas ethylene, a potent plant growth regulator that can affect flowering on most plant species at concentrations as low as 10 ppb.
Paint	Paints, both alkyd and latex, can emit VOCs that could affect plant growth and productivity.
Sealants	Many sealants give off volatile chemicals during the curing phase that are toxic to plants.
Other plastics and rubber	Many plastics contain phthalates, which are used to give plastics flexibility. They can accumulate in plants, are probable carcinogens and have a number of known human health effects.
Wood	The main problem with natural wood products is their likeliness to rot under humid and wet conditions common to plant growth systems.
Aluminum, copper, brass, steel, iron, zinc	Most metals will corrode under high humidity and wet environments. None of these should be used in nutrient or condensed water recovery systems, as they will leach ions to the water and cause toxicity.

Suitable Construction Materials for Plant Production

Stainless steel	Food-grade 316 stainless steel is durable and inert.
Glass	Glass is food-safe and inert.
Teflon	PTFE (polytetrafluoroethylene) is a polymer with excellent chemical resistance and has no off-gassing or leaching properties.
Polypropylene	Unlike polyethylene, polypropylene does not off-gas ethylene or any other plant-toxic chemicals.
Baked enamel	Baked enamel surfaces that effectively coat metals allow use of metals that are generally avoided. For example, brass or copper heat exchanger components that are enamel coated and baked are less likely to leach toxic metals.
Viton	Viton is a polymer that is highly chemical resistant and inert.
Cured silicone	Once cured, silicone adhesives and caulking are suitable for plant growth environments.

TABLE 2.7
Galvanic Compatibility of Metals

	Zinc	Galvanized steel	Aluminum	Cast Iron	Lead	Mild steel	Tin	Copper	Stainless Steel
Zinc									
Galvanized steel									
Aluminum									
Cast Iron									
Lead									
Mild steel									
Tin									
Copper									
Stainless Steels									

Legend:
- Incompatible
- Very low reactivity
- Medium reactivity
- Same material so no reactivity

2.3.6.2 Microbial

Microorganisms can be a huge source of contamination within a plant growth facility. Chapter 8 will provide a thorough examination of this topic and available mitigation techniques.

2.3.6.3 Plant Sources

While plants produce a variety of chemicals that can enter the air or water/ground, one of the most potentially harmful plant source contaminants is the hormone ethylene. Ethylene, a simple molecule with the molecular formula C_2H_4, is a potent plant growth regulator that has implications in flowering, senescence (plant death), fruit ripening, pollination and leaf abscission (leaves falling off), among others. While the presence of ethylene has been well known to cause floral abortion in numerous plant species, there is some current evidence that the addition of ethylene to a cannabis crop can increase flower size and increase production of CBD. Until there is better understanding of the role of ethylene in cannabis production, its removal is preferred. While there are chemical methods for ethylene removal (potassium permanganate), the simplest is to ventilate with fresh air periodically.

2.4 CONTROL SYSTEMS

Control over the various environment parameters can use simple mechanical control methods that operate single parameters like lighting or temperature, or it can use advanced computer control systems that offer data recording, alarm notifications and precise control over all growing parameters in a single system.

2.4.1 MECHANICAL CONTROLLERS

In the most basic of control methods, mechanical timers and bimetallic temperature sensors (thermostats) can very simply provide separate direct control for lighting (on/off) and temperature control through the activation of heaters, fans and vents. For temperature control, this sort of hardware does not provide accurate control with variations from setpoint in the order of 1–2°C. Mechanical lighting timers have been in use for decades, but as with mechanical thermostats, control isn't as precise as modern computer-based systems.

2.4.2 SUBSYSTEM COMPUTER CONTROLLERS

Computers and microcontrollers can provide a much higher degree of control for individual subsystems. Numerous single function hardware controllers are available to control lighting, pH, electrical conductivity, temperature, humidity, pumps and fans. Although these systems work well for the individual hardware they were designed for, the lack of integration with other subsystems is a huge disadvantage in terms of overall environment control. For example, if the irrigation system malfunctions, an integrated system can turn off lighting and reduce air speed to decrease water loss that can result in crop failure.

2.4.3 INTEGRATED COMPUTER CONTROLLERS

To obtain the highest level of systems control and automation, an integrated solution is the most desirable. A fully autonomous control system can manage all the plant management subsystems, including the following.

- Lighting: conventional and multispectral LED lighting systems
- Air management: temperature, VPD, CO_2, ventilation, air speed
- Irrigation: pH, EC, zone delivery, watering schedules, tank mixing

- Data collection: logging of sensors, actuators and their demand setpoints
- Alarm notification: off nominal sensor readings can trigger alarms, respond with mitigation procedures and notify operators of the problem
- Global access: with internet connectivity, monitoring and control can be done from any place with an internet connection

With environment management contained within one system, efficiencies in labor, water use, plant growth, quality, productivity and improved process control are achieved.

BIBLIOGRAPHY

Beytes, C. (ed.). 2021. *Ball redbook: Volume 1, Greenhouses and equipment*, 19th ed. Chicago, IL: Ball Publishing.

Bugbee, B., and O. Monje. 1992. The limits of crop productivity. *Bioscience* 42: 494–502.

Herold, K. E., R. Radermacher, and S. A. Klein. 2016. *Absorption chillers and heat pumps*. Boca Raton, FL: CRC Press.

Jin, D., S. Jin, and J. Chen. 2019. Cannabis indoor growing conditions, management practices, and post-harvest treatment: A review. *American Journal of Plant Sciences* 10: 925–946.

Langhans, R. W., and T. W. Tibbits. 1997. *Plant growth chamber handbook*. North Central Region Publication 340. Iowa: Iowa Agriculture and Home Economics Experiment Station Special Report No. 99.

Potter, D. J., and P. Duncombe. 2012. The effect of electrical lighting power and irradiance on indoor-grown cannabis potency and yield. *Journal of Forensic Sciences* 57: 618–622.

3 Genetics and Plant Breeding of *Cannabis sativa* for Controlled Environment Production

Ernest Small

CONTENTS

DOI: 10.1201/9781003150442-3

WARNINGS

The information in this chapter provides theoretical and practical guidance on acquiring, maintaining and improving plants that are viewed as having abuse or harmful potential, as well as beneficial applications, and are governed by laws of varying severity according to jurisdiction. These regulations are often complex, and even with expert professional guidance, one may unintentionally become entangled in serious legal difficulties. Acquiring materials internationally is governed by particularly strict legislation, and it cannot be assumed that one's local laws correspond to those in foreign locations. Possession of cannabis materials attracts thieves, and security requirements can be very demanding. Living material of cannabis plants is covered—to often uncertain degrees—by intellectual property, copyright and/or plant rights legislation, so there is danger of having to respond to expensive civil lawsuits. Virtually all cannabis strains trace to varieties that were clandestinely developed by underground breeders, and there remains a very strong tradition of continuing such illegal activities. Indeed, the most novel and valuable genetic materials of cannabis are mostly available from the gray (quasi-legal) and illegal cannabis trade, possibly tempting some interested in conducting genetic and breeding research on *Cannabis sativa* to violate laws and risk prosecution in order to acquire much of the best research materials. It cannot be overemphasized that all projects involving cannabis materials—including acquisition, exchange, research, and commercial activities—must be authorized in most countries. Very substantial expenses can be involved in professional research and development programs involving cannabis breeding, and those involved in large-scale business ventures need to carefully evaluate risk/benefit aspects. Hobbyist breeders need to appreciate that while simply crossing strains and selecting oddities is rewarding, the creation of commercially useful varieties typically requires very considerable sophistication and resources. The world's governments have not yet taken up the challenge of preserving invaluable drug forms of cannabis germplasm in long-term genebanks, and while champions are urgently needed to fulfill this role, caution is required because of the associated expenses and complexities.

3.1 EVOLUTION AND AMBIGUITY OF THE WORDS CANNABIS, MARIJUANA AND HEMP

"*Cannabis*" (capitalized and italicized in this manner) is a genus of plants. Most authorities accept that there is just one species of *Cannabis*, *C. sativa*, so *Cannabis* is an abbreviated way of referring to *C. sativa*. Some of the plants of this species are capable of getting people "high," primarily because of the presence of THC. An exact psychopharmacological term for "high" is unavailable (see discussion in Small 2017a). Accordingly, a corresponding word to identify precisely the plants that get people high is lacking (approximate synonyms are euphoric, intoxicant, psychotomimetic and psychotropic; much less exact terms are addictive, drug, hallucinogenic and narcotic). Nevertheless, an exact word to designate the plants and their drug preparations that get people high was universally used until recently. That word is "marijuana," but as explained in what follows, semantic evolution has occurred and current terminology can be ambiguous.

In the past, "cannabis" has been a generalized term (both a noun and an adjective), employed in one or more of the following senses: any or all preparations (drug and non-drug) of *Cannabis* plants, *Cannabis* plants and any related aspects (such as chemical, medicinal, sociological or commercial considerations). In the early 21st century, the word "cannabis" has often been restricted to refer to what has been called "marijuana" in the past, i.e., high-THC plants (with sufficient THC to intoxicate) and associated drug preparations employed for inebriation and/or therapeutics. In some jurisdictions dealing with the subject (such as Canada), CBD is regulated or defined as falling under the term cannabis. Synthetic cannabinoids may also be included under the term cannabis.

In the past, "marijuana" was the principal term designating high-THC biotypes of *C. sativa* and associated drug preparations. For practical purposes, "marijuana" is an exact synonym for the recent, restricted way that "cannabis" is now used by many. The word "cannabis" has tended to replace the word "marijuana" because the latter was (and is) considered by many in the medical and commercial communities to be a pejorative term, much better replaced by the seemingly more scientific and respectable former word. However, the term "marijuana" remains preferred for some audiences (especially recreational users).

"Hemp" is a widespread term applied to dozens of plant species employed as sources of fiber, but particularly to forms of *C. sativa* with limited THC. In the past, "hemp" and "marijuana" have been occasionally interpreted as synonyms. In recent times, hemp plants have become important sources of oilseed and CBD. When hemp is grown for oilseed, it is distinguished as "oilseed hemp" or "hempseed." The key phrase that has been used to distinguish plants authorized for non-euphoric drug uses (both fiber and oilseed) is "industrial hemp." This phrase is employed in Canada and other jurisdictions to refer to plants with no more than 0.3% THC in the dried female reproductive parts, and such plants have been treated legally much less harshly than marijuana plants. However, the term "cannabis" can overlap the marijuana and hemp categories. For example, in Canada, hemp plants employed for CBD extraction are governed by "cannabis" legislation. Accordingly, in consulting literature, legislation and regulations, it is important not to be misled by how these words are being employed.

Until recently, purified CBD has been one of the products produced in controlled environments, but it can be generated much more economically outdoors in field production. Accordingly, this form of "cannabis" is now infrequently commercially generated in controlled environments. For the most part, controlled environment production is based on: 1) strains with cannabinoid profiles that are mostly or almost completely THC (such strains tend to dominate the marketplace); and 2) strains with substantial CBD as well as substantial THC. Occasionally, strains with high concentrations of other non-inebriating cannabinoids than CBD are grown in controlled environments for medical or experimental purposes.

Just what legislative terminology applies to given biotypes of *C. sativa* and associated products often requires expert evaluation. A given plant biotype or one or more of its products may or may not be designated by the terms clarified here. Moreover, regardless of the appropriateness of these terms, independent legislation may apply. In Canada, for example, there is special legislation concerned with "novel organisms," and genetically engineered *Cannabis* falls into this category.

3.2 EVOLUTION OF *CANNABIS SATIVA*

Cannabis sativa is an example of a plant that has been domesticated, i.e., changed genetically by humans to increase its production of valued products. It exists both in the form of wild-growing populations and different selections maintained in cultivation. Moreover, for thousands of years, the different kinds have been distributed over much of the world and genetic interchange has occurred to generate innumerable intergrading variations. The situation is comparable to dogs, in which there are many selected breeds, countless intergrading mongrels and numerous feral populations. Specialists in domesticated plant evolution and classification treat such "crop-weed complexes" in several ways, so there are different nomenclatural systems any one of which may be acceptable, provided that the basis of the names is understood. Unfortunately, much that has been written about the classification of *C. sativa* has been confused and confusing. The presentation in this review is simplified.

The original wild geographical range of *C. sativa* has not been determined with certainly, but was somewhere in central Asia (McPartland et al. 2019). Whether any of the plants growing there at present represent the ancient kinds is uncertain, because of the likelihood of introgression from nearby domesticated plants. *Cannabis sativa* is one of the oldest of crops, and was domesticated thousands of years ago. As illustrated in Figure 3.1, it was separately domesticated in Europe and northern Asia (particularly China), virtually exclusively for fiber, and just occasionally for its edible seeds. Hybridization between the European and Chinese kinds has occurred, but there remain landraces or cultivars that represent the genetic distinctiveness of the two kinds. Regardless, these northern kinds, as well as related wild-growing plants, produce quite limited THC. As discussed later, the fiber selections are particularly unsuitable for breeding drug strains, but the relatively few oilseed selections have some value for this purpose.

Much more pertinent for the potential breeding of cannabis strains, and as also illustrated in Figure 3.1, in parallel to the northern selection of two geographical hemp groups, two southern drug

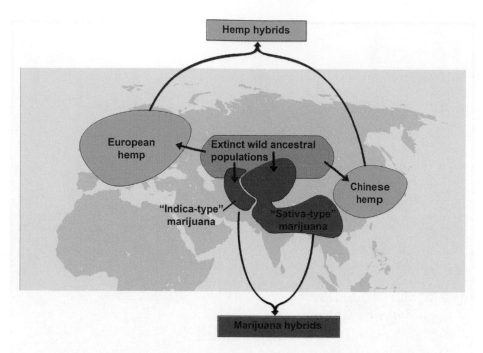

FIGURE 3.1   Approximate postulated geographical locations of ancestral, pre-domesticated *Cannabis sativa* (in orange) and the four principal groups (two groups of low-THC hemp, in green, two groups of marijuana, in red) domesticated more than a millennium ago, and subsequently transported to other parts of the world. Hybridization, mostly during the last century and to a considerable degree in the United States and Europe, has obscured differences between the two hemp groups and between the two marijuana groups.

groups were selected. The history of these two drug groups is complex, but provides critical guidance to understanding the genetics and breeding of cannabis.

For at least the past millennium, cannabis consumption became more firmly entrenched in southern Asia than anywhere else in the world. In Afghanistan and adjacent areas, a class of highly domesticated drug landraces was selected, which coincide with what has come to be known as the "indica-type." By contrast, the kind of significantly less domesticated (albeit more intoxicating) drug plant that was grown in the rest of Asia has come to be known as the "sativa-type." The two kinds are contrasted in Figure 3.2 and Table 3.1. The breeding significance of the indica-type class is discussed later.

Over the last several centuries, the sativa-type has been distributed to much of the world. For outdoor growth, it was often considered superior to the indica-type in producing larger plants with a higher relative content of THC. The domesticated indica-type was extremely poorly adapted to wild existence, while the domesticated sativa-type strains are far hardier and not particularly different from the wild-growing plants growing throughout southern Asia. Not surprisingly, the sativa-type requires much less care in many outdoor circumstances. Aside from the different alleged properties of the indica-type and sativa-type categories, the simple fact that the latter was very widespread geographically, while the former was relatively restricted geographically, doubtless facilitated the much greater historical distribution of the sativa-type throughout the world.

Today, what are advertised as sativa-type strains dominate the cannabis market (both legal and illegal). However, these are often hybrids between the two classes. Most materials (both plants and drug preparations) recognized as sativa-type have been identified as such simply and merely by having a cannabinoid profile with only THC, or sometimes also with only a limited amount of CBD. Conversely, most materials identified as indica-type have been identified as such simply because the cannabinoids include appreciable CBD in addition to substantial THC.

FIGURE 3.2 Interpretations of the ancient indica-type (bottom) and sativa-type (top) cannabis plants.

Source: Prepared by B. Flahey and B. Brookes

TABLE 3.1

Alleged Differences between the Two Ancient Sativa-Type and Indica-Type Classes of Domesticated Marijuana Plants

Group	Sativa-Type	Indica-Type
Early distribution area (see Figure 3.1)	Widespread (southern Asia)	Restricted (Afghanistan, Pakistan, northwestern India and adjacent regions)
Seasonal adaptation	Relatively long (late-maturing), often in semi-tropical regions	Relatively short (early-maturing), adapted to relatively cool, arid regions
Height (under optimal growth conditions)	Relatively tall (2–4 m)	Relatively short (1–2 m)
Habit	Diffusely branched (longer internodes); less dense, more elongated buds	Bushy (short internodes), often conical; very dense, more compact buds
Leaflet width	Leaflets narrow	Leaflets broad
Intensity of leaf color	Leaves lighter green	Leaves dark green
Aroma (i.e., odor and "taste")	Relatively pleasant aroma (often described as "sweet")	Relatively poorer aroma (sometimes described as "sour" and "acrid")
Ease of detachment of heads from secretory glands	Variable	Easily detached
Presence of CBD	Little or no CBD	Substantial CBD
Alleged psychological effects	Relative euphoric: a "cerebral high" promoting energy and creative thought (occasionally panic attacks in inexperienced users, or a drained feeling)	Relatively sedative: physically relaxing, producing lethargy

As pointed out later, the original indica-type represents a syndrome of characteristics that are far superior in most respects for controlled environment production. Tragically, in the latter half of the 20th century, the indica-type was virtually hybridized out of existence by "genetic swamping" (also known as "genetic pollution"; see, for example, Todesco et al. 2016). This phenomenon involves extensive hybridization between a widespread population or species and a much rarer population or species. The result is that the rarer species fails to leave progeny that is "uncontaminated" by the genes of the common species, and so becomes extinct. By contrast, while the more common species may acquire some genes of the rarer group, most of its progeny remain pure. In Asia, particularly in the middle of the 20th century, Westerners introduced the sativa-type class into the range of the indica-type, where spontaneous hybridization genetically swamped the latter. Indica-type material imported into the Western world was extensively hybridized with the sativa-type by underground breeders to generate new strains. The indica-type was viewed as less desirable than the hybrids, and so was not preserved. Unfortunately, when crops are domesticated, they often lose important ancestral genes, and there is evidence that this has happened with respect to cannabinoid-determining genes (Mudge et al. 2018).

3.3 SCIENTIFIC NOMENCLATURE OF GROUPS

The following presentation of terminology is limited to what is required for this chapter. Scientific names for the significant classes of *Cannabis* are listed in Table 3.2, but are avoided in the discussion, for simplicity. Those interested in the complexities of *Cannabis* evolution and classification may consult Clarke and Merlin (2013, 2016), Small (1979a, 1979b, 2017a, 2017b), and McPartland and Small (2020). Because of extensive (indeed, worldwide) hybridization and intergradation among

TABLE 3.2

The Most Important Scientific Names for *Cannabis*

Scientific Name	Circumscription (Interpretation)
C. sativa	The only species of the genus *Cannabis*; includes all plants of *Cannabis*—those used for drugs, other purposes and wild plants
C. sativa subsp. *sativa*	Plants with low (< 0.3%) THC +THC acid, dry weight basis in female inflorescence; includes plants grown as "hemp" and their wild relatives
C. sativa subsp. *indica*	Plants with higher (> 0.3%) THC +THC acid, dry weight basis in female inflorescence; includes plants grown as "marijuana" and their wild relatives
C. sativa subsp. *indica* var. *indica*[1]	Domesticated forms of sativa-type[2] plants
C. sativa subsp. *indica* var. *himalayensis*	Wild forms of sativa-type plants
C. sativa subsp. *indica* var. *afghanica*	Domesticated forms of indica-type plants
C. sativa subsp. *indica* var. *asperrima*	Wild forms of indica-type plants
C. indica	Merely a synonym of *C. sativa* subsp. *indica*
C. ruderalis	Merely wild *C. sativa* subsp. *sativa* (= *C. sativa* subsp. *sativa* var. *spontanea*)

Notes: [1,2] Readers may notice that the scientific name, *C. sativa* subsp. *indica* var. *indica* seems inconsistent with the vernacular name "sativa-type" (shouldn't "var. *indica*" [i.e., variety *indica*] be "indica-type," not "sativa-type"?). The explanation is that the phrases "sativa-type" and "indica-type" arose in the illicit marijuana trade, in ignorance of the naming conventions for scientific names (McPartland 2017). Similarly, one might intuitively deduce that "sativa-type" (which designates the class of *C. sativa* with the highest levels of THC) should correspond with the scientific name *C. sativa* subsp. *sativa* var. *sativa*, but in fact the latter name refers to domesticated forms of hemp, the class with the lowest levels of THC. Unfortunately the vernacular usages of "sativa" and "indica," which are inconsistent with scientific (i.e., Latin) usages, have been the cause of misinterpretations by many authors, but are too established to change. To avoid confusion in this presentation, the vernacular names are employed.

plants of *Cannabis*, classification of groups within the genus and identification of individuals are very difficult.

3.4 THE MEANING OF SOME NON-FORMAL (NON-LATINIZED) GENETIC CATEGORIES

3.4.1 BIOTYPES

A biotype is a group of organisms which share a unique genotype. The concept is sometimes relaxed to mean a group of organisms which share a unique subset of the genotype. The word biotype is used frequently in preference to "strain" in this review because most so-called cannabis strains are unreliable and, as explained in what follows, the strain category is problematic. In biology, the term biotype is often applied to groups which differ genetically only to a very limited extent (although the phenotypic differences may be important). Clones of *Cannabis* are maintained as unique genotypes, and so are biotypes in the narrow sense of the word. Plants generated from cannabis seeds invariably produce at least somewhat different genotypes, but the population may share a unique subset of alleles and so may be a biotype in the wide sense of the term.

3.4.2 CULTIGENS

The word cultigen has been used in various senses to include: 1) the entire set of domesticated plants (excluding the wild plants) of a species; 2) a species that exists only in a domesticated form; 3) distinctive cultivated groups within a species (thus oilseed plants, fiber plants and THC-rich plants can be considered cultigens); and 4) a synonym of cultivar (a mistaken usage).

3.4.3 LANDRACES

Many cultivated plants of *Cannabis* are "landraces" (also spelled land races)—populations domesticated in a locale, typically over long historical periods, substantially by unconscious (non-planned, undeliberate) selection by traditional farmers. These are usually adapted to local stresses, and are often much more variable than modern cultivars. In numerous crops, landraces have provided the raw materials from which cultivars have been selected. Unfortunately, landraces of drug forms of *C. sativa* have been vanishing rapidly and there is little official effort underway to preserve them.

3.4.4 CULTIVARS

Botanists employ Latinized nomenclature to describe plants in various formal categories, such as species and subspecies (Turland et al. 2018). However, since the middle of the 20th century, domesticated selections of plants termed "cultivars," which satisfy certain descriptive and publication requirements, have been the subject of a special—at least partly non-Latinized—code of nomenclature (Brickell et al. 2016). Article 2.3 provides the following definition: "A cultivar is an assemblage of plants that (a) has been selected for a particular character or combination of characters, (b) is distinct, uniform, and stable in these characters, and (c) when propagated by appropriate means, retains those characters." Article 9.1, Note 1 restricts the meaning of cultivar as follows: "No assemblage of plants can be regarded as a cultivar . . . until its category, name, and circumscription has [sic] been published." Webster's Third New International Dictionary (Gove and the Merriam-Webster Editorial Staff 1981, p. 552) provides a more general definition of a cultivar: "an organism of a kind (as a variety, strain, or race) that has originated and persisted under cultivation." Cultivars as defined by the code for cultivated plants can be of quite different nature (e.g., they may be hybrids, clones, grafts [i.e., combinations of species], chimeras [with genetically different tissues] or even plants that are distinct simply because they are infected by a microorganism), but frequently many of the cultivars within a given species differ very little genetically from each other. There are more than 100 recognized cultivars of non-intoxicating forms of *Cannabis* currently grown for fiber and/or oilseed. Only a handful of cannabis strains bred for authorized medicinal usage at present are regarded as cultivars under the code.

3.4.5 STRAINS

There are thousands of so-called strains of marijuana plants (or at least allegedly different strains) that are currently circulated in the black, gray and medicinal marijuana trades (e.g., Figure 3.3). Some authors treat strains as equivalent to cultivars (e.g., Snoeijer 2002), but this is based on confusion between the technical and vernacular (dictionary) meaning of the word cultivar, as previously explained. Article 2.2 of the nomenclatural code for cultivated plants (Brickell et al. 2016) specifically forbids the use of the term "strain" as equivalent to "cultivar" for the purpose of formal recognition. Moreover, this code demands that a number of requirements be satisfied before biotypes can be officially accepted as cultivars, particularly with respect to publication of descriptions. Very few cannabis strains satisfy the descriptive requirements for cultivar recognition, although many *Cannabis* cultivars (mostly grown for fiber or oilseed rather than cannabinoids) do (and by convention are denoted in single quotes: e.g., *C. sativa* 'Debbie'). Some cannabis strains are conceptually identical to *Cannabis* cultivars, and hopefully with the growing medical and commercial importance of cannabis strains, an effort will be made to account for them as adequately as currently done for other domesticated plants. However, unless and until drug selections are professionally recognized as cultivars in the same manner as other important plants, their status is likely to remain chaotic and uncertain. Currently, strain names are highly unreliable (Sawler et al. 2015; Reimann-Philipp et al. 2019; Russo 2019; Schwabe and Mcglaughlin 2019).

FIGURE 3.3 Examples of named strains marketed commercially: the menu of Coffeeshop Smokey, Amsterdam.

Source: Photo by Dominic Milton Trott/*The Drug Users Bible* (CC BY 2.0)

3.4.6 CHEMOVARS

A chemovar (short for chemical variety) is a category defined on the basis of one or a set of chemical constituents. In the case of *C. sativa*, the word is normally employed with specific reference to cannabinoids and/or terpenes to refer to a group of plants distinguishable from other plants by one or more compounds. The plants in an individual chemovar may be genetically related, but not necessarily. Recognition of chemovars can be a way of avoiding biological classification issues. For example, stating that a given chemovar is defined by presence of more than 99% THC in the resin does not require reference to scientific classifications.

3.4.7 CHEMOTYPES

The word chemotype is sometimes employed simply as a synonym of chemovar. However, on the basis of examination of cannabinoid profiles of hundreds of populations, Small and Beckstead (1973a, 1973b) defined the following three principal groups: Chemotype I (drug type), with high THC and low CBD/THC content ratio; Chemotype II (an intermediate group); and Chemotype III (hemp type), with little or no THC and usually a high CBD/THC ratio. Many publications still utilize the three-chemotype scheme, although it has limitations. It has been demonstrated that these three main chemotypes can arise by segregation at a locus (B) within individual F2 progenies of divergent parentage. As noted later in the discussion of cannabinoid genetics, the current interpretation of these chemotypes is that their inheritance is based upon the occurrence, at the B locus, of two codominant alleles, BD and BT, responsible for the presence of CBD and THC, respectively. Other authors have proposed additional chemotypes: Chemotype IV with CBG as the predominant cannabinoid, but also CBD; and chemotype V, with no detectable cannabinoids.

3.4.8 BREEDING LINES

A breeding line is a set of plants bred to become homozygous (genetically uniform) for a particular trait. This facilitates transferring the trait, as needed, to other plants. Professional breeders and breeding companies often guard breeding lines as private intellectual property because of their value.

3.5 HOW TO RECOGNIZE (AND GENERALLY AVOID) RUDERAL (WILD) MATERIAL

Almost all crop plants have been greatly changed genetically from their wild ancestors, in order to minimize the characteristics of the latter which are deleterious or unwanted in cultivation. The wild ancestors frequently maintain genes (alleles) not retained in the domesticates, and so are a source of genes that can be useful for breeding new cultivars, especially for resistance to diseases, herbivores and environmental stresses. Nevertheless, for the most part, professional plant breeders usually use plants that are already domesticated as starting breeding material. They do so because domesticated material already has many desirable genes, whereas wild plants have many undesirable genes which can greatly multiply the work needed to create new cultivars. (Note that Wolter et al. [2019] argue that modern gene editing can much more efficiently transform wild plants into cultivars than can traditional breeding techniques.) This is not to deny the value of wild plants as germplasm, but simply to indicate that for the most part, domesticated material is preferred. Also, since the breeding value of wild germplasm is primarily for resistance to biotic and environmental stresses (which are more substantial and difficult to control outdoors), the wild plants are of much more limited value for controlled environment facilities. So-called *C. ruderalis* (usually merely a designation of northern European wild plants of *C. sativa*) has been used to create autoflowering strains, and this is probably the chief example of employing wild plants in breeding *Cannabis*. Wild plants in North America (mostly the result of escapes from fiber landraces in past centuries) are commonly called "ditchweed," reflecting the lack of respect they are afforded. The scientific classification of wild forms of sativa-type and indica-type cannabis plants is discussed in McPartland and Small (2020; cf. Table 3.2).

In *Cannabis sativa*, the degree of "wildness" is well indicated by several characteristics of the seeds (technically, the so-called seeds are fruits called achenes), as detailed in Naraine et al. (2020) and illustrated in Figure 3.4. To survive, the wild-growing plants must be able to distribute their seeds, and this is reflected by an attenuated seed base which facilitates release of the seeds at maturity; conversely, the seed base of domesticated plants is short and blunt. Wild seeds are also usually smaller than domesticated seeds, generally less than 3.8 mm in length. Wild seeds also have a better-developed camouflagic (spotted and streaked) coating (to avoid being eaten by birds and insects), whereas this layer tends to fall away irregularly in domesticated seeds. Wild seeds have evolved dormancy, i.e., they will not germinate for some time—and when they do so, it is often very sporadically, which makes it difficult to grow them.

3.6 SOURCES OF GERMPLASM FOR BREEDING CANNABIS

"Germplasm" is material that can be used to reproduce or propagate organisms. Plant germplasm is the starting material employed to create new, desirable varieties, so it is critical that the kinds and limitations of available germplasm be appreciated. The need for collecting and preserving *Cannabis* germplasm is outlined by Welling et al. (2015). Backer et al. (2019) reviewed factors that can increase yield of drug forms of *C. sativa*, and noted the importance of analyzing germplasm for the purpose of breeding more productive biotypes.

3.6.1 FIBER HEMP GERMPLASM

Hemp fiber is harvested from the main stalk, hence fiber biotypes are tall—usually over 2 m (Figure 3.5, left). Since the stem nodes tend to disrupt the length of the fiber bundles, thereby

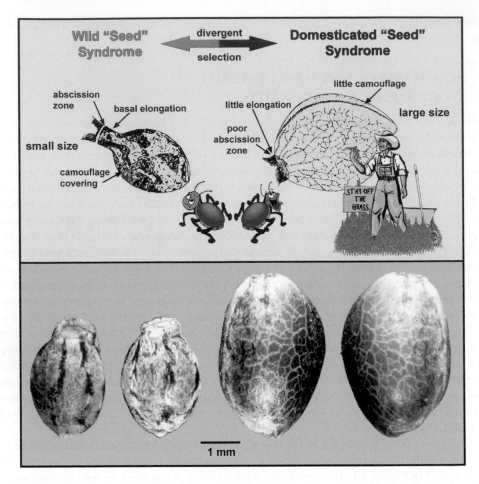

FIGURE 3.4 Seeds (achenes) of wild and domesticated *Cannabis sativa*. Top: Divergent selection for adaptive seed characters between wild and domesticated plants. Bottom left: Wild seeds. Bottom right: Domesticated seeds. The site of attachment of the seeds to the plant is uppermost. Note that the wild seeds are smaller, have a camouflagic persistent covering layer and have an elongated attachment base that facilitates disarticulation.

Sources: Top, prepared by B. Brookes; bottom, photos by E. Small

limiting quality, plants with long internodes have been favored. These plants also often have low genetic propensity for flower production. Also, fiber landraces and cultivars are low in THC, and indeed, relatively low in cannabinoids. These characteristics are very undesirable for cannabis strains intended for controlled environments. Unfortunately from the point of view of breeding new cannabis strains, fiber landraces and cultivars constitute most of the world's lawfully available *Cannabis* germplasm (both in public and private hands).

3.6.2 Oilseed Hemp Germplasm

The seeds of *C. sativa* have been employed for millennia for food and other purposes, but only to a minor degree. Accordingly, there is far less representation of germplasm specialized for oilseed than for fiber. However, in the early 21st century, the hempseed industry has become much more profitable than the fiber hemp industry, and several cultivars have been created that are strikingly similar to the indica-type class of drug plants. Similarly, these hempseed cultivars develop low stature, very compact branching and production of a dense main flowering axis (Figure 3.5, right). As

FIGURE 3.5 The two traditional classes of hemp plants. Left: Tall fiber cultivar 'Petera'. As noted in the text, fiber cultivars are very unsuitable as germplasm sources for breeding drug strains. Right: The oilseed cultivar 'FINOLA', the first modern hemp cultivar developed exclusively for grain. The low stature, autoflowering, limited branching and large compact inflorescence on a single reproductive axis are all characters that have been useful for breeding new hemp oilseed and CBD varieties, and have also been useful for breeding high-THC strains for controlled environments. The breeder, J. C. Callaway, is shown.

Sources: Left photo courtesy of Anndrea Hermann; right photo by Anita Hemmilä, Finola Ltd., permission to reproduce provided by both

detailed by Small (2018), this parallelism is due to the fact that both kinds of plant are designed to maximize flower production—for cannabinoids in the drug form, and for seeds in the oilseed form. The oilseed cultivars produce very little THC, but the fact that they have other very desirable characters means that they deserve to be considered in breeding new drug strains. Seeds of such oilseed cultivars can be purchased (regulations need to be followed).

3.6.3 HIGH-THC GERMPLASM

Most important crops were greatly improved by ancient farmers long before breeding was recognized as a science, and this is so for *C. sativa*. The early evolution of intoxicating plants in southern Asia was detailed earlier. Landraces were distributed and established in Africa and Southeast Asia, and more recently in the Americas, where new local landraces were selected. With the explosion of interest in marijuana that began in the 1960s, clandestine breeders in Western nations created a wide range of strains based on the original Asian landraces. Ironically, law enforcement pressure for decades had the unintended effects of: 1) driving marijuana production indoors where it is harder to locate; 2) increasing knowledge of the technology of indoor production; 3) creating new strains; and 4) increasing potency. Since the early 1970s, cannabis quality and yield efficiency were greatly improved by clandestine breeders and cultivators, especially in the Netherlands and North America (Vanhove 2014). Breeding generated strains that are more productive, faster-maturing, hardier and more attractive to consumers. The cultivation of elite female clones and the use of indoor production techniques that hide plants from the authorities (typically in bedrooms, basements, attics, closets, garages or sheds) was particularly emphasized (Figure 3.6). Growers became able to harvest up to six crops annually, with much greater or faster growth in smaller spaces than achieved previously. It is important to understand that the selection of strains for growth indoors substantially produced biotypes suited for controlled environment production. Notably in the Netherlands and the United

FIGURE 3.6 Short-stature marijuana plants are shown being grown surreptitiously in a confined height-limited indoor space. Illicit breeders have selected such compact strains to avoid detection. Coincidentally, such strains often are well adapted to controlled environments and may constitute very suitable germplasm for breeding.

Source: U.S. government photo (public domain)

FIGURE 3.7 Example prizes for breeding superior cannabis strains. Left: Awards table at the 27th *High Times* Cannabis Cup ceremony in 2014 in Amsterdam. Right: "Dope Cup, Oregon."

Sources: Left, photo by Cannabis Culture (DCC BY 2.0); right photo from WeedStreetWear.com (public domain)

States, there have been competitions for the most impressive strains (Figure 3.7). Numerous strains are currently circulating in the illicit trade, and some of these have been transformed to legality and adopted by the medical and recreational sectors in some nations. Thus, an impressive range of germplasm variability exists for marijuana forms of *C. sativa*, albeit mostly in a state of illegality. Unfortunately, these strains are of very uncertain genetic composition, particularly because of haphazard hybridization by amateur breeders. (See Rahn et al. [2016] for a list of over 600 strains). Even more unfortunate, the original landraces have similarly been subjected to extensive hybridization in recent years, as well as being reduced by law enforcement. High-THC plants still living in the

FIGURE 3.8 High-THC plants growing in the Old World, which represent potentially useful germplasm for breeding. Left: A cannabis landrace growing in the Rif region of northern Morocco. Right: Ruderal (wild-growing, sativa-type) *Cannabis sativa* growing in the Parvati Valley, northern India.

Sources: Left photo by Michel Benoist (CC BY SA 3.0); right photo by Narender Sharma, Blue Particle Solutions (CC BY SA 4.0)

southern Asia (Figure 3.8 right) or northern Africa (Figure 3.8 left), harbor invaluable genes, and are in urgent need of preservation as germplasm.

3.6.4 Public Genebanks

All advanced countries have national genebanks (alternatively spelled gene banks)—mostly collections of seeds maintained in a viable state. A few valuable perennial crops (such as potatoes and orchard trees) are maintained as clones. Most of the larger genebanks are dedicated to crops, and of these, the U.S. National Plant Germplasm System (NPGS) under the control of the U.S. Department of Agriculture is the world's best, and has recently undertaken collection of germplasm of *C. sativa*. Modern public genebanks are wonderful sources of germplasm because they conduct several key activities: 1) they ensure that acquired material is authoritatively identified and collection information is recorded; 2) they examine seeds for purity and viability; and 3) they preserve material under appropriate conditions, regenerating seeds every few years to ensure continuing viability. Frequently, there are associated research programs, often involving characterization of the accessions. Normally, genebanks distribute material to each other and to researchers and breeders. Those who receive seeds are usually required to sign agreements requiring sharing of rights regarding commercial exploitation of the material, and also often limiting distribution of the seeds, as well as adhering to international agreements. Unfortunately, almost all of the representation of *C. sativa* in the world's genebanks is of low-THC material, i.e., hemp landraces and wild collections, which is of limited interest to most breeding of new cannabis strains. Regrettably, governments remain extremely reluctant to invest in high-THC germplasm conservation (Torkamaneh and Jones 2021). Rarely, collections of marijuana strains have been made by governmental organizations for research, and law enforcement purposes, most notably by the National Institute of Health at the University of Mississippi.

3.6.5 Commercial Sources

High-THC germplasm (mostly seeds, occasionally cloned plants) is available from the private sector (Figure 3.9 and Figure 3.10). However, the majority of the most interesting cannabis variants are available only from illegal (black market) or quasi-legal sources (a "gray market" is one that is

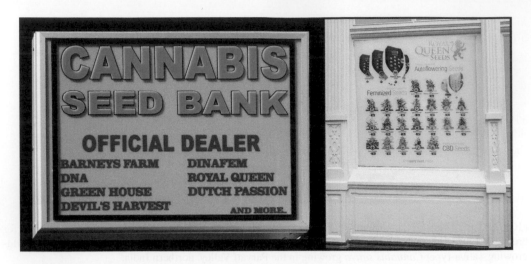

FIGURE 3.9   Store fronts of so-called "seed banks" in Amsterdam.

Sources: Left photo by Anne Jea (CC BY SA 4.0); right photo by Jay Galvin (CC BY 2.0)

FIGURE 3.10   Examples of commercial cannabis germplasm sources. Left: seeds on sale in Amsterdam. Right: Cloned young plants for sale.

Sources: Left photo by Cireremarc (CC BY 2.0); right photo by Harborside Health Centre, California (CC BY SA 2.0)

unauthorized but not significantly discouraged). Some private companies supply seeds and/or seedlings of named strains, both to the public and to authorized professional growers, but generally not when the material is considered to have very high potential value as breeding germplasm, in which case limitations on supplied seeds may be specified. A very restricted range of clonal tissue cultures is currently available from some commercial sources. Tissue cultures represent a potential way of preserving male and autoflowering germplasm that can be difficult to maintain as clones.

An online search of "genebank" turns up a large number of sites maintained by cannabis seed sellers (Figure 3.9), and some private research collections. The largest commercial operations (notably the Sensi Seed Bank) are in the Netherlands. These are not genebanks in the traditional sense, and represent a distortion of how the word has been used. These so-called cannabis genebanks are almost all commercial marijuana seed sellers. The Leafly website (URL depends on country) provides information for thousands of alleged strains, and seed sources for many. There are significant problems with such private sector suppliers. In most jurisdictions, they are considered to be illegal and obtaining seeds from them is a criminal activity, albeit progressively tolerated. Experience has shown that governmental and non-profit institutions are far more likely to conserve significant germplasm, and to do so more competently (better maintenance of genetic purity, better documentation) by comparison with the private sector. The private sector is motivated primarily by the profit potential of its property, and accordingly wishes to keep the most valuable seed stocks out of the hands of competitors, so their most valuable germplasm is almost impossible to obtain. Seeds from private companies is more often than not unreliably documented, of uncertain status (they are almost always hybrids of unknown origin), and often genetically variable or strongly inbred. Feminized seed and autoflowering material are particularly likely to have complex hybrid backgrounds. Companies engage in exaggerated claims in order to inflate the value of the material they market, so (unlike the data from legitimate genebanks) the information provided should not be assumed to be accurate.

3.7   MANAGING A PRIVATE OR PERSONAL GERMPLASM COLLECTION

Whether the objective is breeding, research or production, and whether on a small or commercial scale, there is a need to maintain the starting germplasm for further activities, at least temporarily. Readers are reminded that local laws and regulations need to be followed.

3.7.1   SEEDS

National seed banks have the mandate to store seeds indefinitely (at least for centuries). All seeds lose viability over time, even under controlled storage, and periodic regeneration (typically once every 1–3 decades) to grow a new batch of seeds is very expensive. The advice given here is for short-term storage (1–3 decades), which does not require regeneration. The longevity of *Cannabis* seeds is known to decrease fairly rapidly, to about 75% after two years of storage in a sheltered but otherwise uncontrolled climate. Small and Brookes (2012) experimentally examined the interactions of temperature, humidity and an oxygen-free environment, as they affected seed longevity of *C. sativa*. Progressive lowering of the temperature (from 20°C to −80°C) increased seed longevity; so did progressive lowering of moisture content (from 11% to 4%). A high moisture content (ca. 10%) at room temperature results in fungi attacking and killing the seeds within 18 months. Either reducing the temperature to at least 5°C or reducing the seed moisture content to at least 6% has a huge beneficial effect on maintaining seed viability. Additional reduction of temperature—but not additional reduction of moisture content—has a small supplementary beneficial effect.

The following seed storage procedures are recommended:

- Seed moisture content can be determined as specified by the International Seed Testing Association (www.seedtest.org/en/home.html). In practice, measuring the weight of a sample of seeds before and shortly after drying in a convection oven at 103°C for at least six hours suffices. To avoid sacrificing valuable seeds, substituting a sample of other seeds of *C. sativa* (even dead seeds) that have been stored under the same conditions will usually suffice. Hobbyists are unlikely to have a lab balance appropriate for small seed samples, and can omit this step.
- For storing *C. sativa* seeds in a viable state for up to three years, the moisture content should be limited. Measuring moisture content with small samples is tedious and often impractical, so simply spread them out (single layer) in a well-ventilated warm room for several weeks. High and prolonged heat will greatly shorten viability. Put the dried seeds

in a standard refrigerator (ca. 5°C), in an airtight container in which desiccant is placed to absorb moisture from the seeds and keep them dry. (A desiccant with an indicator that changes color when it becomes damp is recommended.) Individual seed collections can be kept in several small glass jars with the lids perforated, all in a single large glass container in which desiccant is also placed in a similar small glass jar.

- For storing *C. sativa* seeds in a viable state for at least a decade, follow the previously stated procedures but employ a standard freezer (typically reaching −20°C). This should be fitted with an alarm which is triggered should the temperature become excessive.
- For periods of much more than a decade of storage before seed is regenerated, ultra-low-temperature freezers are recommended (these are costly).

3.7.2 Vegetative (Clonal) Cultures

Vegetative reproduction is the only way of guaranteeing that an outstanding biotype, once created or discovered, will be preserved. A given plant can sometimes be maintained for years by a long-day photoperiod, but many photoperiodic plants will eventually flower regardless of light regime (they can still be reproduced by cuttings). Autoflowering biotypes cannot be controlled this way. Removing flower buds can prolong the period before renewal is necessary by cuttings. Most strains are in fact maintained clonally, but the resources and labor necessary are significant.

Some advanced micropropagation technologies are being employed by large commercial operations instead of cuttings to establish new growth, and these cultures represent a means of preserving biotypes (Vassilevska-Ivanova 2019; Monthony et al. 2021). ("Micropropagation" is essentially the "test tube" culture of plant cells, tissues, organs and young plants in aseptic artificial media.) Living material can be preserved as continuously propagated cell cultures (i.e., they are maintained as single cells or proliferating cells not organized into tissues) or as tissue cultures, which can be employed to grow innumerable identical plantlets. Similarly, animal tissues (principally semen, ova [unfertilized eggs] and embryos) of very valuable livestock are now often stored cryogenically in liquid nitrogen, and some large genebanks today similarly conserve apical meristems (growing points or "buds") of some species. Even pollen grains can be maintained long-term under cold storage. Most private firms guard information concerning biotechnological methods for *C. sativa* as intellectual property. However, the U.S. Department of Agriculture has supported research into the conservation of marijuana clones as shoot cultures (Lata et al. 2012). Genetically identical plantlets can be generated rapidly using tissue culture. Pieces of tissue culture have been encapsulated to form "artificial" or "synthetic" seeds, sometimes called "synseeds" (Chandra et al. 2013). These are essentially very young plantlets packaged with fertilizer and water in a gel, and can be planted just like real seeds to produce new plants.

3.8 TRADITIONAL PLANT BREEDING TECHNIQUES AND THEIR APPLICATION TO CANNABIS

Plant breeding is the deliberate creation of new plant biotypes which are superior in one or more desired respects. Generally, the new kinds need to be more competitive or profitable commercially than current offerings, but (especially for ornamental plants) breeders are sometimes content to just create novelties. The underlying genetics of plants are often so complex that only professional plant breeders are competent to avoid critical problems. Nevertheless, amateur or hobbyist breeders— since the mid-20th century mostly conducting their activities under danger of imprisonment—have succeeded in generating a remarkable range of cannabis kinds—and indeed, the fruits of their labor include most of the strains now being grown legally in controlled environments for medical and recreational purposes. There are innumerable guides to cannabis breeding online and in popular cannabis books, and they do not go much beyond providing elementary advice on hybridization and selection. Much more detailed information about breeding is available for hemp (e.g., Ranalli 2004;

Salentijn et al. 2015), some of which is applicable to breeding drug strains, but not much of which pertains to controlled environments. Professional plant breeders typically employ one or more of the following traditional procedures (which in some cases overlap).

3.8.1 INBREEDING

The appearance of a plant with especially desirable feature(s) has for millennia stirred the desire to perpetuate it, but the progeny of sexual organisms naturally diverge from their parents. Inbreeding (restricting reproduction to a very small population with desirable features, or selfing a plant if it tolerates self-fertilization) generates progeny that are especially similar to their parents, including the desired features. This is often a short-term measure, frequently associated with the production of relatively weak plants. Unfortunately, it has been extensively employed by clandestine breeders, with the resulting generation of many clones that are excessively homozygous (genetically uniform and consequently often susceptible to stresses and physiological problems). Careful use of inbreeding to fix characters requires planning and cost/benefit balance of the consequences.

3.8.2 MASS SELECTION

Mass selection (or recurrent mass selection) is probably the oldest and most widely practiced legitimate method of selection for outbreeding crops like *C. sativa*. The seeds of what appear to be relatively desirable plants from a large population are used to propagate the next season's crop, with the result that, over several years, the characteristics judged to have merit tend to become established and uniform. Landraces have evolved in this manner. Continued mass selection is employed to maintain the purity of registered cultivars ("off-types" are rogued away) as, otherwise, domesticated varieties tend to re-acquire wild characteristics.

3.8.3 INDIVIDUAL SELECTION

Individual selection concentrates on evaluating the progeny from a small number of elite plants (those that appear to be champions with respect to desired traits). Once the superior progeny have been identified, they can be used as breeding stock (involving controlled pollination) to generate a supply of plants with superior traits. Because most of the characteristics of value in *C. sativa* can only be evaluated after flowering has occurred, careful records need to be kept so that the breeder will know which seeds to keep and which to discard.

3.8.4 PEDIGREE (COMBINATION) BREEDING

So-called combination or pedigree breeding emphasizes the creation of cultivars by transfer of single genes or gene combinations, achieved by hybridization, backcrossing and selection.

3.8.5 HYBRIDIZATION/SELECTION

Hybridization followed by selection has long been the major tool of plant breeding to increase yield, vigor and uniformity. Indeed, for thousands of years, selection is the basic way that humans have altered organisms to make them more useful. Deliberate hybridization is only about three centuries old. For the professional breeding of most plants (especially crops and ornamentals), the first step is to generate considerable variation (i.e., many diverse plants), which may involve many parents in the initial hybridization, and the planting of many seeds (F1s and backcrosses) so that plants with desired characters can be selected. This can be a slow process (often taking decades), and it requires considerable acreage (usually outdoors) to grow hundreds—or preferably thousands—of plants. This is how new cultivars of hemp continue to be produced, and there is likelihood that in the future, when field cultivation of cannabis strains becomes more popular, they will be produced in the same professional way. To date, most cannabis strains have been produced illicitly, precluding growing large numbers of plants, which would attract law enforcement.

"Hybrid cultivars" are not merely cultivars that are the result of past hybridization between different biotypes (probably all cultivars of *C. sativa* hold this status); as explained in the following, they are the result of a specific marriage of parental strains or cultivars, and must be generated anew each season. They are reminiscent of mules, the progeny of horses and donkeys, renowned for their endurance that exceeds that of their parents but themselves virtually unable to produce offspring. In modern crop hybrid cultivar breeding, the parents can be deliberately inbred, so that when they are mated, their genetic interaction in the F1 hybrid is especially desirable. Compatible combinations generally need to be determined by trial and error. When the plants within an F1 hybrid population are allowed to interbreed, genes determining their superior characteristics segregate and recombine and are so variable that they are of limited advantage to farmers. Accordingly, farmers must purchase certified hybrid seed every growing season from the breeder, who produces the seed by repeatedly crossing a given set of parents. Actually, "pseudohybrid" cultivars of hemp, many of them produced in France, are F2 hybrid populations. Several popular hybrid hemp cultivars are mostly monoecious. These are produced by crossing female dioecious hemp (males are rogued out of the field) pollinated by monoecious hemp.

3.8.6   Heterosis Breeding

Heterosis breeding is a combination of hybridization and selection that takes advantage of particularly desirable genetic partners. Hybridizing distantly related organisms tends to produce heterosis (hybrid vigor). This is the opposite of inbreeding depression (often related to harmful recessive genes being combined from both parents). In heterosis, an F1 hybrid tends to have an increase in some desirable characteristics (such as height, productivity or disease resistance) compared to the mean of the same traits of its parents. Whether hybrid vigor is achieved depends strongly on the interaction of genetic factors from the parents. Up to the present for *Cannabis*, heterosis breeding has been a trial-and-error exercise conducted for hemp, and the generation of hybrid seed requires considerable field plantings. This methodology seems to be very unlikely for controlled environment production, given the present success of clonal production.

3.8.7   Mutation Breeding

Mutation breeding is based on exposing seeds to chemicals or radiation to generate mutants with desirable traits. If a mutated plant is produced, it can be hybridized with normal plants to transfer the altered gene, or simply maintained as a clone. Although most mutations are fatal or undesirable, occasional mutants may be useful. Despite the need to grow and evaluate very large numbers of plants, many cultivars (especially of ornamentals) have been created by this methodology. The technique does not seem to have been used to generate any commercial variety of *C. sativa*.

3.8.8   Polyploidization

Doubling the number of chromosomes of a crop plant has sometimes been found to make it more productive. *Cannabis sativa* normally has a somatic (diploid) number of 20 chromosomes. Doubling of chromosome number is most frequently achieved by exposing meristems (growing points) to a mitotic toxin. (Mitosis is the normal way that cells divide in two, each new cell receiving a full complement of chromosomes). The toxin, usually colchicine, prevents normal division, so that the new cell receives all of the new chromosomes that normally would have been divided between two cells. Tetraploid selections of hemp (i.e., with 40 chromosomes) have been generated frequently, but there are no commercial cultivars. There are also no reports of tetraploid drug forms of *Cannabis* developing more desirable characteristics than their corresponding diploid forms, and polyploidy does not seem to be a useful technique for improving drug aspects (Mansouri and Bagheri 2017; Parsons et al. 2019).

3.9 MOLECULAR (DNA-BASED) PLANT BREEDING TECHNIQUES AND THEIR APPLICATION TO CANNABIS

See Campbell (2019) for an extensive review of the state of molecular breeding techniques applied to *C. sativa*.

3.9.1 MARKER-ASSISTED SELECTION

"Marker-assisted selection" is based on selecting a "marker" that is linked to a trait that a breeder wishes to select. The markers may be morphological, physiological or chemical (especially protein), but are commonly DNA tags identifying particular locations in the genome (see QTL analysis, described next). This indirect selection process can dramatically improve the efficiency of selecting plants with desirable gene combinations. Markers can be employed to transfer single genes or to follow the inheritance of many genes. Genetic markers in *C. sativa* have been found for sex determination (Prentout et al. 2020a, 2020b; Toth et al. 2020). Marker-assisted selection is considered to have excellent potential for *C. sativa* (Pacifico et al. 2006; Nadeem et al. 2018). QTL (quantitative trait locus) analysis examines the relationship between expressed traits and their genetic coding (usually DNA markers). Often times, many QTLs influence a trait, but there is sometimes a simple determining relationship. QTL analysis has great value for breeding crops. QTL marker analysis is reviewed for crops in general by Collard et al. (2005) and Xu et al. (2017). The technique has been utilized only to a limit extent for *Cannabis*, and primarily for hemp—not drug—strains (Campbell 2019).

3.9.2 GENETIC ENGINEERING

Cultivars of many crops today are transgenic, the result of DNA recombination. In transgenic plants (more generally, "genetically modified organisms"), a desired "transgene" (coding for a useful trait) has been identified, isolated, cloned and inserted. Techniques for introducing genes from quite unrelated organisms into given species have been developing since the 1980s. There have been remarkable successes in creating several new kinds of recombinants, although genetically engineered organisms are controversial. An important practical aspect of genetic engineering is the clarification of the genes of a plant under study, and this can lead to manipulation of the genome so that the plant becomes more useful. Modern techniques have made it possible to prepare detailed genetic maps, and in many cases, some of the specific functions of mapped genes have been clarified. In the case of *Cannabis*, this work is in its infancy. In the future, genetic engineering of *Cannabis* may greatly facilitate the production of plants with extremely well-defined characteristics that are very desirable for purposes of industry, medicine and law enforcement (de Meijer 2014). Genetic engineering of *C. sativa* could be relevant to the plant's size, longevity, growth rate, chemical characteristics and tolerance to stresses. In various nations, genetically engineered plants are regulated so as to prevent the possibility of genetically contaminating related crops. In the case of *C. sativa*, pollen from one genetically engineered biotype of a drug, oilseed or fiber class of plant could alter strains or cultivars of the same or the other classes. Possibly more than any other economic crop, because *C. sativa* is heavily regulated and because its pollen is extraordinarily widely distributed by wind, genetically modified forms are likely to encounter societal resistance. However, genetically engineered drug forms that are maintained as non-flowering clones in a controlled environment should be less of a concern, as pollen escape is almost impossible.

3.9.3 GENOMIC EDITING

Gene editing is the use of specialized enzymes to make precise changes in DNA that affect specific traits. The CRISPR/Cas 9 (Clustered Regularly Interspaced Short Palindromic Repeats/CRISPR Associated Protein) system of gene editing has been especially useful for breeding new crops (Wolter et al. 2019). Tailoring the gene affecting a single trait could create a novel biotype that could not be bred by traditional techniques. Conceivably, such a simple change could greatly increase suitability for controlled environments (Folta 2019). Documentation of the genome of *C. sativa* (Van Bakel et al. 2011) stimulated interest in exploiting its genes. Vergara et al. (2017) review genomic information resources for *C. sativa*, and achievements in genome analysis of *C. sativa* are reviewed by Backer et al. (2020), Schilling et al. (2020), Welling et al. (2020), and Hurgobin et al. (2021). The most obvious target for genomic editing is the cannabinoid biosynthetic pathway(s). Some researchers "are aiming to modify chemical synthesis in the cannabis plant by genetically altering its cells to make the desired molecules from shoot to tip, thereby boosting yield" (Dolgin 2019a, p. S5). However, this seems problematical, as the cannabinoids are cytotoxic to plant tissues (Sirikantaramas et al. 2005), which appears to be why they are confined to the secretory trichomes. Commercial companies are claiming successful achievements in creating genomically altered biotypes (Dolgin 2019b).

3.10 KEY ISSUES, PROBLEMS AND CONSIDERATIONS IN CANNABIS BREEDING

The traditional techniques noted previously are routinely carried out in breeding fiber and oilseed cultivars of *C. sativa*, and the DNA-based methods are beginning to be applied. However, the application of both conventional and innovative plant breeding techniques to drug forms is problematical from the viewpoint of hobbyist or amateur breeders. Folta (2019, p. 10) stated:

> Plant breeding requires significant investment in screening genotypes, either in identifying parents to cross or in the analysis of offspring. These activities require access to significant materials, space, labor inputs and other resources, and those may not be realistic in selecting plants specifically for controlled environments.

Indeed, professional crop breeding programs typically extend over at least a decade—often much longer—and usually require cultivation of thousands of plants in large outdoor plots and/or greenhouses, extensive labor and specialized laboratory equipment. The authorized professional development of new drug strains is occurring primarily in the context of very large projects funded by major venture capitalists—especially in biotechnology centers, sometimes associated with university research programs. Continuing restrictions on cannabis make it very difficult for small-scale research and/or development, so it should be understood that the opportunities to engage in sanctioned cannabis breeding at the professional level are very limited.

Amateur breeders commonly let their enthusiasm for (or against) a single characteristic blind them to the necessity of breeding for overall performance. The enormous variation exhibited by dog breeds is instructive. In some cases, breeding has occurred for some particular feature—large size, for example—often with some associated undesirable result, such as shortened life, susceptibility to diseases or behavioral problems. In contrast, breeding for performance (such as hunting, swimming, speed or retrieving) is necessarily based on many interacting considerations (such as intelligence, heart function and lung capacity). In some cases, it may be justified to concentrate on a single character for cannabis, even at the cost of simultaneously acquiring heritable problems. There may be a desire for a rare cannabinoid, for example, and if this is discovered to occur in very high amount in some plants of a population, it could be justified to select/breed for this, at least until the character is stabilized. Professional plant breeders, however, are intimately familiar with the need to address interacting characteristics, albeit the goal may be to optimize a particular trait. It is particularly necessary to be aware that concomitant or pleiotropic deleterious effects may result when selecting for a trait. In evaluating the desirability of the characteristics of given plants, careful observation based on experience should be exercised. It is much cheaper to grow plants outdoors than in controlled environments (in locations where it is legal to do so), and many characteristics are equally valuable for both outdoor and indoor plants, although some are not. Often, very young plants suffice to evaluate critical characteristics, and seedlings can be raised relatively economically indoors.

Aside from facilities to grow plants, assessing cannabinoid content of experimental plants is unavoidable in breeding drug forms of *Cannabis*. Traditionally (and usually illegally), samples of the plants were simply smoked. High-performance liquid chromatography (HPLC) has become the gold standard for determining cannabinoid profiles. Access to the equipment or the cooperation of a laboratory is necessary.

As discussed earlier, some classes of plant should be avoided when breeding cannabis strains, particularly fiber and wild germplasm. Feminized seeds will very likely produce sexually unstable plants, and should be avoided for breeding. As a rule, much less breeding effort is required if one starts with materials that already develops the desired characteristics, and in the cannabis seed world, this means either obtaining seeds from (or tracing to) the Amsterdam market or from technology companies. High-THC Asian and African landraces have been genetically contaminated, as described earlier, but seeds of some have recently become available in the seed trade and deserve to be incorporated in modern breeding efforts.

Assessing heterozygosity can be difficult, but even without sophisticated genetic analysis, progress is possible. *Cannabis* seeds are the result of hybridization normally involving genetically different parents, and they produce plants with genetic representation of both recent parents. They (or any of the possible backcross progeny) can be tested for performance, and the best preserved as clones. However, unless the nature of the seed source is clear (and it usually is not), using seeds of undetermined background is a substantial gamble. Less of a gamble, the most desirable starting breeding materials of *C. sativa* are female clones, especially those that are widely grown. Many of these are outstanding in some desired respects, and indeed they are often hybrids exhibiting heterosis. Since all plants established from a clone share the same genetic makeup, they are extremely uniform in performance. The main disadvantage is that genetic uniformity makes clones very susceptible to the possibility that a new (mutant) disease can become disastrously effective—but should this occur in the case of *C. sativa*, there are many other clones that are likely resistant. Clones can often be induced to produce male flowers (more discussions follow), and the resulting pollen employed to hybridize a different clone. The resulting seeds can be grown, and the plants analyzed for performance, and if justified, new, unique clones produced. While the results are unpredictable, this is a method for generating new clones with considerable heterozygosity. In all cases, generating a novel cannabis strain is not the end of a breeding project. The stability and reliability of performance need to be tested.

3.11 MANAGEMENT OF SEX IN CANNABIS BREEDING

3.11.1 FLORAL STRUCTURE

Cannabis sativa is one of the small minority (ca. 6%) of flowering plants with male flowers (Figure 3.11) and female flowers (Figure 3.12) normally on separate plants. Such plants are termed "dioecious." However, many modern hemp cultivars have been bred to be "monoecious" (with both male flowers and female flowers on the same plants). Before flowering commences, male and female plants cannot be reliably distinguished by appearance, although males tend to be slimmer, taller and more delicate. So-called "pre-flowers" are flowers that often develop first (before inflorescence flowers), in the axils (the crotches where leaves or branches arise) of stem nodes. Genetic markers have been identified which can serve to identify the sex of seedlings and young vegetative plants

FIGURE 3.11 Male plants and their flowers. A: A well-developed plant. B: Flowering branch with male flowers shedding yellow pollen. C: Scanning electron microscope photo showing copious release of pollen from an anther (dehiscence is longitudinal, i.e., along the long axis). Note the line of secretory trichomes along the length of the anther. D: Scanning electron microscope photo showing viable (hydrated) pollen grain (left) and inviable (dehydrated) grain (right).

Sources: Photo A by E. Small; Photo B (public domain) by Erik Fendersson; Photos C–D by E. Small & T. Antle

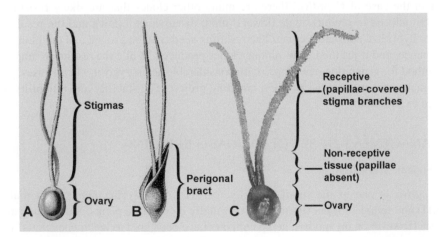

FIGURE 3.12 Female flowers. A: Diagram (public domain) with the perigonal bract removed. B: Diagram (public domain) showing a female flower surrounded by perigonal bract. The perigonal bracts contain the majority of the THC of the buds. C: A live flower

Sources: Diagrams A–B public domain; Photo C by E. Small

(Prentout et al. 2020a, 2020b). A single male flower can produce hundreds of thousands of pollen grains (Figure 3.11C), and larger male plants can bear hundreds of flowers. Typically, the male plants in a population come into flower up to three weeks earlier than the females—an adaptation that promotes outbreeding, but because of overlap in flowering times of males and females, some self-pollination can occur. The male plants die once they shed their pollen. Female flowers occur in clusters which are smaller and relatively obscure compared to those of males. The tiny female flowers consist of a one-compartment ovary and a short apical style with two long thread-like stigmatic branches (Figure 3.12). A perigonal bract (sometimes called a floral bract) subtends each female flower (Figure 3.12B) and grows to envelop the single-seeded fruit that is produced should the flower be fertilized.

3.11.2 GENETIC AND ENVIRONMENTAL CONTROL OF SEX EXPRESSION

Inheritance of sexual expression in *Cannabis* has been studied extensively but is not completely understood. Sexual differentiation in dioecious strains is based on a pair of sex-determining chromosomes, the male being XY, the Y chromosome frequently described as larger. The system produces an approximately 50:50 sex ratio. Sex expression appears to be somewhat determined autosomally (i.e., by chromosomes other than the X and Y), with an X/autosome dosage–type chromosome system (i.e., genes on the autosomes cumulatively influence sex development). Barcaccia et al. (2020) reviewed what is known about the allelic control of sex. Punja and Holmes (2020) speculated that sex determination in *C. sativa* may also be influenced by retrotransposons (a class of mobile DNA sequences that is often widely distributed in genomes).

Whether established from seeds or from clonal stock, it is critical that when commercial cannabis plants flower, they produce only female flowers, since the pollen from any male flowers that develop can fertilize the female flowers, resulting in seed development and the reduction of product quality. Rarely, female plants will spontaneously produce some male flowers, and this may be a fault of particular strains. Environmental stresses during development can cause male flowers to develop on female plants, in which case the problem can be controlled by regulating culture conditions.

3.11.3 MASCULINIZING FEMALE PLANTS TO PROMOTE INBREEDING BY SELF-FERTILIZATION

Sex development can be influenced by hormonal treatments (auxins and ethylene feminize *Cannabis*, whereas gibberellins are masculinizing). The growth regulator Ethrel (with active ingredient ethephon, also has other names) favors the development of female flowers on male plants. Marijuana growers have sprayed Ethrel on young plants to feminize (one might say emasculate) them as much as possible, although it is much simpler to just employ clones. More importantly, male flowers can be induced on female plants by chemical treatments (Mohan Ram and Sett 1982; Lubell and Brand 2018), facilitating the production of feminized seeds, discussed next. However, some biotypes are relatively resistant to being induced to develop flowers of the opposite sex, and some produce pollen of reduced viability (DiMatteo et al. 2020).

3.11.4 PRODUCTION OF FEMINIZED SEED AND FEMALE-PREDOMINANT PLANTS

For well over a century, it has been known that selfing a normally female plant—using pollen from occasional spontaneous male flowers appearing on it or from male flowers induced to develop by manipulating the environment—gives rise to seeds producing only female plants. More detailed methods of seeds feminization will be discussed in Chapter 4. In recent years, the marijuana seed supply industry has employed this (or an analogous) technique to produce "feminized seed." Such seeds produce entirely or mostly female plants, eliminating the need to remove males. Although the resulting plants are extremely inbred, when based on a female plant with exceptionally attractive properties, they will be similar to the mother plant (although not as similar as cloned plants are to their maternal plant). Feminized seed is widely sold online. Feminized seed often produces plants with some male or intersexual flowers, which are very undesirable in breeding programs. As mentioned previously, feminized seed should be avoided in breeding new strains.

3.11.5 DETERMINING THE BREEDING VALUE OF MALES

Cannabis clones, which are females, are the standard material of controlled environment production, so their features are well-evaluated. Although sexually reproducing kinds are available, breeders strongly prefer to begin the creation of new strains by starting with one or more established clonal strains. It is always a gamble to take pollen (whether from sexually reproducing kinds or from clones forced to produce male flowers) and generate hybrid seed using a female clone. Biotechnology may offer ways of predicting successful combinations, but at present, the major way of determining the breeding value of males is progeny testing—that is, by trial and error. The value (technically, the gene combinatorial value) of given males is evaluated by the performance of the progeny (not just the F1 hybrid, but subsequent generations). This necessitates growing out and evaluating large (indeed, very large) numbers of plants, which is generally well outside the capacity of hobbyist breeders. Moreover, while a given plant generated by such a program may turn out to be invaluable and will be preserved as a clone, the original male plant has very likely died and it cannot be further employed (although it is possible to maintain male clones). Despite this discouraging situation, common sense suggests that characters that are valuable in the female clone should be sought in male plants chosen as pollen donors. In the next subsection, such characters are discussed in detail for breeding new female clones for controlled environments, and several of these characters are also apparent in male plants.

3.11.6 HOW TO FERTILIZE FEMALE FLOWERS WITH POLLEN

Fresh pollen is best collected from vigorously growing male plants. Once anthesis (anthers opening to disperse the pollen, as shown in Figure 3.11C) occurs, *Cannabis* pollen gradually loses viability. Initially, the pollen grains may be 90% or more viable. Under good conditions (dry weather), viability after three days typically is reduced to about 50%, and to about 15% after a week (Choudhary et al. 2014). The appearance of pollen under a microscope indicates its viability (note Figure 3.11D, which shows a fresh and a dehydrated pollen grain), and biological stains can also be employed for this purpose.

Transferring pollen from a male plant to the female flower of another plant is simple; ensuring that pollen from other plants do not contaminate the receiving female plant is very difficult. Prodigious quantities of pollen are produced by male plants, and the slightest breeze carries the pollen away, to land on surfaces both nearby and far away. Hemp seed producers in Europe and Canada are required to isolate their outdoor plants by a distance of up to 5 km from other outdoor *Cannabis* plants, in order to produce purebred (pedigreed) seeds (Small and Antle 2003). Commercial controlled growth facilities employ HEPA (high efficient particle air) filters and other technologies not just to exclude pathogenic spores, which can be less than 1 μm (micron) in size, but also to keep pollen (20–25 μm in diameter; Halbritter and Heigl 2020) from contaminating the plants as they are flowering. Also, commercial operations have strict hygienic protocols to keep personnel from importing pollen (and pathogens) to the growth facilities. In a small or personal facility, it is important to keep unwanted pollen-shedding plants well away, to minimize visitors who may have contacted pollen-shedding plants recently and to personally ensure that one is not unintentionally carrying pollen. A change of clothing—which easily carries pollen—may be in order. Fans are universally employed to circulate air in growth facilities, but when plants are being deliberately pollinated, a windy environment is undesirable.

Cannabis clones are female, but some clones are prone to occasionally developing male flowers (Punja and Holmes 2020), so when hybridizing plants, it is important to check that unwanted pollen is not being generated that could interfere with the activities. Note that *Cannabis* clones being maintained in a vegetative state by long-day photoperiods can produce solitary flowers (usually but not always female) on the stems (Spitzer-Rimon et al. 2019), which are easily overlooked. *Cannabis* is not parthenocarpic (producing seeds without fertilization), so if seeds develop, one can be confident that they are the result of pollination. As a check that unwanted pollen has not contaminated a breeding site, a virgin flowering female plant can be kept at the site to ensure it develops no seeds. Also, branches or buds of the experimental plant which was pollinated can be protected temporarily by covering them with a bag for a week (longer coverage can be detrimental). Different kinds of bags for pollinating plants are available from various sources (and are advertised online). Some cultivators spray female inflorescences with water (which will kill pollen) to prepare them for hybridization, but this can induce fungal diseases.

When a female clone is induced to produce male flowers, the pollen may be in very short supply and often much of it is sterile or abnormal, so it is advisable to mark/label the branches with flowers that received the pollen, to facilitate checking on seed development.

Hybridizing two plants is easily accomplished when male flowers of one of them are shedding pollen at the same time as the female flowers of the other has receptive stigmas. Females tend to be in best condition for pollination 2–3 weeks after the first flowers develop. Figure 3.13 shows stigmas at ripe and over-ripe stages. Fertile stigmas are turgid, usually white or off-white (some strains have colored stigmas). Withered stigmas are typically rust-colored or brown. Flowering time of most cannabis strains is determined by photoperiod, which can be controlled in order to coordinate the timing of maturation of pollen of one population with the maturation of stigmas of a second population. Note that to promote outcrossing, male plants of a given population tend to come into flower 1–3 weeks before female plants have receptive stigmas. Most photoperiodic strains can be forced to come into flower within several weeks when exposed to a 12-hour light/12-hour dark regime, but some require light periods as short as nine hours, while a few will flower when the light period is as long as 14 hours. Experimentation may be required to determine the required regimes when hybridizing strains that behave differently. Hybridizing a photoperiodic plant with a day-neutral plant may

FIGURE 3.13 Buds (unfertilized, congested portions of female inflorescences) at appropriate mature and over-mature stage for pollination. Left: Strain Purple Erkle. The whitish appearance of the stigmas indicate that many are receptive to being fertilized by pollen. Right: Strain Bullrider. The brownish appearance of the stigmas indicate that they are over-mature for pollen reception.

Sources: Photos by Psychonaught, released into the public domain

similarly require separate plantings. If fertilization is successful, mature seeds will be available in 4–8 weeks.

Cannabis pollen can be collected, and stored in a relatively viable condition—either for a short or a long period. Branches shedding pollen are shaken over wax or parchment paper until a pile of material is accumulated, forceps/tweezers (or a coarse screen) are employed to get rid of the larger pieces of twigs, leaves, and anthers, and the material is strained (with a fine screen or even just a kitchen strainer). It is then dried (e.g., for two days at room temperature under no or limited light at a low humidity—e.g., 30%), and placed in an airtight (and preferably lightproof) vial. Some hobbyists simply add a desiccation packet from a commercial product to the vial to control moisture. The vial can be kept in a refrigerator (up to several weeks) or a freezer (up to a year). To extract a sample, the container should first be allowed to come to room temperature (otherwise, moisture will condense on the cold pollen grains, killing them). Repeatedly taking the material out of and placing it back into the cold will reduce its viability. Professional preservation of cannabis pollen is also possible by cryopreservation (Gaudet et al. 2020).

Physically transferring pollen can be accomplished in several ways (Clarke [1981] provides basic advice). Moving an entire flowering male plant into the confined space where the female is located is inadvisable, because of the likelihood that pollen will be released in large amounts in the room. (Marijuana seed producers often simply place a male plant in flower among a group of female plants in flower in a room and turn on a fan.) A single flowering branch in a bag is preferable. Flowering branches can be collected from remote locations and maintained (like cut ornamental flowers) with the base in water for several days. Small artist paintbrushes are widely employed to carry samples of pollen to flowering buds. The brushes can be sterilized with alcohol, but immersing in water is sufficient to kill any pollen remaining from previous activities. Hobbyists sometimes employ a "pollinating bladder"—a small rubber bulb with a small exit—which, when loaded with some pollen and squeezed, disperses pollen efficiently over an entire plant or just one bud. Cut-down turkey basters can be used for the purpose.

Safety note: Some people are allergic to pollen. Wearing a respirator (and goggles if very sensitive) may be necessary to avoid pulmonary reactions. Wearing latex or nitrile gloves may be required to avoid dermatitis.

3.12 DIFFERENT NEEDS OF FIELD AND CONTROLLED ENVIRONMENT STRAINS

Aside from *Cannabis*, most commercial plants grown indoors are raised in greenhouses. Most of the crops are high-value ornamentals and vegetables. Also aside from *Cannabis*, in facilities with complete environmental control, there is also some production of vegetables, as well as medicinals (mostly proteins) in genetically engineered plants (mostly tobacco). Generally, only some cultivars of particular species are recommended specifically for indoor growth, and most cultivars are unsuitable or less suitable. In some cases, there has been breeding specifically for controlled growth circumstances. Indeed, as has been described, clandestine marijuana breeders have been doing this for decades. More professionally, it is important to appreciate how the open field conditions and controlled environments provide different stresses that limit production, and accordingly to breed for features that promote efficient growth specifically in controlled environments.

As illustrated in Figure 3.14, plants cultivated in fields—albeit far more protected from environmental stresses than wild plants—must nevertheless withstand both the extremes and the fluctuations of climate (insolation, temperature, wind, rain) and biotic attack (animal pests, weeds, infective microorganisms). In the field situation, there is some possible control of soil water and nutritional status, and of harmful organisms, but far superior control of these and the other stress factors occurs in controlled environments. Folta (2019, p. 6) insightfully concluded "Breeding for controlled environments shifts the focus to a completely different set of plant traits, such as rapid

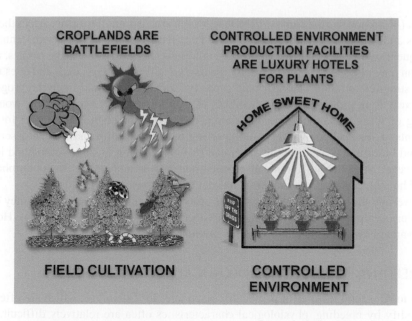

FIGURE 3.14 A contrast of growing cannabis plants in field culture (left) and in a controlled environment (right). In the field, the plants must withstand a wide range of environmental and biotic stresses, and strains suitable for such cultivation necessarily have characteristics that make them less suitable for controlled environments.

Source: Prepared by B. Brookes

growth, performance in low light environments and active manipulation of plant stature." The following sections address these and related subjects.

3.13 BREEDING FOR RAPIDITY OF RESPONSE TO SHORT DAYS; AUTOFLOWERING

Maximizing commercial production in controlled environments requires plants that terminate the vegetative growth phase quickly and rapidly produce a substantial amount of inflorescence. To be competitive commercially, strains need to be harvested in four months or less. Time of flowering is critical to the survival of populations of *C. sativa* grown outdoors, so that both wild and domesticated plants that have grown for many years in a region may be expected to have been selected for an appropriate local flowering time. This limits where the plants can be cultivated outdoors in other locations, but not in controlled environments, since flowering time can be regulated. Potter (2014, p. 35) reported that "a survey of 200 cannabis varieties shows that 88% had a recommended growth period in short day length of 7 to 9 weeks." Populations originating in areas closer to the equator are often naturally programmed to respond more slowly to short days, but this is a slight limitation.

A minority of biotypes of *C. sativa* appear to be programmed to come into flower because of intrinsic developmental reasons, rather than in response to photoperiod. These often have originated at the extremes of the geographical range of *C. sativa*, either at their northernmost locations of survival or near the equator. Biotypes that are indifferent to photoperiod are technically termed day-neutral, more commonly "autoflowering" in the marijuana literature. Day-neutral wild races of *C. sativa* have evolved in the northernmost regions of Eurasia, where they are sometimes termed "*C. ruderalis*." This is adaptive, because the season is so short that the plants might not have time to mature seeds if they relied on photoperiod. Moreover, there are domesticated northern hemp plants of *C. sativa* (best known is the cultivar FINOLA) which are also autoflowering. All of these are low-THC forms and produce relatively small plants. They have been hybridized with cannabis strains and subsequent selections employed to produce autoflowering clones. Near the tropics, autoflowering selection may also have evolved, because the day/night difference is small and does not provide much of a stimulus to retain photoperiodic capacity. Such plants would likely be programmed to grow in a very long growing season regardless of photoperiod, and because they would demand a very long growing season to come into flower, they would be of little use either for controlled environments or for field cultivation. Autoflowering marijuana strains have been described in the underground literature as having been generated by hybridization of short-season and long-season plants. However, such hybridization can produce odd and unpredictable effects on photoperiodic response. The inheritance of autoflowering has not been extensively studied.

Autoflowering strains can be very useful for growing marijuana indoors, since they can be cultivated under continuous light, which induces faster, greater growth (Potter 2014). However, the relative advantages of growing plants under continuous lights requires evaluation.

3.14 BREEDING FOR KEY PHYSIOLOGICAL CHARACTERISTICS

In contrast to structural features of plants (discussed later for *Cannabis*), which are often relatively easy to modify by breeding, physiological characteristics often are relatively difficult. Moreover, controlled environments may often compensate for physiological limitations of given varieties, reducing the need for breeding.

3.14.1 Disease Resistance

Breeders need to keep in mind that *Cannabis* plants are subject to attack by microorganisms (bacteria, fungi, viruses). Bactericides and fungicides are usually not authorized for cannabis, and in any event are considered undesirable because of possible residues. The indica-type marijuana group originated from arid areas as found in Afghanistan and western Turkmenistan, and when strains from this region are grown in high-humidity climates, their dense flowering tops retain moisture and can succumb to "bud mold" caused by *Botrytis cinerea* and *Trichothecium roseum* (McPartland et al. 2000). Unfortunately, since dense buds are considered extremely desirable, they are especially subject to fungi. Control of humidity (which is a significant cost factor) is essential. The danger of fungal spores and bacteria in cannabis products is reduced by the standard practice of irradiation. However, breeding for physiological resistance to microorganisms deserves consideration. For more detailed discussions, please read Chapter 8.

3.14.2 Invertebrate Pest Resistance

Cannabis is quite resistant to insects and other small pests, and controlled growth facilitates their exclusion. Nevertheless, some pernicious greenhouse pests—such as aphids, spider mites, thrips, and whiteflies—are difficult to eradicate once they manage to colonize growth facilities. Biocontrol measures using beneficial insects and other invertebrate predators (such as nematodes and predatory mites) that consume or parasitize invaders, and some organic insecticides, are preferable to employing synthetic pesticides. It should be possible to produce marijuana plants without using any biocides in controlled environments, so that breeding for resistance is not a high priority. For more detailed discussions, please read Chapter 9.

3.14.3 Temperature Tolerance

Optimal temperatures for photosynthesis (which is not necessarily reflective of best temperatures for growth) vary from 25°C to 30°C, depending on variety. Chandra et al. (2011) did not find evidence of differences between drug and fiber varieties, although the former usually originate from hotter home-lands. The growing popularity of LED lighting in controlled environments, which generates significantly less heat than high-intensity discharge lamps, lowers the need for strains to be able to tolerate heat.

3.15 BREEDING FOR SECRETORY GLANDULAR TRICHOMES

3.15.1 BASIC BACKGROUND INFORMATION

Most plant species have very small epidermal appendages termed "trichomes" on the aerial parts, widely considered to be protective against pathogens and arthropod herbivores (Wagner 1991). Trichomes are sometimes termed "hairs," because they are often hair-like, but most biologists reserve the term hair for animals. About 30% of flowering plants possess "glandular trichomes," producing secondary chemicals, usually at the tip of the structure, often in distinctive head-like containers. The substances manufactured are frequently known to serve the plant as protective agents, but are also often immensely useful to humans as natural pesticides, food additives, fragrances and pharmaceuticals. The cannabinoids are produced in specialized tiny secretory trichomes which are almost always multicellular. Classes of glandular trichomes in *Cannabis* have been recognized based on stalk length and size. The so-called stalkless or sessile type, which hardly resembles a hair-like structure, may have a very short stalk that is not visible as it is hidden under the gland head. The stalkless type is often comparatively small, and tends to be common on the leaves. Stalked glands (Figure 3.15), distinguished arbitrarily as long-stalked or short-stalked, tend to be

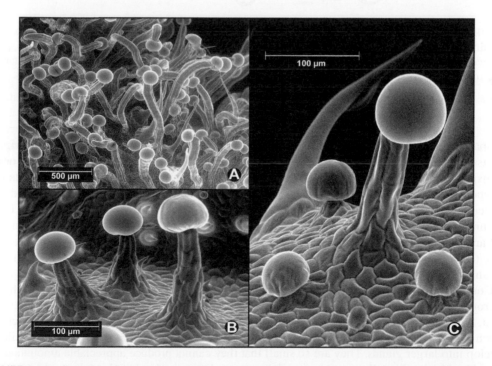

FIGURE 3.15 Scanning electron micrographs of secretory glands of the abaxial (lower or outer) epidermis of perigonal bracts (i.e., the single bract covering each female flower) of high-THC forms of *Cannabis sativa*. (A). Dense concentration of long-stalked glands. (B). Three long-stalked glands. (C). A long-stalked secretory gland (center) around which are three short-stalked multicellular glands. Also shown is a non-glandular hair (a unicellular structure). Resin containing cannabinoids is synthesized in the spherical heads of the glandular trichomes. The perigonal bracts are the most intoxicating plant organ of high-THC forms of the plant.

Source: Prepared by E. Small and T. Antle

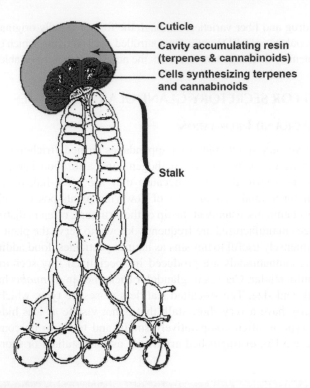

FIGURE 3.16 Diagram of longitudinal section of a long-stalked secretory gland of *Cannabis sativa*. Resin containing cannabinoids is synthesized by the cells in the basal part (shown in red) of the more or less spherical head, and accumulates in the cavity (shown in green) above these cells within the external membrane covering the head. Sometimes the head breaks open and the resin seeps over the adjacent plant tissues.

Source: Adapted from Briosi and Tognini (1894); prepared by B. Brookes

comparatively large and are most noticeable on the abaxial (lower or outer) surfaces of the perigonal bracts and sugar leaves (discussed later). The essential part of the stalked glands is a more or less hemispherical head, sometimes compared in size to the head of a pin. Inside the head at its base are specialized secretory "disk cells," and above these is a non-cellular cavity where secreted resin is accumulated, enlarging the covering sheath (a waxy cuticle) of the head into a spherical blister (Figure 3.16). Most, if not all, of the plant's cannabinoids are synthesized in the secretory disk cells and are confined to the trichomes. (Sirikantaramas et al. (2005) found that cannabinoids are cytotoxic to plant tissues, so it is unlikely that they are located elsewhere.) The resin is a sticky mixture of cannabinoids and a variety of terpenes.

Comparatively small glands with very small heads (sometimes unicellular, typically less than 20 microns in diameter) and very short stalks (sometimes just two cells) often occur over much of the plant. Such glands are often termed "bulbous," in contrast to the larger glands previously described, which are termed "capitate" (meaning head-like). These small glands could simply have failed to develop into larger glands. They are so small that they cannot produce appreciable amounts of the cannabinoids. Additionally, it may be noted that a very distinctive kind of glandular trichome occurs on the male flower's anthers (Figure 3.11C).

3.15.2 BREEDING FOR GLANDULAR TRICHOME DENSITY AND SIZE ON THE PERIGONAL BRACTS

Small and Naraine (2016) reviewed evidence that there has been selection for larger trichome secretory gland heads, as well as greater density of these glands, in the perigonal bracts of some of the most potent cannabis strains. Since most of the plants' cannabinoids are in the stalked secretory glands, and the highest concentrations of these occur on the perigonal bracts, they are an obvious breeding target. Virtually no deliberate breeding for gland size and density has been carried out for *Cannabis*, but the possibility of breeding for trichomes in crop species has been examined by Snyder and Antonious (2009), Glas et al. (2012) and Mishra et al. (2020). Note that this is a quantitative way of improving cannabinoid content, while addressing the proportion of THC in the cannabinoids is a qualitative way (addressed later).

3.15.3 BREEDING FOR SUGAR LEAF SECRETORY TRICHOMES

"Sugar leaves" are very small, frequently unifoliolate (with just one leaflet) leaves that occur among the flowers in the inflorescence. The smaller sugar leaves are richly covered with secretory glands, but they are less dense on the larger ones, and so contribute proportionately lesser amounts of cannabinoids to the buds. Indeed, many cannabis producers trim away ("manicure") the larger sugar leaves to increase the THC content (a selling criterion). This requires a great deal of labor, and trimming machines are often employed, at least to do an initial manicuring. The removed leaves are relatively high in cannabinoids, and are salvaged for extracts. It might seem that a solution to the need for trimming is to breed for reduction or elimination of the sugar leaves, but almost certainly because they are part of the bud, they provide essential photosynthate for its development. It would make more sense to breed for increased concentration and size of the glandular trichomes on the sugar leaves. This would not only eliminate the need for trimming, but would increase the cannabinoid production of the plant.

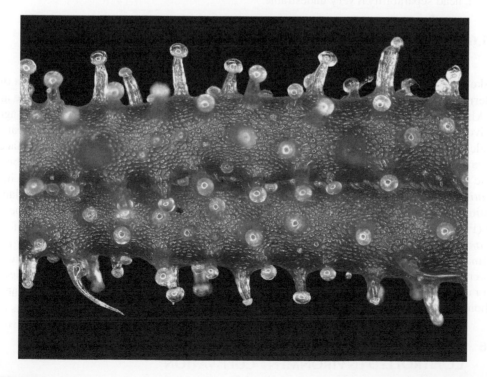

FIGURE 3.17 Edge view of a sugar leaf (small leaf within a bud) covered by long-stalked secretory glandular hairs.

Source: Photo by E. Small

3.15.4 Breeding for Leaf Secretory Trichomes

In past times, although leaves are relatively low in THC, the foliage—as well as the inflorescence—was employed to a much greater extent for drugs than accepted in today's marketplace. In theory, mutations might be found that make the foliage as well-covered with large secretory trichomes as the perigonal bracts. However, as noted earlier, the principal market for controlled environment cannabis is buds. The breeding strategy for increasing foliage cannabinoids makes sense for the cannabinoid extraction industry, which is more efficiently served by field-grown plants.

3.15.5 Breeding for Reduced Separability (Brittleness) of Glandular Trichome Heads

Most of the plant's cannabinoids occur in the heads of the stalked capitate glandular trichomes. The narrowed portion of the top of the stalk, just below the base of the head (note Figure 3.16), has been characterized as an "abscission layer" by some biologists, but this is misleading. In no way is the so-called "abscission layer" of cannabis stalked trichomes comparable to the abscission zone at the base of the foliage of deciduous trees, or at the base of fruits that abscise at maturity. Just why stalked glandular trichomes develop a constricted area just beneath the gland heads is unclear adaptively. In the living state, the gland heads always burst immediately when touched, but do not readily fall off from the living plant. When the plant is dried, however, the gland heads do fall off very readily at the constricted zone when agitated. This facilitates harvesting the heads for hashish preparation, and some strains may have been selected for ease of harvesting the heads for making intoxicating preparations. Contrarily, for preparing buds for commercial sale, this is very undesirable because the inevitable agitating that occurs when handling the buds can result in the loss of the THC-containing heads. Clearly the fragility of the stalk is an important breeding criterion and, in general, head separability is very undesirable.

3.15.6 Unicellular Trichomes: Can They Be Eliminated or Transformed into Glandular Trichomes?

Unlike the multicellular glandular (i.e., gland-tipped) trichomes discussed previously, most of the epidermal surfaces of *C. sativa* are covered by unicellular, non-glandular trichomes. These are of two kinds: cystolithic and simple, both of which are thought to be mechanical defenses against herbivores (Bar and Shtein 2019). "Cystolithic" trichomes (Figure 3.18A–C) have small particles of calcium carbonate embedded in the base, and are mostly on the adaxial ("upper") surfaces of the leaves. This feature tends to make the plant unpleasant to chew and less palatable to herbivores. These trichomes are quite stiff, and can cause skin irritation and dermatitis in people who handle *C. sativa* extensively without suitable protective clothing. Also common on much of the plant are additional unicellular hairs, simple hairs that lack the basal stony concretions and tend to be slimmer (Figure 3.18B). These are silicified, also tending to increase the unpalatability of the foliage. In *Cannabis*, such hairs are present on both surfaces of the leaves, but in young leaves, they can be dominant on the abaxial ("lower") surface. All of these hairs represent a waste product from the grower's perspective, and they probably decrease photosynthesis. Selection of mutants decreasing or eliminating these non-glandular trichomes—or preferably, transforming them into glandular trichomes—could be highly advantageous.

3.16 BREEDING FOR PLANT ARCHITECTURE FOR CONTROLLED ENVIRONMENT CULTIVATION

In 1934, the French entomologist August Magnan calculated that bee flight was aerodynamically impossible: bees should not be able to fly because of the haphazard way their wings flap. His analysis was misguided, because it assumed that bees fly on the same aviation principles that enable

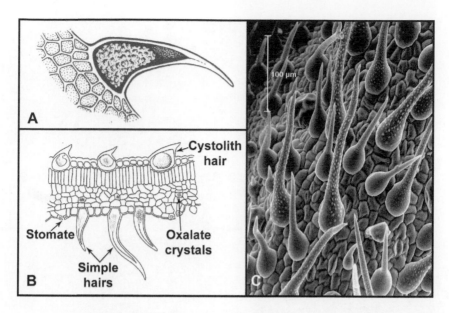

FIGURE 3.18 Non-secretory unicellular types of cannabis trichomes ("hairs"). A: Diagram of a single cystolithic hair (calcium carbonate concretions are embedded in the base). B: Section of leaf showing cystolith hairs on adaxial ("upper") surface and simple hairs on abaxial ("lower") surface. C: Scanning electron micrograph of adaxial ("upper") surface of a young leaf of *Cannabis sativa*, showing protective cystolithic trichomes arrayed in one direction.

Sources: Diagrams A–B public domain images; Photo C by E. Small and T. Antle

airplane flight. The lesson is that it is hazardous to interpret abilities from appearance. Nevertheless, there is considerable evidence in crop plants that performance is very strongly determined by plant architecture, and so structural features provide superb guides for breeding better cannabis strains. Indeed, there is a model of plant architecture that remarkably exhibits the kinds of structural features that are exceptionally appropriate for controlled environment growth facilities for cannabis plants. This is the indica-type syndrome discussed earlier, and illustrated in Figure 3.19. The architectural features of the indica-type archetype that are particularly suited to controlled environments are: limited height, strongly reduced internodes resulting in very short branches and condensed inflorescences, and large leaves (leaflets). These features are discussed in detail in Small (2018), and summarized in the following.

3.16.1 Breeding for Stem Features (Height, Branching, Internodes)

Most strains (excluding daylength insensitive kinds) can be made to grow large simply by manipulating the photoperiod to keep the plants vegetative and allowing them to grow for long periods under luxuriant conditions. Generally for controlled environments, there is little motivation to breed large plants—either very tall or wide (with strongly spreading branches). The lower leaves of tall plants may not receive sufficient light. Plants with spreading branches can make handling (tying branches to supports, trimming away damaged branches, harvesting) difficult.

Stems, including branches, are necessary to subtend leaves and flowers. However, reducing the proportion of stem tissue is a way of increasing the efficiency of crop production (i.e., increasing the "harvest index") for a desired product (unless, of course, the economic product is stem tissue). Plants with a reduced proportion of stem tissues have superior yields because some of the reserves normally dedicated to vegetative growth are redirected toward the harvestable product. As reviewed by Small (2018), increasing the harvest index by reducing stem tissue was a large part of the "Green

FIGURE 3.19 A contrast of the tall, diffusely branched "sativa-type" and the short compact "indica-type" cannabis classes, as they existed in the past. As discussed in the text, the indica-type syndrome of domesticated characteristics is consistent with the group representing an extremely advanced cultigen of *Cannabis sativa* selected in past times, and a model for selecting comparable features for controlled environment growth facilities.

Source: Prepared by B. Brookes

Revolution" in the middle of the 20th century, which produced the most significant advances in productivity of many of the world's most important grain crops. Strategically, the dwarfed plants of the Green Revolution proved to be most productively grown in high densities employing fast growing cycles. Indeed, plants with limited (or at least short) branching are naturally superior to strongly branching plants for the purpose of fully and uniformly occupying a field and maximally utilizing solar irradiation. This is obviously desirable for optimizing production, a very important goal in modern plant breeding that will become increasingly significant. This points the way to how *Cannabis* is best grown in controlled environments. The indica-type of plant (Figure 3.20) is a model of desirable size and branching characteristics.

"Dwarfism" of the plants of the Green Revolution—and indeed, of the indica-type of *Cannabis*—is not merely small size: it is a growth syndrome caused by shortening of the internodes, and is most frequently controlled by mutations controlling gibberellins (Small 2018). The mutations responsible for dwarfism have been identified in several crops, but not yet in *Cannabis*. Regardless, dwarf (i.e., indica-type) forms of drug strains are widely grown and these represent valuable germplasm for breeding. Folta (2019, p. 8) noted "New methods in gene editing may target gibberellin receptors or synthesis mechanisms, creating custom dwarfing varieties."

3.16.2　Breeding for Foliage Features

Crop architecture governs reception of light for photosynthesis. Leaves are incredibly efficient solar radiation collectors: a cannabis leaf absorbs over 90% of received incident light (personal

FIGURE 3.20 Contrasting patterns of development of the vegetative stem of *Cannabis sativa* (female plants). Left: Strong, diffuse branching pattern (correlated with long internodes) and considerable height, typical of most well-developed, open-grown plants of *Cannabis sativa*, and inappropriate for controlled environments. Right: Short height and very condensed branching (correlated with short internodes) of an "indica-type" landrace from Afghanistan, representing an architecture particularly appropriate for controlled environments.

Sources: Photos by E. Small

observation). When plants are grown at high densities, sunlight will only penetrate a canopy for a certain depth. In tall cultivars, this means that the lower leaves photosynthesize inefficiently for lack of light, and if light intensity falls below the compensation point, they may even drain rather than contribute to photosynthate accumulation (Bilodeau et al. 2019). To optimize a crop's efficiency of energy capture on an area basis, presumably there is an ideal layout of foliage (canopy structure) which is related to leaf dimensions. Since controlled environments normally employ overhead lighting, canopy architecture best suited for field plants and for controlled environment plants probably differ. The study of crop canopy suitability for controlled environments is limited, but common sense suggests that the degree of light transmission through the canopy is an important breeding criterion.

There are two solutions to adapting cannabis plants to the limited light of controlled environments: 1) physically manipulate the form of the plants (i.e., trim the height and/or train the branches) to limit the density of the canopy—as often practiced by illicit marijuana producers; or 2) select plants that grow naturally to an appropriate size and shape to maximize use of artificial light (which usually involves inducing flowering with short days). In both approaches the goals are: shortening the light-intercepting canopy and distributing leaf exposure so that most of the foliage is well lit and growth is optimized. In approach (1), effort is required to physically arrange branches to maximize exposure to the lights, while in approach (2), the natural architecture of the plants achieves this. Approach (1) typically requires using twist ties to attach branches to a horizontal trellis wire or cord framework, trimming as necessary and removing lower fan leaves. Sometimes stem leader buds are destroyed to induce proliferation of branches. Growth regulators could be used to induce desired plant forms, but applying chemicals to cannabis plants is unwise. There are several reasons why reducing labor is desirable: 1) it increases efficiency and reduces costs; 2) working under intense lights in a hot room is very unpleasant; and 3) manipulation by laborers can contaminate the plants with microorganisms and pollen.

As detailed by Small (2018), in numerous crops, the leaves of domesticated forms are larger than is the case in related wild species. This is likely due to the greater photosynthetic capacity of large leaves, the result of selection by humans to be more productive in a given limited area. Indeed, this pattern seems to be true for the three classes of domesticated *Cannabis* (fiber, oilseed, and marijuana), all of which tend to have larger leaves and leaflets than do wild *Cannabis* plants. This is particularly well illustrated by the indica-type class representing advanced domesticated cannabis, discussed earlier. The leaves of indica-type plants have leaflets that are larger, but especially wider (Figure 3.21). *Cannabis* leaves sometimes have as many as 13 leaflets, and it might seem that breeding for leaflet number might be advantageous. However, as leaflet number increases, the leaflets become crowded and overlap. Wider leaflets rather than more leaflets appears to be a better strategy for increasing the photosynthetic capacity of individual leaves. Indeed, on the basis of modeling considerations for tomato leaves, Sarlikioti et al. (2011) concluded that for a given leaf area, bigger but fewer leaflets were better at intercepting light than more but smaller leaflets. These considerations strongly suggest that breeders of cannabis strains maintain and accentuate plants with large leaves with wide leaflets. The clones cultivated in controlled environment growth facilities usually have large leaves with impressively wide leaflets.

FIGURE 3.21   A contrast of leaves of "sativa-type" (left; narrow leaflets) and "indica-type" (right; wide leaflets) marijuana plants.

Source: Photo by Transmitdistort (CC BY 3.0)

3.16.3 BREEDING FOR INFLORESCENCE FEATURES (BUDS, COLAS, COLOR)

3.16.3.1 Terminology

In botany (including horticulture, agriculture and gardening) a "bud" is a meristem (the growing point of a part of a plant, producing a stem, flower, leaf or other organ). The word bud has been distorted in "marijuana language" to mean a tight cluster of flowers in the female inflorescence (flowering branch system) of *Cannabis*. In this review, the context should be sufficient to indicate which sense is meant. The term "inflorescence" refers to: 1) a group or cluster of flowers on an ultimate branch; and/or 2) the entire branching system bearing flowers (excluding "pre-flowers" discussed earlier). When the flowers are fertilized and develop fruits, the fruit-bearing branching systems are termed "infructescences." In many marijuana strains, the ultimate flowering stems have been selected to develop very congested buds. The term "cola" is sometimes used as a synonym of "bud" but is more usually applied to a large branch (especially the central stem) bearing many buds close together.

3.16.3.2 Why Mother Nature Doesn't and Humans Do Produce Buds

Well-developed wild plants of *C. sativa* produce female reproductive structures mostly on the ends of long branches, so that the resulting seeds are well separated. Moreover, the flowers and seeds mature sequentially. These developmental features represent survival strategies to prevent: 1) a large standing crop of ripened seeds on the plants; and 2) concentrations of seeds at particular locations on the plants, both of which would be extremely attractive to herbivores, particularly birds. Humans, by contrast, prefer crops to develop large standing concentrations of simultaneously maturing forms of the desired product (seeds in the case of oilseed hemp, flowers in the case of drug plants). Buds are popular as a sales item because they are usually a reliable indicator of high-grade marijuana (it is impossible to judge the quality of manicured marijuana without smoking it or measuring THC content). Buds are too large to smoke directly (they can be placed on large vaporizers), so they are broken up into a tobacco-like consistency, often using an herb grinder.

3.16.3.3 Selection for Bud Location and Size

There are two contrasting architectural strategies that need to be examined: centralization of the reproductive parts in a large compact axis (Figure 3.22A–B), or development of the reproductive parts in several smaller compact structures on different branches (Figure 3.22C). Production of a centralized axis carrying all or most of the buds greatly facilitates harvesting (this is the strategy being employed for hempseed cultivars). When buds are scattered on different branches, they tend to mature at different times, necessitating collection at different times. (This is like harvesting tomatoes from indeterminate (vine) forms as they ripen over several weeks. Some growers harvest the terminal inflorescence of cannabis plants first, since it matures relatively quickly.) Marijuana growers sometimes damage the meristem (growing point) of the main stem to cause the plant to branch and develop more buds, but once again, there is considerable labor involved.

Just as there is a market for standard-size (i.e., large) tomatoes and miniature ("cocktail") tomatoes, large and small buds can be marketed. When the plants are of the dwarf (indica-type) class, they strongly tend to have short internodes, which in the buds results in a quite compact but relatively large form, considered desirable. However, the denseness of flowers can retain moisture, leading to fungal diseases unless humidity is carefully controlled. In any event, the form and size of the buds needs to be evaluated as breeding criteria.

3.16.3.4 Selection for Bud Color

Anthocyanin pigments can produce various red, purple and pink coloration of the buds, and other pigments can result in a yellow or golden hue. Some strains have buds with different colors in the bracts, stigmas and sugar leaves (Figure 3.23). (Immature stigmas are usually naturally white,

FIGURE 3.22 Desirable and undesirable distribution of flowers for production of buds. A, B: Development of heavily congested main axis inflorescences ("colas")—an ideal architecture for controlled environments as this produces buds with very limited branching. C: Very diffuse distribution of small buds on highly branched plants.

Sources: Photo A by WeedStreetwear (public domain); B by Erik Fenderson (public domain); C by M. Martin Vicente (CC BY 2.0)

FIGURE 3.23 Colorful buds.

Sources: Photos from WeedStreetWear.com (public domain)

becoming reddish or brown with age). Humans are fond of color, and mutations producing interestingly colored buds have market value; when they occur, they are often worth preserving.

3.17 BREEDING FOR CANNABINOIDS

Although there has been a substantial cannabinoid extract market for THC and CBD, supplied by authorized controlled environment producers, the supply is increasingly coming from outdoor growers. In practice, this means that plants specialized for CBD production are inappropriate for controlled environment commercial production. For cannabis products based on extracts, there is no particular need to create strains with a particular balance of THC and CBD, since these compounds can simply be combined in desired proportions. The commercial controlled environment production of buds is largely for strains that are dominated by THC and—to a lesser extent—strains with both THC and CBD. There are very few producers of other cannabinoids, but since this niche is still experimental and very high value, the plants are grown in controlled environment facilities. Accordingly, there is a need to breed strains dedicated to production of the minor cannabinoids in controlled environments.

3.17.1 THE FIVE FUNDAMENTAL DETERMINANTS OF CANNABINOID DEVELOPMENT

3.17.1.1 Environmental Modification

Environmental stresses are known to influence relative (i.e., qualitative) cannabinoid development, but the effects appear to be minor (Small 2017a). For example, Namdar et al. (2019) found that under LED fixtures, the ratio of cannabigerolic acid (CBGA) to THCA was 1:2, changing to 1:16 when grown under high-pressure sodium light. Quantitative production of course varies with size of plant, but is also determined by adequacy of several growth requirements (Vanhove et al. 2011; Backer et al. 2019).

3.17.1.2 Stage of Maturity

Cannabinoids do change quantitatively as plants mature (generally, they increase at flowering) and sometimes qualitatively (e.g., CBC tends to present in substantial amounts in juvenile plants). However, very young plants can usually be employed to evaluate qualitative cannabinoid profiles, as there is relatively limited cannabinoid plasticity (De Backer et al. 2012; Welling et al. 2018).

3.17.1.3 Site (Organs)

Cannabinoids develop virtually exclusively in the plant's trichomes (they are suspected to occur in laticifers, which are internal ducts), but the location on the plant may influence development. For example, Potter (2009) recorded a greater presence of CBC in the small (non-stalked) secretory glands of the foliage than in the large (stalked) glands of the inflorescence. Similarly, Bernstein et al. (2019) found different ratios of cannabinoids in the foliage compared to the inflorescence. Reports of cannabinoids in parts of the plant lacking trichomes (roots, seeds, pollen) are likely the result of contamination.

3.17.1.4 Quantitative Inheritance Genes

The absolute quantity of cannabinoids produced by an individual plant or by a population (on an area basis) depends on growth and developmental traits (such as size and proportion of tissues constituted by secretory glands), which: 1) are probably determined polygenically and 2) are unrelated (at least substantially) to cannabinoid biosynthetic pathways; and 3) may be especially subject to environmental modification. On the whole, there is very limited information available regarding the inheritance of quantitative genetic factors, some of which may be pleiotropic (i.e., with correlated effects). Probably most hobbyist cannabis breeders do not appreciate the importance of distinguishing quantitative and qualitative aspects of cannabinoid inheritance.

3.17.1.5 Qualitative Inheritance: Genes Controlling Biosynthetic Pathways

Qualitative inheritance of the cannabinoids is determined by genes controlling their biosynthetic pathways. These genes determine the relative amounts of particular cannabinoids. The possibility of linkage to quantitative genes (i.e., correlated multiple effects) needs to be kept in mind.

3.17.2 BIOSYNTHESIS AND KEY CONTROLLING GENES OF THE MAJOR CANNABINOIDS

A pentyl side chain has the formula C_5H_{11}. The biosynthetic pathways of the major cannabinoids with pentyl side chains (CBC, CBD, CBG and THC) were established during the 1990s. The first event in the pentyl cannabinoid biosynthesis is the production of cannabigerol (CBG), produced by condensation of a phenol-derived olivetolic acid and a terpene-based geranylpyrophosphate catalyzed by the enzyme geranylpyrophosphate:olivetolate geranyltransferase. From CBG, Δ^9-THC, CBD and CBC are synthesized, each by a specific synthase enzyme. An outline of the biosynthesis of the two most important cannabinoids, THC and CBD, is shown in Figure 3.24. For more complete analyses of cannabinoid biosynthesis, see Flores-Sanchez and Verpoorte (2008), Van Bakel

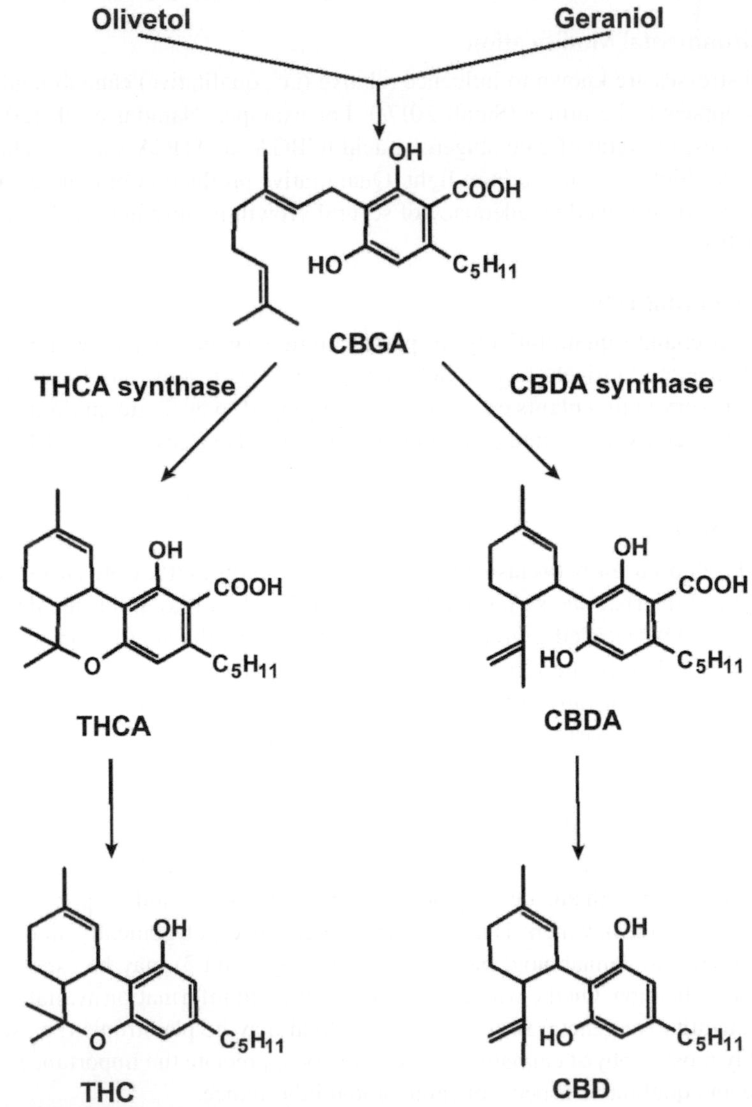

FIGURE 3.24 Biosynthetic pathway of THC and CBD, the predominant cannabinoids of *Cannabis sativa*.

Notes: CBGA = cannabigerolic acid, THCA = tetrahydrocannabinolic acid, CBDA = cannabidiolic acid (the carboxylated forms of CBG, THC and CBD, respectively); decarboxylation (conversion of THCA to THC and conversion of CBDA to CBD) is not part of the biosynthetic pathway, but occurs spontaneously with aging and/or heat

et al. (2011), Gagne et al. (2012), Sirikantaramas and Taura (2017), Laverty et al. (2020), Wenger et al. (2020) and Singh et al. (2021).

De Meijer et al. (2003) found evidence that inheritance of the key cannabinoids THC acid and CBD acid is determined by the allelic status at a single locus (referred to as B), and that THCA development in *C. sativa* is under the partial genetic control of codominant alleles. Allele B_D is postulated to encode CBDA synthase, while allele B_T encodes THCA synthase. This genetic model holds that plants in which CBDA is predominant have a B_D/B_D genotype at the B locus, plants in which THC is predominant have a B_T/B_T genotype and plants with substantial amounts of both THCA and CBDA are heterozygous (B_D/B_T genotype). De Meijer and Hammond (2005) found that plants accumulating CBG have a mutation of B_D (which they term B_0) in the homozygous state that encodes for a poorly functional CBD synthase; and De Meijer et al. (2009) selected a variant of this that almost completely prevents the conversion of CBG into CBD.

The hypothesis that the enzymes that produce THCA (THCA synthase) and CBDA (CBDA synthase) from the same precursor compound, cannabigerolic acid, are controlled exclusively by two alleles of the same gene was challenged by Weiblen et al. (2015), who found that THCA synthase and CBDA synthase are encoded by two separate but linked regions. THC-predominant plants simply have a non-functional copy of CBDA synthase, so they convert all cannabigerolic acid into THCA. Other evidence also indicates that other genes control the pathways to THCA and CBDA (Van Bakel et al. 2011; Onofri et al. 2015). It has been speculated (see Dolgin 2019b) that cannabinoid qualitative expression may also be influenced by retrotransposons (a class of mobile DNA sequences that is often widely distributed in genomes).

Regardless of the complexities of genic control of the cannabinoids, progeny of hybrids and backcrosses between marijuana (high-THC) and hemp (low-THC) parents segregate dramatically for THC production, facilitating transfer of desired characters from the latter to cannabis strains.

3.17.3 "Minor" Cannabinoids

Geographical biotypes have been found with one or more rare cannabinoids in unusually high presence, which is probably the result of genetic drift (change in population genetics occurring in small populations simply by haphazard survival of unusual plants). CBC (cannabichromene) is a frequent minor constituent of highly intoxicating strains of *C. sativa*, especially from Africa, and strains high in CBC have been selected for medicinal experimentation. CBC is present in substantial amounts in juvenile plants and declines with maturation. De Meijer et al. (2009) found plant variants in which CBC persisted into maturity and noticed that this is associated with a reduced presence of perigonal bracts and secretory glands. CBG (cannabigerol) rarely dominates the resin of *Cannabis*. Some geographical races with minor or trace amounts of cannabinoids have been described, notably for CBGM (cannabigerol monomethyl ether) in some northeastern Asian populations, CBDV (cannabidivarin) in some populations from central Asia and THCV (tetrahydrocannadivarin) in some collections from Asia and Africa. Regardless of source, plants with unusually high concentrations of any of the more than 100 infrequent cannabinoids have potential economic value, and need to be considered for preservation as clones. GW Pharmaceuticals, centered in the United Kingdom, maintains clones with high amounts of some of the minor cannabinoids. However, it needs to be remembered that infrequent natural cannabinoids can potentially be economically synthesized.

3.18 BREEDING FOR TERPENES

3.18.1 Background

The characteristic odors of *Cannabis* plants are due to their essential oil. Essential oils are complex mixtures of organic (hydrocarbon) chemicals and particularly include terpenes and oxygenated compounds such as alcohols, esters, ethers, aldehydes, ketones, lactones, phenols and phenol ethers. Terpenes typically dominate essential oils. Terpenes are made up of units of isoprene: $CH_2 = C(CH_3) - CH = CH_2$. Monoterpenes consist of two isoprene units, while sesquiterpenes consist of three. "Terpenoids" are related compounds, although the term is often used as a synonym of terpenes. Many terpenes are extremely odoriferous, detectable by smell at very low concentrations. The terpenes of *Cannabis* are manufactured in the same epidermal glands (secretory glandular trichomes) in which the cannabinoids of *Cannabis* are produced. The cannabinoids and terpenoids make up the resinous secretion of the glands, but the former are odorless. Indeed, the cannabinoids and terpenoids have a parental biosynthetic precursor in common (pyrophosphate). Unlike the terpenes, the cannabinoids are odorless. Terpenes may account for about 1% of marijuana but can comprise as much as 10% of the secretory glands (Potter 2009). Terpene biosynthesis is discussed by Andre et al. (2016), Allen et al. (2019), Booth et al. (2017, 2020) and Booth and Bohlmann (2019).

3.18.2   THE PRINCIPAL *CANNABIS* TERPENES

The composition of essential oils has been found to vary considerably among strains and cultivars of *C. sativa* (Elzinga et al. 2015; Hazekamp et al. 2016; Lynch et al. 2016; Mudge et al. 2019). The two most common terpenes in the plant world are α-pinene and limonene, and both are present in the essential oil of *C. sativa*. Other common terpenes of *Cannabis* include myrcene, linalool, beta-caryophyllene, caryophyllene oxide, nerolidol and phytol. Some terpenes in *Cannabis* are quite pleasant in odor: limonene is fruity (lemons are rich in this chemical), while linalool has a rather sweet smell. *Cannabis* essential oils with high sesquiterpene concentrations tend to smell bad, while oils with high monoterpene percentages (but a low α-humulene or caryophyllene oxide concentration) tend to have pleasant smells. Depending on biotype, monoterpenes represent 48—92% of the volatile terpenes, and sesquiterpenes represent 5—49%. Monoterpenoids usually make up most of the essential oil of *Cannabis*. The aroma of *C. sativa* is particularly due to the monoterpenes pinene and limonene, which frequently comprise over 75% of the volatiles and often dominate "headspace" odor near the plant. The monoterpenes evaporate relatively faster than other components, so the composition of essential oil actually in the harvested plant (and capable of being extracted) may differ from the volatiles released around the fresh plant. Consequently, the odor of the living plant is not necessarily indicative of the relative composition of the plant's essential oil or of the odor of the dried plant.

3.18.3   ECONOMIC IMPORTANCE OF *CANNABIS* TERPENES

As is well known, "aroma" is a combination of taste and odor (that is, what people perceive as "taste" or "flavor" is strongly determined by odor). It has become traditional to judge the quality of marijuana by its aroma (Figure 3.25), and many consumers are convinced that a strong aroma indicates desirable potency. Moreover, there has been widespread speculation that terpenes interact with cannabinoids to produce or accentuate therapeutic effects (Russo 2011), and some consumers prefer certain terpenes for their alleged healthful value. An unpleasant odor often does not disqualify material from being consumed by humans, and indeed the odor of some popular marijuana strains is quite objectionable. Nevertheless, illicit marijuana breeders for the last half century have been creating strains with distinctive aromas, with a tendency to select more pleasant odors.

3.18.4   THE WISDOM OF BREEDING FOR TERPENES

The tobacco industry has added aroma ingredients to its products for many years, and the marijuana industry currently commonly similarly adds aroma to products based on extracts (especially for electronic vaporizers and for edibles). However, regulations in some jurisdictions prevent adding materials to any form of cannabis. Adding substances to medical marijuana may require demonstration of efficacy or safety. In the long term, it would seem that if genuine medical benefit is gained

FIGURE 3.25 Smelling a marijuana sample to judge its merit—a widespread ritual. The odor is an important commercial criterion, but of uncertain future marketplace value.

Source: Photo by My 420 Tours (CC BY SA 4.0)

from adding given terpenes to medical cannabis, there should not be an objection. Current regulations generally do not permit the addition of terpenes to marijuana simply for attractive aroma, in the manner that lemon and pine aromas are widely added to detergents. Nevertheless, the prospects seem reasonable that "flavored" or "aromatized" herbal marijuana will become popular since this is already the case for extracts. Given that there are no known unique terpenes in *Cannabis*, and that far cheaper sources of the same terpenes can simply be added to cannabis drugs, breeding *for* terpenes is questionable. Nevertheless, since the primary cannabis sales product is still buds and aroma is a critical consumer criterion, breeding *against* objectionable aromas anticipated to reduce sales remains necessary.

3.19   ACKNOWLEDGMENTS

Brenda Brookes skillfully assembled and enhanced the illustrations for publication. Creative Commons Licenses employed in this article: CC BY 2.0 (Attribution 2.0 Generic): http://creativecommons. org/licenses/by/2.0/; CC BY 3.0 (Attribution 3.0 Unported): http://creativecommons.org/licenses/ by/3.0/; CC BY SA 2.0 (Attribution ShareAlike 2.0 Generic): https://creativecommons.org/licenses/ by-sa/2.0/; CC BY SA 3.0 (Attribution ShareAlike 3.0 Unported): http://creativecommons.org/

REFERENCES

Allen, K. D., K. McKernan, C. Pauli, J. Roe, A. Torres, and R. Gaudino. 2019. Genomic characterization of the complete terpene synthase gene family from *Cannabis sativa*. *PLoS One* 14(9): e0222363. https://doi.org/10.1371/journal.pone.0222363.

Andre, C. M., J. F. Hausman, and G. Guerriero. 2016. *Cannabis sativa*: The plant of the thousand and one molecules. *Frontiers in Plant Science* 7. https://doi.org/10.3389/fpls.2016.00019.

Backer, R., G. Mandolino, O. Wilkins, M. A. ElSohly, and D. L. Smith. 2020. Editorial: Cannabis genomics, breeding and production. *Frontiers in Plant Science*. https://doi.org/10.3389/fpls.2020.591445.

Backer, R., T. Schwinghamer, P. Rosenbaum, et al. 2019. Closing the yield gap for cannabis: A meta-analysis of factors determining cannabis yield. *Frontiers in Plant Science*. https://doi.org/10.3389/fpls.2019.00495.

Bar, M., and H. Shtein. 2019. Plant trichomes and the biomechanics of defense in various systems, with Solanaceae as a model. *Botany* 97: 651–660.

Barcaccia, G., F. Palumbo, F. Scariolo, A. Vannozzi, M. Borin, and S. Bona. 2020. Potentials and challenges of genomics for breeding *Cannabis* cultivars. *Frontiers in Plant Science*. https://doi.org/10.3389/fpls.2020.573299.

Bernstein, N., J. Gorelick, and S. Koch. 2019. Interplay between chemistry and morphology in medical cannabis (*Cannabis sativa* L.). *Industrial Crops and Products* 129: 185–194.

Bilodeau, S. E., B. S. Wu, A. S. Rufyikiri, S. MacPherson, and M. Lefsrud. 2019. An update on plant photobiology and implications for cannabis production. *Frontiers in Plant Science* 10: 296. https://doi.org/10.3389/fpls.2019.00296.

Booth, J. K., and J. Bohlmann. 2019. Terpenes in *Cannabis sativa*—from plant genome to humans. *Plant Science* 284: 67–72.

Booth, J. K., J. E. Page, and J. Bohlmann. 2017. Terpene synthases from *Cannabis sativa*. *PLoS One* 12(3): e0173911. https://doi.org/10.1371/journal.pone.0173911.

Booth, J. K., M. M. S. Yuen, S. Jancsik, L. L. Madilao, J. E. Page, and J. Bohlmann. 2020. Terpene synthases and terpene variation in *Cannabis sativa*. *Plant Physiology* 184: 130–147.

Brickell, C. D., C. Alexander, J. J. Cubey, et al. (eds.). 2016. *International code of nomenclature for cultivated plants*, 9th ed. Leuven, Belgium: International Society for Horticultural Science.

Briosi, G., and F. Tognini. 1894. *Anatomia della canapa (Cannabis sativa L.). Parte prima: Organi sessuali*. Milan, Italy: Istituto Botanico Della R. Universita de Pavia, Series 2–3, 91–209.

Campbell, B. J. 2019. *Plasticity, allelic diversity, and genetic architecture of industrial hemp (Cannabis sativa L.)*. PhD thesis. Fort Collins: Colorado State University. https://mountainscholar.org/bitstream/handle/10217/197447/Campbell_colostate_0053A_15680.pdf?sequence=1&isAllowed=y.

Chandra, S., H. Lata, I. A. Khan, and M. A. ElSohly. 2011. Temperature response of photosynthesis in different drug and fiber varieties of *Cannabis sativa* L. *Physiology and Molecular Biology of Plants* 17: 297–303.

Chandra, S., H. Lata, I. A. Khan, and M. A. ElSohly. 2013. The role of biotechnology in *Cannabis sativa* propagation for the production of phytocannabinoids. In *Biotechnology for medicinal plants—micropropagation and improvement*, ed. S. Chandra, H. Lata, and A. Varma. Berlin: Springer-Verlag, 123–148.

Choudhary, N., M. B. Siddiqui, S. Bi, and S. Khatoon. 2014. Effect of seasonality and time after anthesis on the viability and longevity of *Cannabis sativa* pollen. *Palynology* 38: 235–241.

Clarke, R. C. 1981. *Marijuana botany: An advanced study: The propagation and breeding of distinctive cannabis*. Oakland, CA: Ronin Publishing.

Clarke, R. C., and M. D. Merlin. 2013. *Cannabis: Evolution and ethnobotany*. Los Angeles: University of California Press.

Clarke, R. C., and M. D. Merlin. 2016. *Cannabis* domestication, breeding history, present-day genetic diversity, and future prospects. *Critical Reviews in Plant Sciences* 35: 293–327.

Collard, B. C. Y., M. Z. Z. Jahufer, J. B. Brouwer, and E. C. K. Pang. 2005. An introduction to markers, quantitative trait loci (QTL) mapping and marker-assisted selection for crop improvement: The basic concepts. *Euphytica* 142: 169–196.

De Backer, B., K. Maebe, A. G. Verstraete, and C. Charlier. 2012. Evolution of the content of THC and other major cannabinoids in drug-type cannabis cuttings and seedlings during growth of plants. *Journal of Forensic Sciences* 57: 918–922.

De Meijer, E. P. M. 2014. The chemical phenotypes (chemotypes) of *Cannabis*. In *Handbook of cannabis*, ed. R. G. Pertwee. Oxford: Oxford University Press, 89–110.

De Meijer, E. P. M., M. Bagatta, A. Carboni, et al. 2003. The inheritance of chemical phenotype in *Cannabis sativa* L. *Genetics* 163: 335–346.

De Meijer, E. P. M., and K. M. Hammond. 2005. The inheritance of chemical phenotype in *Cannabis sativa* L. (II): Canabigerol predominant plants. *Euphytica* 145: 189–198.

De Meijer, E. P. M., K. M. Hammond, and M. Micheler. 2009. The inheritance of chemical phenotype in *Cannabis sativa* L. (III): Variation in cannabichrome proportion. *Euphytica* 165: 293–311.

DiMatteo, J., L. Kurtz, and J. D. Lubell-Brand. 2020. Pollen appearance and in vitro germination varies for five strains of female hemp masculinized using silver thiosulfate. *HortScience* 55: 547–549.

Dolgin, E. 2019a. A boosted crop: Genetic engineering could enable cannabinoids of pharmaceutical interest to be produced on an industrial scale. *Nature* 572: S5–S7.

Dolgin, E. 2019b. Genomics blazes a trail to improved cannabis cultivation. *PNAS* 116: 8638–8640.

Elzinga, S., J. Fischedick, R. Podkolinski, and J. C. Raber. 2015. Cannabinoids and terpenes as chemotaxonomic markers in *Cannabis*. *Natural Products Chemistry & Research* 3: 181. https://doi.org/10.4172/2329-6836.1000181.

Flores-Sanchez, I. J., and R. Verpoorte. 2008. Secondary metabolism in cannabis. *Phytochemistry Reviews* 7: 615–639.

Folta, K. M. 2019. Breeding new varieties for controlled environments. *Plant Biology* 21: 6–12.

Gagne, S. J., J. M. Stout, E. Liu, Z. Boubakir, S. M. Clark, and J. E. Page. 2012. Identification of olive-tolic acid cyclase from *Cannabis sativa* reveals a unique catalytic route to plant polyketides. *PNAS* 109: 12811–12816.

Gaudet, D., N. S. Yadav, A. Sorokin, A. Bilichak, and I. Kovalchuk. 2020. Development and optimization of a germination assay and long-term storage for *Cannabis sativa* pollen. *Plants* 9(5): 665. https://doi.org/10.3390/plants9050665.

Glas, J. J., B. C. J. Schimmel, J. M. Alba, et al. 2012. Plant glandular trichomes as targets for breeding or engineering of resistance to herbivores. *International Journal of Molecular Sciences* 13: 17077–17103.

Gove, P. B., and the Merriam-Webster Editorial Staff. (eds.). 1981. *Webster's third new international dictionary of the English language unabridged*. Springfield, MA: Merriam-Webster Inc.

Halbritter, H., and H. Heigl. 2020. Cannabis sativa. In *PalDat—a palynological database*. www.paldat.org/pub/Cannabis_sativa/303779;jsessionid=CFD8FF05DEEC1726860869524F45B6F0.

Hazekamp, A., K. Tejkalová, and S. Papadimitriou. 2016. Cannabis: From cultivar to chemovar II—a metabolomics approach to cannabis classification. *Cannabis and Cannabinoid Research* 1(1): 202–215.

Hurgobin, B., M. Tamiru-Oli, M. T. Welling, M. S. Doblin, A. Bacic, J. Whelan, and M. G. Lewsey. 2021. Recent advances in *Cannabis sativa* genomics research. *New Phytologist* 230: 73–89.

Lata, H., S. Chandra, Z. Mehmedic, I. A. Khan, and M. A. ElSohly. 2012. In vitro germplasm conservation of high Δ^9-tetrahydrocannabinol yielding elite clones of *Cannabis sativa* L. under slow growth conditions. *Acta Physiologiae Plantarum* 34: 743–750.

Laverty, K. U., J. M. Stout, and M. J. Sullivan. 2020. A physical and genetic map of *Cannabis sativa* identifies extensive rearrangements at the THC/CBD acid synthase loci. *Genome Research* 29: 146–156.

Lubell, J. D., and M. H. Brand. 2018. Foliar sprays of silver thiosulfate produce male flowers on female hemp plants. *HortTechnology* 28: 743–747.

Lynch, R. C., D. Vergara, S. Tittes, et al. 2016. Genomic and chemical diversity in *Cannabis*. *Critical Reviews in Plant Sciences* 35: 349–363.

Mansouri, H., and M. Bagheri. 2017. Induction of polyploidy and its effect on *Cannabis sativa* L. In *Cannabis sativa L.: Botany and biotechnology*, ed. S. Chandra, H. Lata and M. A. ElSohly. Berlin: Springer-Verlag, 365–383.

McPartland, J. M. 2017. *Cannabis sativa* and *Cannabis indica* versus "Sativa" and "Indica." In *Cannabis sativa L.: Botany and biotechnology*, ed. S. Chandra, H. Lata and M. A. ElSohly. Berlin: Springer-Verlag, 101–120.

McPartland, J. M., R. C. Clarke, and D. P. Watson. 2000. *Hemp diseases and pests: Management and biological control*. Wallingford, Oxon: CABI.

McPartland, J., W. Hegman, and T. Long. 2019. *Cannabis* in Asia: Its center of origin and early cultivation, based on a synthesis of subfossil pollen and archaeobotanical studies. *Vegetation History and Archaeobotany* 28: 691–702.

McPartland, J. M., and E. Small. 2020. A classification of endangered high-THC cannabis (*Cannabis sativa* subsp. *indica*) domesticates and their wild relatives. *PhytoKeys* 177: 81–112.

Mishra, A., P. Gupta, S. S. Dhawan, and R. K. Lal. 2020. Adaptability and stability pattern of a genetic character—trichome diversity in menthol mint (*Mentha arvensis* L.) *Journal of Medicinal and Aromatic Plant Sciences* 42: 121–126.

Mohan Ram, H. Y., and R. Sett. 1982. Induction of fertile male flowers in genetically female *Cannabis sativa* plants by silver nitrate and silver thiosulphate anionic complex. *Theoretical and Applied Genetics* 62: 369–375.

Monthony, A. S., S. R. Page, M. Hesami, and A. M. P. Jones. 2021. The past, present and future of *Cannabis sativa* tissue culture. *Plants* 10(1): 185. https://doi.org/10.3390/plants10010185.

Mudge, E. M., P. N. Brown, and S. J. Murch. 2019. The terroir of *Cannabis*: Terpene metabolomics as a tool to understand *Cannabis sativa* selections. *Planta Medica* 85: 781–796.

Mudge, E. M., S. J. Murch, and P. N. Brown. 2018. Chemometric analysis of cannabinoids: Chemotaxonomy and domestication syndrome. *Scientific Reports* 8: 13090. https://doi.org/10.1038/s41598-018-31120-2.

Nadeem, M. A., M. A. Nawaz, M. Q. Shahid, et al. 2018. DNA molecular markers in plant breeding: Current status and recent advancements in genomic selection and genome editing. *Biotechnology & Biotechnological Equipment* 32: 261–285.

Namdar, D., D. Charuvi, V. Ajjampura, et al. 2019. LED lighting affects the composition and biological activity of *Cannabis sativa* secondary metabolites. *Industrial Crops and Products* 132: 177–185.

Naraine, S. G. U., E. Small, A. E. Laursen, and L. G. Campbell. 2020. A multivariate analysis of evolutionary divergence among feral, fiber, oilseed, dioecious/monoecious and marijuana variants of *Cannabis sativa* L. *Genetic Resources and Crop Evolution* 67: 703–714.

Onofri, C., E. P. M. de Meijer, and G. Mandolino. 2015. Sequence heterogeneity of cannabidiolic- and tetrahydrocannabinolic acid-synthase in *Cannabis sativa* L. and its relationship with chemical phenotype. *Phytochemistry* 116: 57–68.

Pacifico, D., F. Miselli, M. Micheler, A. Carboni, P. Ranalli, and G. Mandolino. 2006. Genetics and marker-assisted selection of the chemotype in *Cannabis sativa* L. *Molecular Breeding* 17: 257–268.

Parsons, J. L., S. L. Martin, T. James, G. Golenia, E. A. Boudko, and S. R. Hepworth. 2019. Polyploidization for the genetic improvement of *Cannabis sativa*. *Frontiers in Plant Science* 10: 476. https://doi.org/10.3389/fpls.2019.00476.

Potter, D. J. 2009. *The propagation, characterisation and optimisation of Cannabis sativa L. as a phytopharmaceutical*. PhD thesis. London: King's College.

Potter, D. J. 2014. A review of the cultivation and processing of cannabis (*Cannabis sativa* L.) for production of prescription medicines in the UK. *Drug Testing and Analysis* 6: 31–38.

Prentout, D., O. Razumova, H. Henri, M. Divashuk, G. Karlov, and G. A. B. Marais. 2020a. Development of genetic markers for sexing *Cannabis sativa* seedlings. *BioRxiv*. https://doi.org/10.1101/2020.05.25.114355.

Prentout, D., O. Razumova, B. Rhoné, et al. 2020b. An efficient RNA-seq-based segregation analysis identifies the sex chromosomes of *Cannabis sativa*. *Genome Research* 30: 164–172.

Punja, Z. K., and J. E. Holmes. 2020. Hermaphroditism in marijuana (*Cannabis sativa* L.) inflorescences—impact on floral morphology, seed formation, progeny sex ratios, and genetic variation. *Frontiers in Plant Science* 11: 718. https://doi.org/10.3389/fpls.2020.00718.

Rahn, B., B. J. Pearson, R. N. Trigiano, and D. J. Gray. 2016. The derivation of modern cannabis varieties. *Critical Reviews in Plant Sciences* 35: 328–348.

Ranalli, P. 2004. Current status and future scenarios of hemp breeding. *Euphytica* 140: 121–131.

Reimann-Philipp, U., M. Speck, C. Orser, et al. 2019. Cannabis chemovar nomenclature misrepresents chemical and genetic diversity; survey of variations in chemical profiles and genetic markers in Nevada medical cannabis samples. *Cannabis and Cannabinoid Research* X: 1–16. https://doi.org/10.1089/can.2018.0063.

Russo, E. B. 2011. Taming THC: Potential cannabis synergy and phytocannabinoid-terpenoid entourage effects. *British Journal of Pharmacology* 163: 1344–1364.

Russo, E. B. 2019. The case for the entourage effect and conventional breeding of clinical cannabis: No "strain," no gain. *Frontiers in Plant Science* 9: 1969. https://doi.org/10.3389/fpls.2018.01969.

Salentijn, E. M. J., Q. Zhang, S. Amaducci, M. Yang, M. Luisa, and L. M. Trindade. 2015. New developments in fiber hemp (*Cannabis sativa* L.) breeding. *Industrial Crops and Products* 68: 32–41.

Sarlikioti, V., P. H. V. de Visser, G. H. Buck-Sorlin, and L. F. M. Marcelis. 2011. How plant architecture affects light absorption and photosynthesis in tomato: Towards an ideotype for plant architecture using a functional-structural plant model. *Annals of Botany* 108: 1065–1073.

Sawler, J., J. M. Stout, K. M. Gardner, et al. 2015. The genetic structure of marijuana and hemp. *PLoS One* 10(8): e0133292. https://doi.org/10.1371/journal.pone.0133292.

Schilling, S., C. A. Dowling, J. Shi, et al. 2020. The cream of the crop: Biology, breeding and applications of *Cannabis sativa*. *Authorea*. https://doi.org/10.22541/au.160139712.25104053/v2.

Schwabe, A. L., and M. Mcglaughlin. 2019. Genetic tools weed out misconceptions of strain reliability in *Cannabis sativa*: Implications for a budding industry. *Journal of Cannabis Research* 1: 1–16.

Singh, A., A. Bilichak, and I. Kovalchuk. 2021. The genetics of *Cannabis*-genomic variations of key synthases and their effect on cannabinoid content. *Genome* 64: 490–501.

Sirikantaramas, S., and F. Taura. 2017. Cannabinoids: Biosynthesis and biotechnological applications. In *Cannabis sativa L.: Botany and biotechnology*, ed. S. Chandra, H. Lata, and M. A. ElSohly. Berlin: Springer-Verlag, 183–206.

Sirikantaramas, S., F. Taura, Y. Tanaka, Y. Ishikawa, S. Morimoto, and Y. Shoyama. 2005. Tetrahydrocannabinolic acid synthase, the enzyme controlling marijuana psychoactivity, is secreted into the storage cavity of the glandular trichomes. *Plant Cell Physiology* 46: 1578–1582.

Small, E. 1979a. *The species problem in Cannabis: Science and semantics. Volume 1, science.* Toronto: Corpus.

Small, E. 1979b. *The species problem in Cannabis: Science and semantics. Volume 2, semantics.* Toronto: Corpus.

Small, E. 2017a. *Cannabis: A complete guide.* Boca Raton: CRC Press, Taylor & Francis.

Small, E. 2017b. Classification of *Cannabis sativa* in relation to agricultural, biotechnological, medical and recreational utilization. In *Cannabis sativa L.: Botany and biotechnology*, ed. S. Chandra, H. Lata and M. A. ElSohly. Berlin: Springer-Verlag, 1–62.

Small, E. 2018. Dwarf germplasm: The key to giant *Cannabis* hempseed and cannabinoid crops. *Genetic Resources and Crop Evolution* 65: 1071–1107.

Small, E., and T. Antle. 2003. A preliminary study of pollen dispersal in *Cannabis sativa*. *Journal of Industrial Hemp* 8(2): 37–50.

Small, E., and H. D. Beckstead. 1973a. Common cannabinoid phenotypes in 350 stocks of *Cannabis*. *Lloydia* 35: 144–165.

Small, E., and H. D. Beckstead. 1973b. Cannabinoid phenotypes in *Cannabis*. *Nature* 245: 147–148.

Small, E., and B. Brookes. 2012. Temperature and moisture content for storage maintenance of germination capacity of seeds of industrial hemp, marijuana, and ditchweed forms of *Cannabis sativa*. *Journal of Natural Fibers* 9(4): 240–255.

Small, E., and S. G. U. Naraine. 2016. Size matters: Evolution of large drug-secreting resin glands in elite pharmaceutical strains of *Cannabis sativa* (marijuana). *Genetic Resources and Crop Evolution* 63: 349–359.

Snoeijer, W. 2002. *A checklist of some cannabaceae cultivars. Part A: Cannabis.* Leiden: Division of Pharmacology, Amsterdam Center for Drug Research.

Snyder, J. C., and G. F. Antonious. 2009. Trichomes—importance in plant defence and plant breeding. *CAB Reviews: Perspectives in Agriculture, Veterinary Science, Nutrition and Natural Resources* 4: 1–16.

Spitzer-Rimon, B., S. Duchin, N. Bernstein, and R. Kamenetsky. 2019. Architecture and florogenesis in female *Cannabis sativa* plants. *Frontiers in Plant Science*. https://doi.org/10.3389/fpls.2019.00350.

Todesco, M., M. A. Pascual, G. L. Owens, et al. 2016. Hybridization and extinction. *Evolutionary applications* 9: 892–908. https://doi.org/10.1111/eva.12367

Torkamaneh, D., and A. M. P. Jones. 2021. Cannabis, the multibillion dollar plant that no genebank wanted. *Genome*. https://doi.org/10.1139/gen-2021-0016. Epub ahead of print. PMID: 34242522.

Toth, J. A., G. M. Stack, A. R. Cala, et al. 2020. Development and validation of genetic markers for sex and cannabinoid chemotype in *Cannabis sativa* L. *Global Change Biology Bioenergy* 12: 213–222.

Turland, N. J., J. H. Wiersema, F. R. Barrie, et al. (eds.). 2018. *International code of nomenclature for algae, fungi, and plants (Shenzhen Code) adopted by the Nineteenth international botanical congress Shenzhen, China, July 2017.* Regnum Vegetabile 159. Glashütten: Koeltz Botanical Books.

Van Bakel, H., J. M. Stout, A. G. Cote, et al. 2011. The draft genome and transcriptome of *Cannabis sativa*. *Genome Biology*. https://doi.org/10.1186/gb-2011-12-10-r102; http://genomebiology.com/content/pdf/gb-2011-12-10-r102.pdf.

Vanhove, W. 2014. *The agronomy and economy of illicit indoor cannabis cultivation.* PhD thesis. Belgium: Ghent University. https://biblio.ugent.be/publication/4256480/file/4256507.pdf.

Vanhove, W., P. Van Damme, and N. Meert. 2011. Factors determining yield and quality of illicit indoor cannabis (*Cannabis* spp.) production. *Forensic Science International* 212: 158–163.

Vassilevska-Ivanova, R. 2019. Biology and ecology of genus *Cannabis*: Genetic origin and biodiversity. In vitro production of cannabinoids. *Genetics and Plant Physiology* 9: 75–98.

Vergara, D., H. Baker, K. Clancy, et al. 2017. Genetic and genomic tools for *Cannabis sativa*. *Critical Reviews in Plant Sciences*. http://doi.org/10.1080/07352689.2016.1267496.

Wagner, G. J. 1991. Secreting glandular trichomes: More than just hairs. *Plant Physiology* 96: 675–679.

Weiblen, G. D., J. P. Wenger, K. J. Craft, et al. 2015. Gene duplication and divergence affecting drug content in *Cannabis sativa*. *New Phytologist* 208: 1241–1250.

Welling, M. T., L. Liu, T. Kretzschmar, R. Mauleon, O. Ansari, and G. J. King. 2020. An extreme-phenotype genome-wide association study identifies candidate cannabinoid pathway genes in Cannabis. *Scientific Reports* 10: 18643. https://doi.org/10.1038/s41598-020-75271-7.

Welling, M. T., L. Liu, C. A. Raymond, O. Ansari, and G. J. King. 2018. Developmental plasticity of the major alkyl cannabinoid chemotypes in a diverse *Cannabis* genetic resource collection. *Frontiers in Plant Science* 9: 1510. https://doi.org/10.3389/fpls.2018.01510.

Welling, M. T., T. Shapter, T. J. Rose, L. Liu, R. Stanger, and G. J. King. 2015. A belated green revolution for *Cannabis*: Virtual genetic resources for fast-track cultivar development. *Frontiers in Plant Science* 7: 1113. https://doi.org/10.3389/fpls.2016.01113.

Wenger, J. P., C. J. Dabney III, M. A. ElSohly, et al. 2020. Validating a predictive model of cannabinoid inheritance with feral, clinical, and industrial *Cannabis sativa*. *American Journal of Botany* 107: 1423–1432.

Wolter, F., P. Schindele, and H. Puchta, 2019. Plant breeding at the speed of light: The power of CRISPR/Cas to generate directed genetic diversity at multiple sites. *BMC Plant Biology* 19: 176. https://doi.org/10.1186/s12870-019-1775-1.

Xu, Y., P. Li, Z. Yang, and C. Xu. 2017. Genetic mapping of quantitative trait loci in crops. *The Crop Journal* 5: 175–184.

4 Cannabis Propagation

Max Jones and Adrian S. Monthony

CONTENTS

4.1 INTRODUCTION

Commercial production of cannabis requires a reliable supply of high-quality planting material to meet production targets, minimize insect/disease/viral pressure and maintain the quality and consistency of the end-product. For smaller controlled environment operations, this may require the production of hundreds of plants per cycle, while larger facilities may require tens of thousands or more. In contrast, outdoor or seasonal greenhouse production may require hundreds of thousands or even millions of plants, or seeds, to be ready by a specific time of year.

DOI: 10.1201/9781003150442-4

Cannabis is a versatile plant with many options available for propagation, including seed, vegetative stem cuttings, grafting, layering and micropropagation using plant tissue culture methods. Each of these approaches has specific advantages, challenges and limitations, making the ideal propagation strategy dependent upon the production system, the scale of the operation and the end-use of the plant. To establish the ideal strategy for a specific production system, understanding the underlying biology of the cannabis plant and the various propagation strategies that are available is paramount. In this chapter, an overview of propagation techniques and their respective applications will be provided, based on recent literature and experience. However, it is also important to recognize that this is a rapidly developing field, especially with respect to micropropagation, and additional reading may be necessary to stay up to date with new developments.

4.2   SEED-BASED PROPAGATION

In order to develop a propagation strategy for any plant, it is critical to understand the biology of the species. Although cannabis biology is discussed in Chapter 1, some aspects that are of particular relevance to propagation strategies will be presented here. Specifically, although cannabis is a highly variable species, it is predominantly a dioecious (separate male and female plants) short-day annual. Typical of most annual plants, cannabis produces an abundance of seed that can be used for propagation, with a single plant capable of producing hundreds or thousands of seeds depending on its size. Since cannabis is primarily dioecious, it is considered an obligate out-crosser, meaning that it produces seed exclusively through cross-pollination (this is mostly accurate, but self-pollination occurs in monoecious cultivars and when hermaphroditic flowers develop). A major implication of this is that seedling populations generally display a high degree of genetic and phenotypic diversity.

While this diversity may be acceptable for small-scale operations, it is often not suitable for large-scale production, depending on the specific end-use. For many larger operations, this variability presents challenges: both in cultivation and meeting government-mandated quality assurance requirements. For example, from an agronomic perspective, variation in the maturation date requires multiple passes of selective harvesting rather than harvesting all of the plants at once. Selective harvesting adds cost and complicates scheduling, making it highly undesirable. From a quality assurance (QA) perspective, this variability leads to unacceptable phytochemical inconsistency within a batch. That being said, for some applications, such as outdoor production for extraction purposes, this variability may be acceptable at scale since the cost of seed-based propagation is relatively low and the extraction process will homogenize the end-product. Though seed is not suitable for all operations, it is viable for others and is an important approach to propagate cannabis. Furthermore, as plant breeders develop more uniform seed-based cultivars, the viability of seed-based propagation will likely increase and be more widely adopted.

While production of seed from drug/medicinal cannabis is largely unregulated and lacks formal standards, this is not the case for hemp. In contrast, hemp has been legal in Canada and other jurisdictions for decades, but is heavily regulated, and there are well-developed protocols and requirements as outlined in the Industrial Hemp Regulations (SOR/2018–145). For example, in Canada, hemp can only be cultivated from a list of government-approved cultivars and can only be planted from pedigree seed produced using strict guidelines and inspections overseen by the Canadian Seed Growers' Association (CSGA). This is largely required in hemp to ensure that the THC level remains below 0.3%, but it also ensures varietal and mechanical purity, as well as high seed quality. While similar regulations are not established for drug/medicinal cannabis, it would be advisable to use these standards as a starting point to develop best practice to ensure genetic/ mechanical purity and seed quality of drug/medicinal cannabis. The specifics of these recommendations are presented in what follows, but more detail can be found in the CSGA's Circular 6 (CSGA 2019).

4.2.1 Production of Seed from Open Pollinated Cultivars

The simplest approach to seed production is open pollination, in which plants of a specific cultivar are grown together and permitted to randomly pollinate within the population. While this is commonly employed to produce hemp seed, it is generally not used for the production of medicinal/drug-type cannabis, due to the high degree of genetic and phenotypic diversity present within the resulting seed. A key requirement for seed production using open pollination is large population sizes. This would have been difficult to maintain during cannabis prohibition and likely is a contributing factor to the relative lack of open pollinated cultivars of drug/medicinal cannabis in comparison to hemp.

Despite the current lack of modern open pollinated cultivars of drug/medicinal cannabis, there is significant interest in collecting and maintaining landraces, which are often from open pollinated populations, for future breeding efforts (see Chapter 3). As prohibitions of drug/medicinal cannabis are lifted, open pollinated cultivars could become an important method for cannabis propagation, especially for lower-cost outdoor production systems where some variability is acceptable. Open pollinated cultivars may even be preferred by some producers to maintain a higher degree of genetic variability within their field. While the breeding and development of new cultivars is beyond the scope of this chapter, there are several factors that are required to maintain and propagate open pollinated cultivars.

The first factor that needs to be considered is the population size needed to maintain the genetic diversity present within the cultivar/landrace. If the population of plants is insufficient, it will lead to genetic erosion—and important traits that define the cultivar could be lost. As such, for open pollinated cultivars and landraces, it is important to maintain sufficient population sizes to ensure the full genetic compliment is preserved within the population over multiple generations. The precise number of individuals in a population required for this has not been determined, and it will depend on the genetic diversity present with a given cultivar/landrace. However, it is likely that hundreds of plants should be used at a minimum.

The second major consideration for maintaining an open pollinated cultivar is ensuring the genetic purity of the stand and resulting seed. As a wind-pollinated species, pollen can travel long distances, with evidence of pollen traveling many kilometers (Cabezudo et al. 1997). This makes contamination from nearby fields a real concern for maintaining genetic purity of seed. While official isolation distances for drug/medicinal-type cannabis seed production have not been determined, it is reasonable to assume that requirements for hemp seed production would be applicable. The general isolation requirements required by the CSGA for industrial hemp include that there be no other cannabis plants within 100 m of the field, and a density of less than 10 plants/Ha beyond that point. For registered seed production, the CSGA requires a 4,800 m isolation distance from cultivation sites of any other cultivar, while for certified seed, an 800 m isolation distance is required. While pollen can potentially travel further than these isolation distances (Cabezudo et al. 1997), the amount of cross-contamination using these recommendations will be minimal. These isolation distances should be adopted for seed production in medicinal/recreational cannabis, and must include isolation from hemp, as well. However, it should also be noted that these requirements were developed for outdoor seed production, and if seed is being produced in controlled environments where outdoor air is filtered, it may be possible to reduce the isolation distance accordingly.

Another consideration for seed production regulated by the CSGA is the land use history. Specifically, it is required that the land being used for seed production has not been used for hemp cultivation in the last three years for registered seed and two years for certified seed. This requirement helps ensure that volunteer hemp plants from previous crops do not establish and introduce other genetics into the population. This is less likely in medicinal/drug-type cannabis, since males are generally removed to avoid seed production, but should be taken into consideration regardless. Additional care should be taken in areas where feral cannabis populations are known to exist.

4.2.2 Production of Seed Using Controlled Pollination

While open pollination is the simplest and cheapest approach to seed production, it is generally not used for cannabis due to the variability present within the resulting seed population. Instead, it is more common to produce seed by crossing two specific cultivars or individual genotypes (F1 hybrid seed). Ideally, the parental cultivars/lines used in the cross have been selected based on their performance in previous test-crosses to determine their combining ability before being used for commercial propagation. The process of conducting test-crosses and assessing combining ability is outside the scope of this chapter, but once the parental lines are identified they can then be used to produce large numbers of seed using controlled pollination.

Controlled pollination can be further divided into two general categories. The first approach is to produce a cross between two distinct seed-based populations to include the full genetic diversity of each. To achieve this, one cultivar will be assigned as the male, while the other will be selected as the female parent. Seedlings of both cultivars are then germinated, and plants are sorted based on sex. Ideally, this is done using molecular markers such that they can be selected at a young age and planted strategically (Hesami et al. 2020), but can also be done using morphological characteristics at a later stage. Once the plants are sexed, the females from the male parental line will be discarded, while the males from the female parent are discarded. From there, both cultivars are planted together to allow the male plants to pollinate the females randomly. In general, fewer plants from the male line are required and one row of males can be planted for every four rows of females. In addition to the isolation distances and other requirements discussed in the seed-based propagation section (Section 4.2), it is important to monitor the field to remove any males/females that were missed, or any hermaphroditic individuals. The end-result of this process will provide a cross between two cultivars, but will include the full genetic background of both. While this approach has applications for breeding programs, it would generally result in high levels of variability and is not commonly used for propagation.

A variation of the described method is the production of feminized seeds from this type of cross so that there are no males present in the progeny. The specifics of producing feminized seed will be discussed in what follows, but the general process requires that the pollen is derived from a female plant that has been induced to go hermaphroditic and produce male flowers. By doing this, the pollen that is produced is genetically female and all seed resulting from it will be female. The general method to do this is the same as previously outlined, except that the "male" parents used for the cross are actually female plants that are treated to induce anatomically male flowers. Another minor modification is that a higher ratio of "male":female plants may be required as feminized pollen appears to be less viable than true male pollen (DiMatteo et al. 2020).

The much more common approach to seed production is the selective pollination between two individual genotypes to produce F1 hybrid seed. As with the approach previously outlined, it is ideal to conduct test-crosses to assess combining ability prior to commencing larger-scale propagation. Hybrid seed is produced in much the same manner as the previous approach, except that the male and female lines are initially selected from a seedling population and clonally propagated to produce the plants used for seed production such that all seed has the same mother and father. The genetic and phenotypic variability of seed produced through controlled pollination of individuals will depend on the heterozygosity of the parental lines, but would generally be less than the previous methods—and, in some cases, can be very uniform. This is especially true if the parental lines have been inbred, in which case the resulting F1 hybrid seed can be nearly genetically identical. As with the previously described approach, this can be done crossing a specific male with a specific female, or feminized seed can be produced using two different female lines, with one used as the pollen donor. While breeding is outside the scope of this chapter, it should be noted that there is some evidence that the outcome will vary depending on which genotypes are used as the mother and father, and this should be considered and tested before implementation. Hybrid seed can also be produced using hand pollination as described in Chapter 3, which may be especially useful to ensure good seed set in feminized seed production.

4.2.3 Production of Feminized Seed

As with many dioecious crops, only one of the sexes produces an economically important product in cannabis. In the case of cannabis, only the female produces sufficient amounts of cannabinoids and males are not commercially useful for this application. Furthermore, female plants that are pollinated and set seed produce lower yields of cannabinoids. Seeded plants are of much lower value and often not marketable as dried flowers; they can be used for extraction, but the overall yield and value of seeded cannabis flowers is significantly lower and undesirable. To avoid pollination and fertilization of female cannabis plants, it is imperative that no male plants are present. The labor-intensive and costly process of identifying and removing male plants from a seedling population can be eliminated using vegetative propagation (see Section 4.3), or producers can use feminized seed as outlined in what follows.

To produce feminized seed, genetically female plants are treated with various chemicals or environmental stressors to induce the formation of anatomically male flowers (Ram and Jaiswal 1972; Ram and Sett 1982b, 1979). While these flowers appear similar to true male flowers and produce viable pollen, they are still genetically female. Since cannabis sex determination is largely analogous to the XX/XY system in humans, this means that the XX pollen can only contribute the female sex determination and all seeds that are produced by crossing a female (XX) with a hermaphroditic female (XX) are female (XX).

Although sex determination in cannabis is similar to the human XX/XY system, it is much more plastic and hermaphroditism is a common and not fully understood phenomenon. This can lead to unwanted male flower production and seed set, despite the use of all female plants, and a number of considerations should be taken when making feminized seed. For example, male flowers can be induced using environmental stress, especially through disrupting the photoperiod, and higher temperatures, but this approach essentially selects for genotypes that are predisposed to producing hermaphroditic flowers in stressful conditions. Using stress to induce hermaphroditic flowers may lead to unwanted pollination in a commercial setting and should be avoided.

The alternative approach to induce male flowers is through the application of various chemicals that act as or inhibit the activity of specific plant growth regulators (PGRs) (Ram and Jaiswal 1972; Ram and Sett 1982b, 1979). Since this approach relies on manipulating PGRs rather than inducing hermaphroditism from a stress response, a PGR-based approach is likely to act independently of a plant's natural propensity to form hermaphroditic flowers, therefore making it the preferred approach. Some examples of compounds that can induce male flowers on female plants include gibberellic acid and a variety of ethylene inhibitors (2-aminoethoxyvinyl glycine, silver nitrate, silver thiosulphate [STS], etc.). Ethylene is a gaseous plant growth regulator associated with sex determination in a variety of plant species (Dolan 1997), and ethylene inhibitors are the preferred class of chemicals for inducing hermaphroditic flowers in cannabis. This class of compounds induced hermaphroditic flowers by inhibiting the action of ethylene, either through the suppression of key enzymes involved in ethylene biosynthesis, or by inhibition of the action of ethylene receptors. As would be expected, the application of ethylene to a male plant can induce the formation of female flowers and could be used to produce a mixture of male (XY) and super male (YY) seeds, should an application for this become of interest in cannabis cultivation (Ram and Sett 1982b).

Among ethylene inhibitors, silver nitrate and STS are the two most well documented for this application and will be the focus of this section. Initial studies found that applying drops of silver nitrate or STS to the growing tip of female cannabis plants on a daily basis for five days induced the formation of male flowers and viable seed (Ram and Sett 1982a). However, silver nitrate demonstrated some toxicity to the plant in the form of tip dieback and produced fewer male flowers than STS. As such, STS is the preferred compound.

Since these initial studies demonstrating the efficacy of STS, the application method has evolved, and most producers now spray the entire plant to saturation with an STS solution. In a recent publication evaluating STS treatments across four hemp genotypes, it was found that all of them responded, but the degree of conversion was dependent on genotype (Lubell-Brand and Brand 2018). At 0.3 mM concentrations of STS, the conversion ranged from 42–91% of the flowers in the terminal inflorescence being male, but at 3 mM, the percentages ranged from 95–100%. As such, it appears that either concentration would be sufficient for male flower production, but the higher level is more reliable, especially in cases when the genotypic response is not known. It is also important to recognize that in this study, the solution was applied three times on a weekly basis beginning on the day of switching to a short-day photoperiod. Although further refinement of this method could lead to a reduction in the number of applications or an alteration of the application interval, the method as it currently stands remains a viable approach to produce feminized seed. The following is a protocol based on this study outlining the specific steps required.

4.2.3.1   Plant Preparation
- Healthy plants of the genotype of the desired "male" parent should be selected and maintained in the vegetative state (under long days).
- If either parent is a day-neutral (a.k.a. autoflowering), the treatment should be timed to coincide with the start of flowering. In a cross between two day-neutral plants, they should be planted such that the flower development is synchronized or that the "male" parent initiates flowering first.

4.2.3.2   Mixing the STS Solution
- STS is generally considered to be unstable and is most often made fresh for each application. However, research evaluating its biological activity has found no loss in efficacy over three months when stored in plastic or glass at 2°C (Cameron et al. 1985). While the recommendation of using only fresh STS solutions may be overly cautious, this is still recommended as best practice and may be a reasonable precaution, given the value of this process.
- The STS solution is produced using 0.1 M stock solutions of silver nitrate and sodium thiosulfate. Unlike the STS solution, these are both known to be relatively stable and can be used for many months when stored in the dark at low temperatures (refrigerated). To make the 0.1 M solutions, mix 16.987g/L of silver nitrate and 15.811 g/L of sodium thiosulfate in separate containers using distilled water.
- To produce the STS solution, slowly mix the 0.1 M stock solution of silver nitrate with the 0.1 M stock solution of sodium thiosulfate at a ratio of 1:4 by volume. The final concentration of silver in this solution is 20 mM. This is then diluted with distilled water to obtain the desired concentration (i.e., 1 part solution to 6.66 parts distilled water to achieve 3 mM final solution). Tween 20 is also added at a rate of 0.1% to act as a surfactant (i.e., 10 ml/L).

4.2.3.3   Application of the STS Solution
- This solution is then sprayed onto the entire plant until saturation at the time of switching to short-day conditions (or at the initial signs of flowering for day-neutral genotypes), and repeated weekly for three weeks.
- Pollen from the developing male flowers can be collected and used to produce seed as you would with typical pollen, but greater care may be warranted to ensure adequate pollination (see the following).

While feminized seed production is relatively straightforward and has been well established, there are important factors to consider when generating feminized seeds and using them for propagation or in long-term breeding programs. The induction of male flowers on a female plant provides an opportunity to self-pollinate, which can have significant benefits for developing inbred lines and hybrid seeds in a breeding program. However, it is important to recognize that this can also result in inbreeding depression, and for most propagation systems, it should be avoided. As such, it is important to consider what the intended cross is and only use seed collected from the appropriate plant. There is also a common belief that feminized seed has a higher tendency to produce hermaphroditic plants, especially if multiple generations of feminization are used. Although this has not yet been demonstrated in the academic literature, it could be the result of inbreeding or improper breeding strategies which could result in heritable epigenetic changes (Eccleston et al. 2007), causing an increased tendency of feminized seed to produce hermaphroditic flowers. Until this is better understood, it would be wise to conduct trials with the parental lines of interest to determine whether this issue arises, before implementing this method at a production-scale.

Another issue that has been identified with feminized pollen is that it demonstrates morphological abnormalities that may make it less effective than true male pollen. In a recent study comparing five hemp genotypes induced to produce male flowers, all five genotypes produced irregular-shaped pollen (DiMatteo et al. 2020). The induced male flowers also tended to produce smaller pollen grains with lower germination rates than the true male plants. It has been well established that feminized pollen can produce feminized seed; these results suggest that extra care should be taken to ensure good pollination, including a higher reliance on hand pollenating, or a higher ratio of "male" to female plants should be used.

The final consideration when considering the use of STS (or any compound) is the potential human health and environmental impacts. Ionic silver is not exceptionally toxic to humans, but is extremely toxic to aquatic plants and animals (Howe et al. 2002). Concentrations of 1–5 µg/L are known to kill a variety of animals, including insects, amphibians and fish. The toxicity of silver ions varies with the chemical form and the availability of free silver ions. While silver thiosulphate is less toxic than silver nitrate, it is important that containment strategies are considered before initiating any application of the compound. This is relatively easy in an indoor setting where plants can be contained and any runoff can be kept out of the natural environment, but presents some major challenges for larger scale production in a field setting. Prior to initiating such a program, producers need to ensure that they are following their local regulations regarding application and containment to avoid negative environmental impacts.

4.2.4 Harvesting, Storing and Cultivation of Seed

What is often referred to as a cannabis "seed" is in fact an achene: a type of fruit (Figure 4.1). However, for the purposes of this chapter, the term seed will be used. Cultivation practices for seed production are similar to those developed for producing dried inflorescence or extracts. Since best practices for cannabis cultivation are discussed in the other chapters, the following section will focus on some of the key differences and considerations when the objective is seed production. For larger-scale seed production, a lot can be learned from the extensive experience with hemp seed. When cultivating cannabis for dried inflorescences or extracts, unfertilized inflorescences are harvested at a point that maximizes cannabinoid content. In seed production, this is different: harvest date is determined based on seed maturity. In general, cannabis seeds mature approximately 4–6 weeks after pollination, but the specific time to maturity will vary by genotype and environmental conditions. Another important consideration is that in some cases, pollination will occur over an extended period of time and seed maturity will not be uniform. This is particularly true when relying on passive pollination and can be minimized by removing males once the bulk of pollination has been achieved or by using hand pollination to synchronize the event. However, cannabis inflorescences are indeterminate and mature at different rates within the plant, so despite these precautions, seed maturity will be somewhat variable.

4.2.4.1 Harvest

Due to the variable time to seed maturity and influence of environmental parameters, harvest should be based on developmental signs rather than calendar date. Mature seeds are typically gray-brown

FIGURE 4.1   *C. sativa* cv. Finola seeds (a) and germinated seedling (b). Scalebar = 1 mm

and often have a darker mottled pattern (Figure 4.1). While mottling usually occurs with maturity, some genotypes do not produce this pattern, making it an unreliable indicator unless you are familiar with the genotype. Upon maturation, the calyx will start to dry and the seed will readily fall from the plant (shattering), leading to potential seed loss if it is not harvested with care. For smaller-scale indoor production, seed can be collected over a period of time or whole plants can be manually harvested and carefully processed to minimize seed loss. For larger outdoor seed production, harvesting is done using a combine when seeds first start to shatter (Baxter and Scheifele 2000). At this point, approximately 70% of the seed has reached full maturity, but further delaying harvest will result in increased seed loss, reduced quality and a more difficult harvest for the machinery. In situations with limited drying capacity, harvest should begin earlier and be dried in batches to minimize shattering losses. For more details on large-scale production of hemp seed, which could be adapted for other cannabis types, a number of government fact sheets are available, and one developed in your region (or a similar environment) should be consulted.

At harvest, seed can contain from 10–30% moisture (Baxter and Scheifele 2000), which is not suitable for long-term storage. Seed moisture content for hemp needs to be reduced to 9% or less for safe storage, and harvested plants should be dried immediately using forced air at room temperature to maintain quality. Unlike cannabis harvested for medical or recreational purposes, there is no need for slow drying or curing, and higher aeration levels can be used. By the time the seed has reached 9% moisture or less, the plant material will be dried and can be carefully crushed to release the seed. Seed can then be further cleaned using a standard forced air seed cleaner.

4.2.4.2   Storage

Seed moisture recommendations for storing hemp seed is 9% or less, as previously mentioned. This should also be combined with low temperatures to preserve germination rates and plant vigor. Previous work has found that cannabis seeds (hemp and drug type) with 11% moisture content stored at 20°C had 0% germination rates following 18 months of storage (Small and Brookes 2012). The decline of seed viability was slowed by either reducing seed moisture content or storage temperature. Reducing the moisture content to 6% or less prevented any significant decline in germination over 66 months, even at 20°C. Lowering the storage temperature to 5 °C prevented decline in germination in most treatments, but some decline was still observed at 11% moisture content.

Further reduction in storage temperature prevented germination rate decline even at 11% moisture content. Based on these results, the ideal storage conditions for cannabis seed is 5–8% seed moisture content at temperatures of −20 °C or below; allowing germination rate to be maintained for five years or more. In situations when these conditions are not available, ensuring appropriate moisture levels will help maintain germination rates. However, these recommendations are based on germination percentage and it is possible that seed vigor will decline with time in storage. As such, it is recommended that seed be stored at both low temperature and appropriate moisture content to maintain germination rates, as well as plant vigor.

4.2.4.3 Germination

While some wild forms of cannabis demonstrate seed dormancy, this is not the case for domesticated cannabis seeds, which will readily germinate once they are fully mature (Small 2016). In cases when less domesticated genetics (i.e., wild germplasm, some landraces, etc.) are being cultivated, various treatments such as cold stratification may be beneficial, but for most applications, this is not needed. There is a general practice of pre-soaking, or even pre-germinating, seeds prior to planting to increase germination speed and uniformity, but this is generally not necessary. This practice may damage the emerging radicle during transplant, in addition to being less efficient at a large scale. If increased speed and uniformity are desired, seed priming could likely be a better approach, although seed priming has not yet been studied in cannabis.

For germination, seeds should be planted in a growth medium with a good balance of water-holding capacity, drainage, air-filled porosity and other standard factors. The ideal medium will depend on the specific production system, but some soilless potting mixes developed for plant propagation are well suited. Seeds can germinate in light or dark conditions (Small 2016), and it is best to ensure they are covered in the medium to provide good anchorage for root development. Planted seeds should be maintained in standard cultivation conditions (see also Chapters 5, 6 and 7 for further information) and provided with sufficient water to assure a moist, but not saturated medium. Germination does not need fertilizer, but fertilization can begin 1–2 weeks after germination.

Seeds will generally germinate within the first few days, and the cotyledons will emerge from the media within the first week (see Figure 4.2a). The first set of true leaves will develop with a single leaflet with opposite arrangement (Figure 4.2). The following leaf will generally have three leaflets, and this will increase as the plant grows (Figure 4.2c–d). Initial growth will be slow, but the growth rate will increase rapidly as the plant develops and care should be taken not to overwater during these early, slow stages of growth. Generally, seeds can be planted at relatively high density (i.e., 48 cell flats, Figure 4.2b), but should be transplanted as the growth rate begins to increase. For further information on optimal growing conditions in controlled environments see Chapters 5, 6 and 7.

4.3 VEGETATIVE PROPAGATION

The alternative to using seed is vegetative propagation, in which elite individuals are clonally propagated to maintain their genetic and phenotypic characteristics. This is a common approach in many horticultural crops that do not grow true to type from seed (e.g., apples and most perennial fruit) or for dioecious species when a specific ratio of males to females is desired (e.g., hops, kiwi, dates). For cannabis, clonal propagation is often preferred to ensure uniformity within and between batches for a consistent flavor/aroma, phytochemical profile, harvest date and other important characteristics. This approach also ensures that all plants are female and avoids the need to cull males, although the use of feminized seed can also address this issue.

While there have been some non-photoperiodic (a.k.a. day-neutral or autoflowering) cultivars developed, most are short-day plants that flower when exposed to long, uninterrupted, dark periods. This is important as it allows plants to be maintained as vegetative plants under long-day conditions (i.e., 16/8 hours). As with most plants, vegetative propagation is more successful from vegetative tissues, and maintaining cannabis in this state allows them to be used as mother plants to use for

FIGURE 4.2 Seed-based propagation of cannabis depicting: a) a recently germinated seedling; b) Seedling flats showing typical planting density; c) and d) larger seedlings following transplant; e) A field of seed-based cannabis plants during flowering; and f) an areal view of fields established from seed.

Sources: All images were provided by Avicanna, Inc. from its cultivation site in Santa Marta, Colombia

vegetative propagation for extended periods of time. Proper maintenance and training of mother plants is reviewed in Chapter 7, so this section fill focus on the cloning process itself.

4.3.1 STEM CUTTINGS

Rooted stem cuttings are by far the most common approach to vegetative propagation in cannabis. During this process, apical cuttings with about three fully developed nodes are removed from the mother plant, treated with an auxin based rooting powder, inserted into a rooting medium

FIGURE 4.3 Cutting rooting process. From up, middle left to lower right: cuttings in water, trimmed extra leaves and stems, cuttings dipped into rooting hormone powder then inserted to growing medium in plug tray.

Source: Photo by Youbin Zheng

(Figure 4.3) and maintained in a high-humidity environment to allow root development. In order to take successful cuttings, it is first important to understand the biological basis of plant growth and development during this process, as this will inform important downstream decisions and aid in troubleshooting problems. Once a familiarity with this has been developed, the importance of knowing how to take a cutting comes into play.

One of the most consequential differences in the biology of a whole plant compared to a stem cutting is where they derive their energy. During regular growth, a plant is photoautotrophic and actively uses photosynthesis to fix carbon and store energy in the form of carbohydrates. When a stem cutting is taken, much of the energy used to form roots comes from the stored carbohydrates found within the stem and leaves of the cutting, making photosynthesis less important. As a consequence of this difference, rooting success is largely related to the health and well-being of the mother plant from which the cutting was taken, as the mother plant's health directly affects the availability of energy reserves and nutrients needed for root development. When poor rooting is encountered, it is often worth looking at mother plant maintenance, as well as the actual rooting procedure, to ensure that the cuttings are healthy and have the resources needed to complete the process of root formation. Despite the importance of mother plant health in the rooting process, this has largely been ignored in cannabis research, and optimal conditions are not fully known. There are some general principles that can be applied in cannabis based on research in other species. For example, moderately low levels of nitrogen are ideal to prevent overly succulent growth; however, adequate amounts of nitrogen are still required to maintain plant health and ensure good rooting (Rein et al. 1991). Other nutrients that are known to play important roles in rooting success include phosphorous, zinc and calcium, and particular attention should be given to those nutrient levels in the mother plant. Another important difference between stem cuttings and whole plants is that cuttings do not yet have roots and therefore are extremely susceptible to wilting at low relative humidity. Additionally, a lack of roots limits their ability to obtain nutrients from the medium (if provided) and adds to the importance of mother plant nutrition to support this new growth from the cuttings.

Together, these differences have important consequences with practical implications related to protocol development and troubleshooting. For example, light levels used for rooting cuttings are generally much lower than during plant growth. This reduces water loss through evapotranspiration, but requires adequate carbohydrate and nutrient reserves to be present in the tissue. Another common practice is to trim the leaves to reduce the surface area and water loss. However, this practice also has the effect of reducing available carbohydrates and nutrients, and can be counterproductive in some cases (Caplan et al. 2018). Leaves are a major source of auxin, the plant growth regulator that leads to root formation, and removal of leaves from cuttings can also reduce auxin supply and impact rooting. As such, it is important to consider the costs and benefits of such practices before deciding the best method for any given production system. By understanding some of these factors, better decisions can be made to improve rooting and address problems as they occur. The following section will provide a more detailed overview of how stem cuttings are taken in cannabis.

4.3.2 STEM CUTTINGS IN CANNABIS

Cannabis is relatively easy to propagate using stem cuttings, which makes this a viable approach to propagate the plant at a commercial scale. Rooting rates over 90% are achievable, given healthy stock plants and an ideal environment, although some cultivars may be less prolific. Findings from a recent study suggest that the rooting response of cannabis may be genotype dependent (Campbell et al. 2021), further contributing to the growing body of literature which shows that growth requirements in cannabis can be heavily influenced by genetic background (Page et al. 2021; Monthony et al. 2021b). Protocol optimization may vary based on the specific cultivar; however, the environment, available resources and production preferences outlined in the following section will provide a general overview of best practices.

The first decision to be made is from where to take the cutting. In many species, there is a marked difference in the performance of cuttings taken from different locations within the plant. Some people believe that this is also true for cannabis, although a recent study found that there was no difference in rooting stem cuttings taken from different locations within a plant (Caplan et al. 2018). Location had no major effect in this study, but another study reported that cuttings with a larger stem diameter (2.9–3.2 mm) performed better than those with smaller stem diameters (Cockson et al. 2019), and this could lead to false conclusions that position affects performance if stem diameter variation within the plant is not taken into consideration. Overall, it is generally recommended that healthy apical cuttings with about three fully developed nodes are selected from an actively growing vegetative plant (Caplan et al. 2018). Lower sections of the branch can also be used if plant material is in short supply, but this is suboptimal, as it will result in less uniform plants. One notable exception to this is when cuttings are taken to replace an aging mother plant, in which case lower branches are likely to have a lower mutational load than branches near the plant's apex and therefore provide new mothers with better genetic fidelity.

When taking the cutting it is important to ensure that all surfaces and tools are cleaned and disinfected before starting and regularly between cuttings to prevent the spread of disease. This can be done using ethanol, rubbing alcohol (isopropyl alcohol), bleach or other products, and is an important preventative measure. Using a sharp blade, the cutting should be taken by making a 45° cut immediately below a node. Ideally, cuttings should be inserted into the media immediately, but can be stored at low temperatures and high humidity for several days if needed (e.g., in disposable plastic sandwich bag with moist paper towel stored in a refrigerator).

Once the cuttings have been obtained, there are a few commonly performed procedures that can improve rooting success. These include stem wounding, removal of leaves and trimming leaves. Stem wounding is often done in difficult-to-root species, and has also been reported to increase the success rate in cannabis and reduce the time for roots to develop (Campbell et al. 2019). In this particular study on stem wounding, using a scalpel to scrape the epidermal tissue from the bottom 5 cm of the stem increased rooting from less than 40% to about 80% in two of three cultivars, making it worth the extra time and effort it entails. However, in most cases, rooting rates are much higher without wounding, and it is questionable if the benefits of wounding would outweigh the extra time/ cost. As such, stem wounding does appear to improve rooting success, but should be reserved for difficult-to-root cultivars or situations when rooting is otherwise problematic. In another study, it was found that cuttings rooted best when all three leaves were left on the cutting and were not trimmed (Caplan et al. 2018). However, there are cases when trimming leaves may still be necessary, such as when leaves come into contact with the growth medium after the cuttings are inserted, as this contact can provide an opportunity for disease. Some growers may still choose to trim leaves to reduce overlap between cuttings, as overlap can also facilitate the spread of disease. The choice to trim or not to trim may also be affected by the growth habit of the specific cultivar as some have much longer petioles and a broader growth habit where this becomes more problematic. Taken together, the aforementioned strategies bring situational benefits depending on the challenges faced in a given production facility, and their integration into a propagation protocol must be decided upon based on a cost/benefit specific to your circumstances.

Once the cuttings have been prepared, they are generally treated with an auxin-based rooting powder/gel prior to being inserted into the desired rooting media. As previously mentioned, auxin is a natural plant growth regulator produced by leaves and shoot meristems. In a natural setting, this compound is transported from the shoots toward the roots and provides the plant with a system to direct growth and development. When a stem cutting is removed from the plant, the auxin accumulates at the base and promotes root initiation. Cannabis can be rooted naturally; however, the application of various synthetic auxins can increase the response, resulting in quicker and more prolific rooting. The use of these compounds is standard practice for cannabis and most vegetatively propagated species (Caplan et al. 2018). Despite the importance and widespread use of exogenous auxin to improve rooting in cannabis, no studies have been published that compared concentration, formulation or type of auxin. In some cases, good results have been obtained using commercial preparations of 0.1% indole-3-butyric acid (IBA; usually recommended for herbaceous or softwood cuttings) or 0.2% IBA (recommended for semi-hardwood) rooting powders, but many studies report relatively low rooting success, even with these treatments. It is possible that other auxin formulations would provide better results, especially for cultivars that are relatively slow or difficult to root. For example, other species often benefit from a mixture of IBA and 1-naphthaleneacetic acid (NAA) rather than using IBA on its own. Additionally, liquid formulations of auxin improve rooting in many species. Further research is needed to provide empirically based recommendations for which type/formulation of auxin should be used, but in situations when rooting is slow or less successful, alternative auxin formulations would be a logical approach to improve the system.

After the cuttings have been treated with rooting powder/gel, they are ready to be planted into a rooting medium. Choice of rooting media can play an important role in early rooting success of cuttings and should not be overlooked. Rockwool has been recently highlighted as one of the preferred media for root establishment (Campbell et al. 2021; Lubell-Brand et al. 2021). In a study of four commonly used rooting media, Campbell et al. (2021) report that rockwool offered a 7–13-fold improvement over the second most effective rooting medium. However, choice of rooting medium is also influenced by the production system that plants are being propagated for. For example, for many hydroponic production systems, cuttings may be rooted in rockwool or oasis cubes, while transplants intended for outdoor systems would more likely be rooted in a peat-based soilless potting mix. Regardless, a medium that provides a balance of water-holding capacity to ensure the cuttings have adequate access to water and good drainage to provide gas exchange is ideal. Growing media are often pre-soaked with water or a weak nutrient solution (i.e., electrical conductivity of 0.7 dS × m^{-1} and a pH of 5.2), and in some cases a preventative beneficial microbe such as *Trichoderma harzianum* is also added. In general, there is a tendency to overwater cuttings, so attention should be paid to ensure the medium is moist, but not waterlogged. Existing research has shown the importance of rooting media in promoting rooting in other species, but to date, few studies have comparing rooting media types in cannabis, and this represents an important area for future exploration.

Another consideration at this point is the planting density of the cuttings. While it is tempting to plant as many plants as can fit in a tray to maximize production efficiency, this can lead to the spread of disease and result in plant loss. It is better to ensure that the cuttings are spaced apart such that there is not too much leaf overlap. As previously mentioned, this is also a common reason for producers to trim leaves, as it will enable higher densities.

Once the cuttings are inserted in the medium, they need to be placed in an appropriate environment to facilitate root initiation and development. As mentioned previously, one of the most important factors is to provide a high-humidity environment to avoid desiccation. This can be achieved using various approaches including plastic humidity domes, covering benches with plastic film or using mist/fog benches. This decision may also influence the selection of growing media; for example, propagation in a mist bed where water is regularly added into the system may be better suited to media with higher drainage than a humidity dome where water may not be added at all. Regardless of the system used for this, the objective is to provide a high relative humidity (~95%) for the first 4–7 days and then slowly reduced to ambient levels over the next week or two by slowly opening vents on the humidity dome, reducing misting frequency, etc.

Light quality, intensity, and photoperiod are also important considerations when rooting cuttings. While these conditions have not been fully optimized, light intensities of 100–200 µmol·m^{-2}·s^{-1} are often used and considered appropriate, but many herbaceous species perform better with slightly higher levels, around 300 µmol·m^{-2}·s^{-1}. In a recent study comparing different spectra at 200 µmol·m^{-2}·s^{-1}, Moher (2020) found no major differences, suggesting that this is not a critical consideration and that fluorescent or LED systems are equally effective. Typically, long-day photoperiods are used, ranging from 16 hours to continuous light, but there have been no comparative studies on this aspect.

Temperature is another critical factor for successful vegetative propagation. As with many other parameters, this has not been fully optimized, but temperatures of about 24 °C are generally used. While most studies and environmental controls only consider air temperature, the temperature of the rootzone is arguably more important for rooting cuttings. It would be preferable to use a temperature probe inserted into the medium to regulate temperature during this phase. In general, the ideal conditions are that the rootzone is 2–3°C warmer than the air temperature to promote faster rooting response and relatively slow shoot growth. While this is generally not reported in the cannabis literature, one study did find that rootzone heating increases rooting by day 14 from 12.5% (25.6°C) to 66.7% (27.8°C), and increases a qualitative rooting index measured at day 28 from 0.40 to 0.68 on a scale from 0–1 (Cockson et al. 2019). Based on this, it is evident that rootzone heating is beneficial for cannabis cuttings and should be implemented if possible. In general, even if rootzone heating cannot be implemented, it is preferable to base environmental controls on rootzone temperature. This may result in quicker shoot growth than desired, but this is usually preferable to slow rooting due to suboptimal temperatures.

One of the main issues during vegetative propagation is disease control, especially damping off diseases caused by *Pythium* spp., *Phytophthora* spp. and *Fusarium* spp. In the case of cannabis, this is particularly problematic due to the lack of certified control agents and consumer preferences for cannabis products that have not been treated with pesticides and fungicides. As such, the best approach to mitigate this issue is through preventative measures to eliminate the primary source of inoculum. As mentioned previously, this starts with using proper hygiene and sanitation protocols to ensure that all tools, materials, substrates and water are free of pathogens. The second common source of pathogens is from the mother plants themselves, and it is important that they are maintained as healthy plants and steps are taken to control pests, disease and viruses (see Chapter 8 and Chapter 9 for further detail). Other than these preventative measures, control agents are limited and will depend on the specific local regulations. Due to these restrictions, many producers have opted for the use of beneficial microbes such as *T. harzianum*, but their efficacy has not been demonstrated for cannabis, and it is not known if there is a benefit or not.

4.4   USE OF IN VITRO TECHNOLOGIES FOR CANNABIS PROPAGATION

While traditional vegetative propagation offers an affordable approach to propagate genetically and phenotypically uniform plants from elite genetics, it has a number of drawbacks that are specifically problematic for large-scale production. The primary issue is that the mother plants used as a source of cuttings are vulnerable to insect pests, bacterial/fungal pathogens and viral infection. As a result, mother plants often serve as the primary inoculum and can dramatically increase disease pressure during production. Although insects and bacterial/fungal pathogens can be controlled to some degree with biological and chemical controls, viral infections cannot be controlled through these means, and preventative measures are required. Some other issues with traditional vegetative propagation include the amount of space required to maintain mother plants, often accounting for 10–15% of commercial floor space, the potential for genetic mutations to occur and a general decline in vigor over time.

An alternative approach for clonal propagation is through micropropagation, which uses plant tissue culture techniques to mass propagate plants. During the process of micropropagation, plant tissues are surface sterilized to eliminate insects, bacteria and fungi, and are further propagated under axenic conditions to maintain sterility. As a result, propagules produced through micropropagation are free of insects and microbial pathogens, providing reliably clean starting material to reduce the primary inoculum and disease pressure during production. While micropropagation on its own does not eliminate viral infections, more specialized techniques using meristem culture can be used to produce virus-free plants from infected individuals. These can then be further propagated in vitro to produce clean planting materials free of insects and microbes, in addition to being virus-free. Since micropropagation is conducted in small vessels and requires lower light levels than maintaining mother plants, it has the potential to greatly reduce the floor space required for propagation and specifically for maintaining large numbers of elite genetics. While the mutation rate in tissue culture is often higher than during regular growth, low temperature culture and cryopreservation techniques can be used for preservation of genetics and provide a superior approach for long-term maintenance.

Despite the advantages of micropropagation, it requires specialized equipment and expertise, and is generally more expensive than traditional propagation techniques. For many crops, micropropagation can be up to ten times as expensive as traditional methods. However, it is important to note that there is no need to select one or the other: the most efficient approach is likely a combined system, as we see in other crops. A good example of this hybrid propagation platform is used in the seed potato industry, where the initial plant material (nuclear stock) is provided as virus-indexed plants produced through micropropagation. The nuclear stock is then multiplied, first as mini-tubers in the greenhouse and then again in isolated fields where infection is unlikely. During this multi-step propagation system, the plants are regularly tested for important pests, diseases and viruses to ensure they are below a critical threshold. The end-result is that the potato industry reaps most of the benefits of micropropagation (i.e., lower insect, disease and viral pressures) at an affordable cost.

In a similar manner, there are several options for the propagation of cannabis. These include the exclusive use of traditional propagation through stem cuttings (lowest cost but highest insect/disease/virus pressure), a hybrid model whereby mother plants are initiated through micropropagation and then further multiplied in controlled environments using stem cuttings with regular testing (higher cost, but lower insect, disease and virus pressure), or exclusive use of micropropagation (highest cost, but lowest insect, disease and virus pressure). The optimal approach will undoubtedly depend on many factors and should consider in the cost of micropropagation vs. traditional methods, the value of floor space required for maintaining mother plants, the resources and expertise availabilities and the level of insects, diseases and viruses in your facility.

While the specific role of plant tissue culture and micropropagation in any given production facility will vary, these techniques provide important tools for the industry and will play a major role moving forward. The following section will provide a general summary of the current state of these techniques and their application. For a more thorough review of recent academic work, several reviews have recently been published (Monthony et al. 2021c). However, it is important to note that many of these techniques are still in their infancy and will continue to be refined and improved in the near future.

4.4.1 Micropropagation

A general definition of micropropagation is the use of plant tissue culture techniques to propagate plants, but this can take many forms each with their own advantages, challenges, and limitations. Despite this variation, the process is generally divided into several key steps. Initially divided into four stages, a fifth stage (stage 0) was added in recognition of the importance stock plant maintenance plays in the process. The following is a general summary of each stage of micropropagation, with cannabis-specific protocols presented below:

> *Stage 0: Stock plant selection and maintenance.* As with any propagation technique, the identification and maintenance of parent plant material has a significant impact on the success of the protocol. Stock plants should be well nourished, receive adequate light levels, and be relatively insect/pathogen free. In general, plants should be maintained in a similar fashion as mother plants used for cuttings (See Chapter 7). While this is the ideal situation, it should be noted that micropropagation is often used to rescue plants that are struggling for a number of reasons including insects, disease, abiotic stress or viral infection. In these cases, suboptimal stock plants must be used, but initial success rates and vigor will suffer accordingly.
>
> *Stage 1: Establishment of axenic cultures.* During this stage, explants from stock plants are surface disinfected to eliminate microbial contamination and establish them in vitro. The objective of this step is to eliminate microbes from the surface of the plant, while minimizing tissue damage. As highlighted previously, the success in this stage is highly dependent on the health/vigor of the stock plants. Additionally, the environment in which they are grown will have a significant impact on the success of this stage and the specific protocol that is used. For example, explants collected from plants growing outdoors will general have much higher microbial load than those collected from plants growing in a greenhouse or indoors, and will require a more aggressive sterilization process. Once established in vitro, nodal explants generally produce an initial flush of shoots, followed by sporadic growth while the tissues acclimate to in vitro conditions. Growth patterns of other explant types (i.e., leaf tissue, isolated meristems, etc.) may vary, but given that nodal explants are the more common for cannabis, this will be the focus.

Stage 2: Multiplication. This is arguably the most important stage of micropropagation as it is where the number of plants is increased. The true power of micropropagation is the exponential multiplication that occurs during this stage. For example, with a multiplication rate of 10, where one shoot can produce ten new shoots each cycle, a single shoot can produce one million plants within six cycles. Assuming a cycle of two months, this means that within a single year, one nodal explant could produce 1,000,000 plants; however, this assumes that none are lost to contamination or other causes. Multiplication can be achieved through various means, including shoot proliferation through nodal cultures (the most common and commercially adopted), de novo shoot organogenesis or somatic embryogenesis, as will be discussed in more detail in what follows.

Stage 3: Rooting and stem elongation. In this stage, plantlets produced in stage 2 are transferred to a medium to induce root development and sometimes to allow shoot elongation. In vitro rooted plants can then be transferred to the greenhouse (stage 4). Alternately, stages 3 and 4 can be combined and unrooted plantlets from stage 2 can be directly transferred to the greenhouse and rooted ex vitro. The latter option is often preferred for commercial production, provided that survival rates can be maintained as it eliminated one step in vitro and reduces labor costs.

Stage 4: Acclimatization. Often a bottleneck in micropropagation systems, this is the step in which the plantlets are transferred from tissue culture to their intended growing conditions (i.e., greenhouse, growth room, etc.). Success during this stage is highly linked to the quality of plants produced in earlier stages, and this should be considered when problems occur.

4.4.2 Micropropagation of Cannabis Through Shoot Proliferation

Micropropagation through shoot proliferation is a process in which new plants are produced from existing meristems rather than de novo regeneration from non-meristematic cells. This approach is widely used for commercial micropropagation in many crops, as it is relatively easy and is thought to minimize the chances of genetic mutation through somaclonal variation. The latter point is important, as de novo regeneration can result in higher mutation rates and an increased chance of off-type plants. While off-types have not been reported in cannabis, very few studies have tested for this— and based on other plants, it should be expected to some degree. To date, this is by far the most common approach to micropropagate cannabis, although it is worth noting that the terminology in existing literature is often confusing, with some articles referring to this process as regeneration. Other challenges that rises from existing micropropagation methods are that multiplication rates reported in the literature are highly variable, many published methods are not fully developed (i.e., lacking some stages) and there have been challenges replicating published protocols. The following subsections will provide a summary of the five stages of micropropagation for shoot proliferation in cannabis, as well as a more specific sample protocol (Figure 4.4).

4.4.2.1 Stage 0

Stock plants should be maintained in the vegetative state and cared for as you would a standard mother plant (see Section 7.3, Stock Plants). For induction into tissue culture, it is best to cultivate them in an indoor or greenhouse setting to minimize microbial contamination and allow for shorter disinfection times, but explants can also be taken from outdoor plants if needed.

4.4.2.2 Stage 1

The first decision to make when establishing plants in tissue culture is the selection of explants. For shoot proliferation, nodal explants are the most common, as they have existing meristems capable of developing into new shoots. From our experience, larger apical explants with 2–3 nodes work well,

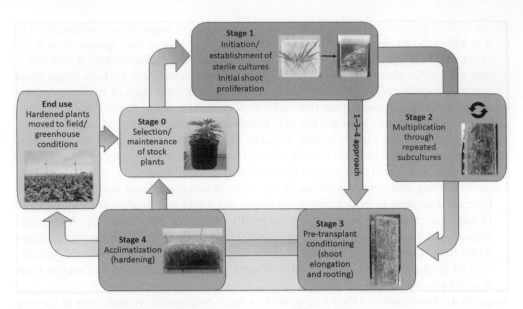

FIGURE 4.4 The five-stage micropropagation process for tissue culture. The arrow connecting stages 1 and 3 indicates the 1–3–4 approach (whereby stage 2 is skipped), which is common in *C. sativa* micropropagation methods. Inclusion of stage 2 allows for repeated subcultures of in vitro plants (indicated by the circular arrows), therefore facilitating large-scale multiplication or long-term germplasm storage. Stages 3 and 4 are connected to indicate that these stages are often combined into a single step for commercial propagation.

Source: Adapted from Monthony et al. 2021c

while smaller single-node explants have a lower survival rate and are slower to initiate new growth. As such, explants with two or more nodes are ideal for culture initiation (see Figure 4.5).

While there are many different solutions that can be used to surface disinfect cannabis explants, the most common is sodium hypochlorite (bleach) with the addition of a surfactant to break the surface tension of the solution. Explants are typically treated for 7–10 minutes in a 10% commercial bleach solution with the addition of 0.01% Tween 20 as a surfactant. This is followed by three five-minute rinses in sterile deionized (or distilled) water. Cannabis plants cultivated indoors tend to be relatively clean and contamination rates are low, but concentrations and incubation times may need to be adjusted for plants coming from outdoors or those with heavier microbial loads. A one-minute ethanol wash is also often included prior to the bleach solution and is thought to be more effective at killing some insects. The latter is an important consideration, as thrips and mites can survive the sterilization process and are capable of moving between vessels. While thrips and mites can cause damage on their own, the problem is compounded in a tissue culture setting, as they can carry microbes with them and cause catastrophic losses due to contamination.

Following surface sterilization, the explants are cultured in sterile culture vessels containing induction medium (Figure 4.5). While many different vessels can be used, we prefer to use culture tubes with a single explant per tube for this step to prevent spread of contamination and use space efficiently. Lata et al. (2016) recommend Murashige and Skoog (MS) medium supplemented with 0.5 µM thidiazuron (TDZ) or meta-topolin (mT) for shoot proliferation at this stage, and we have found that either of these options work well, but further improvements in this media are likely to be developed. Additionally, 1 mL/L of Plant Preservative Mixture™ (PPM) is often added at this stage to reduce contamination. Following inoculation, cultures should be closely monitored for contamination over the next couple weeks and discarded immediately (re-surface sterilization can be attempted, but is generally difficult and would only be advisable in situations when the explant is irreplaceable).

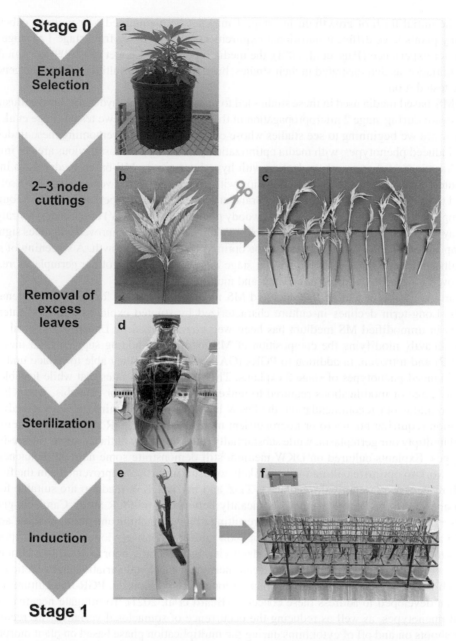

FIGURE 4.5 Induction of vegetatively propagated nodes from *C. sativa* mother plants (stage 0) into stage 1 of tissue culture: a) Selection of healthy starting material; b) and c) 2–3 node cuttings are taken and excess tissue such as leaves are removed to reduce contamination; d) nodes are sterilized using bleach and rinsed with distilled water; e) and f) individual 2–3 node explants are each transferred to capped culture tubes containing sterile induction medium and the racks are moved to controlled growth conditions under artificial lighting.

4.4.2.3 Stage 2

During stage 1, there is generally an initial flush of growth from the parent material, and many explants will form multiple shoot clusters (MSCs). While several articles have reported these numbers as multiplication rates for micropropagation, it should be noted that these reported rates only refers to the initial growth observed in stage 1 of micropropagation. This is an important distinction in that the initial flush of growth during stage 1 may not be maintained over multiple subcultures and many plants have different nutritional requirements as they move from stage 1 to stage 2 culture. In our experience (Page et al. 2021), the media reported by Lata et al. (2009) are suitable for culture initiation as demonstrated in their studies, but not for stage 2 multiplication of the genotypes we have tested it on.

The MS-based media used in these studies led to high rates of hyperhydricity, basal callusing and culture death during stage 2 micropropagation in the five cultivars that we tested (Page et al. 2021). Only now are we beginning to see studies whose objectives include overcoming these undesirable culture-induced phenotypes, with media optimization presenting the most obvious area of improvement. One successful approach to dealing with hyperhydricity in MS-based media, is to increase agar content to 1% (w/v), which in combination with the use of 0.2 μm vented culture vessel lids is reported to prevent hyperhydricity (Lubell-Brand et al. 2021). In our experience, we have found that switching from MS to Driver and Kuniyuki woody plant medium (DKW) basal salts is a straightforward way to reduce hyperhydricity and improve explant health; these improved outcomes signal that the composition of MS basal salts may not be optimal for cannabis growth. A screening of several basal salts confirmed that DKW was best for stage 2 micropropagation of our germplasm, resulting in improved long-term culture maintenance and multiplication.

We are not alone in finding that standard MS media is unsuitable for stage 2 maintenance of explants. Long-term declines in culture characterized by stunted explant height and interveinal chlorosis in unmodified MS medium has been well characterized by Lubell-Brand et al. (2021). Only by heavily modifying the composition of MS medium by adding supplemental mesos (Ca, Mg and P) and nitrogen, in addition to PGRs (GA and mT) were they able to reduce undesirable culture-induced phenotypes of stage 2 explants. This example illustrates that while feasible, there is a wide range of modifications required to make MS media work for cannabis. As such, DKW medium remains our recommendation: the DKW basal salt with vitamins provides a ready-to-use formulation requiring no macro or micronutrient modifications or PGR supplementation to maintain and multiply our germplasm, while substantially reducing many of the negative culture-induced phenotypes. Explants cultured on DKW medium still demonstrate some nutrient deficiencies and ongoing work to optimize culture media is likely to result in further improvements in the future.

With respect to plant growth regulators, TDZ and mT in DKW medium are suitable for plant growth and development, but are not significantly better than no PGR at all. Cannabis generally exhibits a high degree of apical dominance, resulting in low branching and multiplication, and even with the addition of cytokinins, the multiplication rates are relatively low. Prolonged exposure of explants to cytokinins often results in hyperhydricity and generally poor growth. In addition, there are concerns that prolonged PGR exposure can increase somaclonal variation and genetic instability, which is undesirable for the maintenance of long-term germplasm. PGR-free culture methods have been developed to address these concerns (Beard et al. 2021). To mitigate negative culture-induced phenotypes, as well as reducing the occurrence of somaclonal variation, we recommend cycling shoots on and off of cytokinins during the multiplication phase based on plant morphology, or alternatively only using PGRs during the initiation phase and then eliminating them from the culture medium during stage 2.

4.4.2.4 Stage 3

Although stage 2 of micropropagation can be challenging and the multiplication rates are often low, explants will generally root well on basal medium or medium supplemented with IBA (Table 4.1). A critical factor in rooting success is the health and quality of the plants produced during stage 2 of micropropagation. Specifically, hyperhydricity is well known to inhibit root development in many species, and it is a common problem in cannabis micropropagation. The use of 0.2 μm vented lids has also been reported to reduce hyperhydricity during in vitro rooting with IBA (Lubell-Brand et al. 2021). If problems are encountered during the rooting phase, it is often best to evaluate the quality of plants being produced and focus on optimizing stage 2 rather than stage 3 micropropagation.

TABLE 4.1

Recommended Culture Conditions for Stages 1–4 of the Five Stages of Cannabis Micropropagation

Stage	Recommendations	
1	**Basal salts:** DKW with vitamins **Gelling agent:** 0.6–1% (w/v) agar **Carbon source:** 3% (w/v) sucrose **PGRs:** 0.5 µM TDZ or 0.5 µM mT (optional)	**Additives:** 1 mL/L PPM to reduce contamination **Culture Vessel:** one explant per capped culture tube
2	**Basal salts:** DKW with vitamins **Gelling agent:** 0.6–1% (w/v) agar **Carbon source:** 3% (w/v) sucrose **PGRs:** 0.5 µM TDZ or 0.5 µM mT (optional; we recommend cycling between PGR and PGR-free media every subculture)	**Additives:** 1 mL/L PPM (optional, based on contamination rates) **Culture Vessel:** four explants per GA-7 (or similar) or ten explants per We-V box 0.2 µm vented lids to reduce hyperhydricity (optional)
3	*For Rooting In Vitro Directly from Stage 2* **Basal salts:** DKW with vitamins **Gelling agent:** 0.6–1% (w/v) agar **Carbon source:** 3% (w/v) sucrose **PGRs:** 2.5–5 µM IBA	**Additives:** 1 mL/L PPM (optional, based on contamination rates) **Culture Vessel:** four explants per GA-7 (or similar) or ten explants per We-V box 0.2 µm vented lids to reduce hyperhydricity (optional)
4	*For Ex Vitro Rooting from Stage 2 Explants (Combined Stage 3 and Stage 4 Approach)* **Soilless media:** rockwool, Oasis, etc. **Culture Vessel:** Plug-trays with clear propagation domes or open trays in mist beds *For Explants Rooted in Stage 3 and Being Acclimatized Ex Vitro* **Soilless media:** rockwool, Oasis, or peat- or coir-based potting mix. **Culture Vessel:** Plug-trays with clear propagation domes or open trays in mist beds	**PGRs:** IBA based rooting hormones (talc- or gel-based) **Nutrients:** Supplementation with half-strength fertilizer.

Abbreviations: Driver and Kuniyuki woody plant medium (DKW); weight per volume (w/v); thidiazuron (TDZ); meta-topolin (mT); Plant Preservative Mixture™ (PPM); indole-3-butyric acid (IBA); plant growth regulator (PGR)

4.4.2.5 Stage 4

Acclimatization can present a bottleneck in many micropropagation systems. This process represents a significant transitionary period for the plants, moving from an in vitro environment with high humidity, low light and supplemental carbohydrates into a situation with higher light levels, lower humidity and no supplemental sugars. In general, the key to this process is to make the transitionary period a gradual transition. This is done by carefully washing the medium from the roots (to remove residual sugars), planting them in potting mix and placing them in a high-humidity, low-light environment. From this point, the humidity should be gradually decreased while slowly providing higher light levels. This environment can be created using mist beds or humidity domes, and slowly decreasing the misting frequency or opening the dome. Alternately, this can be done using a programmable controlled environment chamber to precisely control these factors. Regardless, the objective is to slowly adjust these factors to match the ultimate growing conditions over the course of approximately two weeks, in a manner very similar to stem cuttings.

As with stage 3 micropropagation, cannabis acclimatizes well to ex vitro conditions, provided that the quality of the plantlets are good and over 95% survival is realistic. As is the case with rooting, if the plants are hyperhydric or in poor health, the survival rate will be much lower, and the best way to address this is by focusing on the upstream micropropagation protocols to produce healthier plants rather than focusing on stage 4 micropropagation itself.

An alternative approach to stage 3 and stage 4 is to combine them into a single step and directly rooting shoots in the greenhouse/growth room (Table 4.1). In this case, the same principles are used to slowly transition from the in vitro setting to the growth room: well-developed healthy shoots without roots are removed from culture, dipped in a rooting powder/gel and planted in a mist bed/ humidity dome. This approach is attractive as it eliminates a step in vitro, thereby reducing overall costs. From our limited experience, cannabis responds well to this approach and it is a viable alternative that has also been explored by other groups (Murphy and Adelberg 2021; Mestinšek-Mubi et al. 2020). However, this method has yet to be implemented at a large scale.

Sample Protocol

Stage 0

- Select healthy plants maintained in the vegetative stage as shown in Figure 4.5a.

Stage 1: Surface Sterilization and Initiation

- Cut healthy branches from the mother plants and cut them into 2–3 node explants, as shown in Figure 4.5b–c.
- Rinse explants under running water for approximately 5–10 minutes to remove debris.
- Remove large leaves and petioles from the explants so as to reduce the surface area upon which contaminants may remain (Figure 4.5b–c).
- In a flow bench (from here forward), place explants into vessel with 10% commercial bleach (sodium hypochlorite) and 0.01% Tween 20 and agitate for seven minutes. If contamination is an issue with your material, increase the time or concentration incrementally to identify a suitable method (Figure 4.5d).
- Transfer all explants into a pre-sterilized vessel containing distilled water (deionized or reverse osmosis [RO] water can also be used). Agitate for five minutes, and repeat two more times.
- Transfer onto culture tubes containing pre-sterilized initiation medium with one explant per vessel (Figure 4.5e–f and Table 4.1).
- Place in culture room under long-day photoperiod ($\geq$ 16 hours).
- Monitor for contamination and plant health.

Stage 2: Multiplication

- When shoots have developed from the initial explants (approximately four weeks), remove them from the vessel and dissect into 2–3-node explants.
- Transfer explants into multiplication medium (Table 4.1) on a slight angle approximately 1 cm deep. Multiple explants can be cultured in a single vessel (i.e.,. 4/GA-7, 10/We-V), but it is recommended to avoid overcrowding.
- Close and seal the vessels, and place them in the culture room under a long-day photoperiod.
- Repeat this stage until the desired number of plants are produced. Note that the explants in the initial subcultures will respond sporadically, but this should become more uniform after several cycles.

Stage 3: In Vitro Rooting

- During stage 2 of micropropagation without cytokinins, we typically observe approximately 40–50% rooting. These plants can be directly used for stage 4 micropropagation.
- For rooting any unrooted shoots, transfer whole shoots onto root induction medium after making a fresh cut at the base of the plant (Table 4.1).

- Similar to stage 2 micropropagation, several explants can be cultured per vessel, but over-crowding should be avoided.

Stage 4: Acclimatization

- Remove rooted explants from the culture medium and carefully rinse them under running water to clean the roots.
- Transplant them into pre-moistened potting mix and place them in a high-humidity environment. This can be achieved using humidity domes, mist/fog beds, controlled environment chambers, etc.
- Over the next two weeks, start to provide the plants with half-strength fertilizer and slowly reduce the humidity (slowly open humidity domes, decrease mist frequency, etc.).
- Once new growth is observed, plants can be transplanted and treated as typical plants.
- Note that rooted plants can also be transferred into rockwool or other growing media by splitting the medium in half and carefully placing the plantlet inside.

4.4.3 MODIFIED MICROPROPAGATION THROUGH SHOOT PROLIFERATION

Although micropropagation through shoot proliferation using nodal explants (as described in Section 4.4.2) is common in many species, several authors have noted that the multiplication rate in cannabis is relatively low (Page et al. 2021; Murphy and Adelberg 2021; Wróbel et al. 2020). A number of factors contribute to the low multiplication rates, including a lack of multiple shoot cluster (MSC) formation (the formation of several shoots from the base of the explant) in many genotypes, strong apical dominance that results in very little branching and the poor performance of single node explants that necessitates the use of multi-node explants. As a result, each two-node explant typically results in a single shoot which usually fails to produce multiple shoots, or a branched growth habit that would provide more material for increase multiplication. In many species, MSC formation and branching can be induced by the addition of cytokinins, but from our experience, they offer no considerable improvement in branching of cannabis cultures.

To address this issue, (Wróbel et al. 2020) developed a modified version of shoot proliferation in which nodal explants are cultured as described previously, but after some initial growth, the shoot tip is removed. By removing the shoot tip, apical dominance is broken, and the lateral branches start to develop, resulting in a more branched growth habit. The lateral shoots are then used as explants and subcultured. By doing this, they were able to increase their multiplication rate from an average of 0.9 × (losing plants) to 3.0 × and increase the average shoot height from 0.9 cm to 2.2 cm. While a multiplication rate of 3.0 is still relatively low for commercial micropropagation, it is a significant improvement and could form a basis for future methods. However, it should be noted that this approach requires an extra step in vitro that would increase labor costs and risks of contamination, so these factors should be considered before adopting this modification. A similar hedging method is presented by Murphy and Adelberg (Murphy and Adelberg 2021). This method consists of the repeated removal and subsequent ex vitro rooting of shoot tips from in vitro grown explants. These in vitro explants' bases have been allowed to root in the culture medium and are maintained as in vitro "micro" mother plants over a 12-week period. This hybrid in vitro/ex vitro model produces comparable multiplication rates as the in vitro method proposed by Wróbel et al. (2020). Hybrid systems, which rely on the repeated removal of shoot tips to generate ex vitro clones, are gaining popularity. A retipping method proposed by Lubell-Brand et al. (2021) similarly uses shoot tips from in vitro plants for the production of ex vitro clones. These ex vitro clones are subsequently be retipped for further plant multiplication, yielding competitive multiplication rates. The apparent growing popularity of these systems is likely owing to the favorable trade-off between the benefits of tissue culture (i.e., rejuvenation of plants, safe and reliable storage of germplasm), while also avoiding the increased labor associated with the long-term production of clones in vitro.

A similar, unpublished approach has been adopted to a limited degree in our lab. In this method in vitro, rooted explants are repeatedly coppiced for the harvest of nodal explants. A flush of shoots is subsequently generated from nodes at the intact base, and these form the shoots which can be used for future subcultures. Although the multiplication rate using this approach cannot be directly compared to traditional shoot proliferation methods, since the mother plant provides many flushes, it seems from our experience to be a viable alternative and often produces higher-quality shoots. This method is loosely modeled after how cuttings are taken from mother plants in a greenhouse or growth room, further supporting the emerging trend of maintaining in vitro "micro" mother plants for cloning. One of the challenges with this approach is that the culture medium will dry out and/or deplete over the course of time and needs to be replenished. To achieve this, we have successfully used a variety of liquid culture systems including a temporary immersion rocker system and a thin layer liquid culture system. While there are undoubtedly improvements to be made, we have had good success using basal DKW medium without any plant growth regulators for the mother plants, while the removed shoots can be cultured on standard multiplication medium, basal DKW, or root induction medium, depending on their final destination.

4.4.4 MICROPROPAGATION USING FLORAL TISSUES

Another approach to cannabis micropropagation that has the potential to increase multiplication rates is the use of floral tissues rather than vegetative tissues. The potential for this method was first highlighted by Piunno et al. (2019), where floral tissues from flowering plants growing in the greenhouse or indoor production facility were used to establish plants in culture. This technique worked on two of the three cultivars tested, and could be done using immature or mature inflorescence tissues. The use of floral tissues to establish cultures has significant applications for propagating elite germplasm and developing more efficient breeding strategies, but also has potential to improve micropropagation techniques in general.

The use of floral tissues for micropropagation could represent a substantial improvement over vegetative tissues as the inflorescence of cannabis is a highly branched compound raceme (Spitzer-Rimon et al. 2019). Due to this floral architecture, reproductive tissues contain a high density of meristems that can each potentially develop into new shoots. Recent work has demonstrated that in vitro cannabis plants respond to photoperiod in a similar manner as regular plants, and that flowering can reliably be induced under short-day conditions (12/12) (Moher et al. 2020).

Subsequent work has exploited this to provide a floral micropropagation system based on floral reversion (Monthony et al. 2021a). In this study, established in vitro shoots were transferred to short-day conditions (12/12) to induce flowering (Figure 4.6a). Once the shoots were flowering, immature floral explants of different sizes (single floret vs. pairs) were cultured in long-day conditions on a variety of media containing BA or mT (Figure 4.6b–c, Figure 4.7a, Figure 4.7c). The result of the study found that single florets resulted in the highest multiplication rate, although pairs performed comparably on the same DKW-based medium supplemented with 1.0 µM mT with the added benefit of being easier to dissect and having a higher likelihood of reversion (Figure 4.7). Based on these findings, the end goal of the user should determine whether single florets (best for maximizing multiplication) or pairs (best for low labor and high multiplication) should be used. Since shoots produced from the floral explants can then be transferred back to short-day conditions to induce flowering, this approach can be used repeatedly for stage 2 micropropagation (Figure 4.4). Based on this study, the multiplication rate using this system is projected to be approximately 18.2 × with single florets, which is much greater than what has been reported for micropropagation using vegetative shoots. The multiplication potential over ten tissue culture cycles using this method (both single florets and pairs) and other published vegetative methods is compared in Figure 4.7c. Although this method has great potential, it should be noted that flower induction requires a separate area with photoperiod control and this extends the time for each cycle. Below is a sample protocol for floral based micropropagation.

FIGURE 4.6 Stages of floral reversion-based micropropagation: a) image of in vitro flowering plant grown under short-day (12/12) conditions as a source of explants; b) representative explant consisting of two florets; c) a petri plate showing four floral explants; d) an example of young shoot emerging from a floral explant; e) a petri plate with many shoots developing from floral explants and ready to be subcultured; and f) a rooted shoot derived from floral reversion ready to transfer to the greenhouse.

4.4.5 MICROPROPAGATION THROUGH DE NOVO REGENERATION

De novo regeneration is the production of new shoots or somatic embryos from non-meristematic tissues such as leaves, hypocotyls or roots. Regeneration can occur directly from existing cells or through an intermediary callus phase, but in either case, the process requires the de-differentiation and subsequent re-differentiation of somatic cells. This is fundamentally different than micropropagation through shoot proliferation whereby new plants are derived from pre-existing meristems and

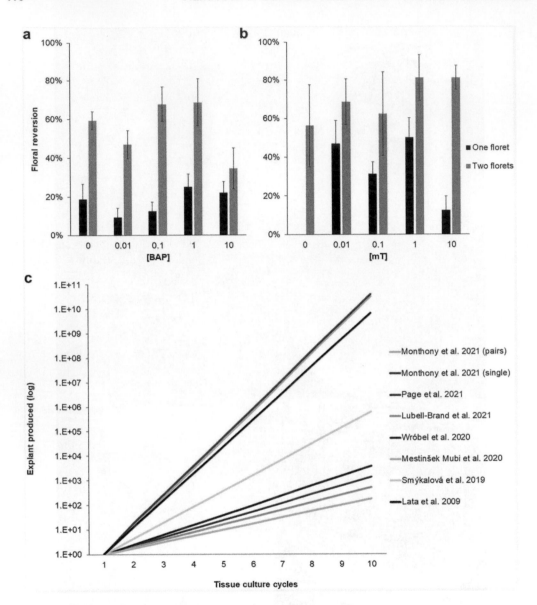

FIGURE 4.7 Percentage floral reversion in single and pairs of florets at different BAP and mT concentrations expressed in μM (a and b) and the multiplication potential of existing vegetative methods compared with single and pairs (c). The interaction between Floret and [BAP] or [mT] were not significant as determined by an analysis of variance. However, 1 μM mT treatment showed the highest percentage reversion of the tested PGRs.

Source: Monthony et al. (2021a)

at no stage do cells need to go through the de- and re-differentiation processes. In addition to being distinct developmental processes, this difference has a number of practical implications.

One of the main advantages of de novo regeneration is that theoretically, any plant cell can develop into a new plant. Since a plant is made up of millions of such cells, the theoretical potential for multiplication is dramatically greater than relying on the limited number of meristems present in the plant. While this is outside the scope of this chapter, de novo regeneration also provides the basis for more advanced biotechnological breeding tools such as transgenics and genome editing.

For these applications, a single cell can be manipulated and used to regenerate an entire plant that contains the new or modified gene. However, it is widely accepted that de novo regeneration—especially when there is an intermediary callus phase—results in significantly higher rates of mutation and can present challenges for large-scale propagation of true-to-type plants. From a commercial perspective, shoot proliferation is often preferred, both to minimize the rates of mutation and because it is technically simpler.

While there have been several reports of de novo regeneration in cannabis, most researchers report low rates of success and a high degree of genotypic variability (Monthony et al. 2021c). One exception to this is an article from 2010 that reported that 96.6% of leaf derived callus produced shoots at a rate of 12.3 shoots per explant (Lata et al. 2010). However, in a recent attempt to replicate this study across ten different genotypes, no regeneration was observed (Monthony et al. 2021b). For a more detailed review of this area, the reader is directed to a recent review (Monthony et al. 2021c), but based on existing published literature, there is no well-developed de novo regeneration system available that works across a wide range of genotypes. Together with the increased potential for somatic mutations, this approach is not currently recommended for commercial propagation.

4.5 OTHER APPLICATIONS OF IN VITRO TECHNOLOGY IN CANNABIS

In addition to propagation, in vitro technologies provide a number of other applications that can be integrated into the production system. These techniques can play an important role in rejuvenating mother plants, improving plant breeding efficiency, preserving cannabis genetics and establishing a clean plant program. While this chapter will not go into great depth on these topics, the following subsections provide a brief summary of some of these areas.

4.5.1 REJUVENATION OF MOTHER PLANTS

Cannabis is commonly propagated through stem cuttings from mother plants for many years. Over time, many producers have observed a general decline in plant health and cannabinoid content. While this has not been quantified, and the underlying mechanism is not known, it is generally thought to result from genetic mutations or epigenetic changes that accumulate as the plant matures. In general, young juvenile plants/tissues produce more vigorous cuttings, while root development becomes more challenging with maturity.

Despite maintaining cannabis mother plants in a vegetative state, based on morphological characteristics, it appears that they lose juvenility over time. Important concepts with respect to plant juvenility are that it is not uniform within a plant, with some tissues being physiologically more mature than others, and that it is at least in part reversible and various treatments can rejuvenate plants. Among the methods that can rejuvenate plants is the use of micropropagation.

Micropropagated plants are well known to revert to a more juvenile state in many plant species. In cannabis, this is widely thought to be the case, and micropropagated plants often demonstrate more juvenile growth characteristics. While further research is needed to explore this phenomenon and unravel the underlying mechanism, it appears that micropropagation is a good approach to rejuvenate declining mother plants and may reverse this serious issue.

4.5.2 IMPROVED PLATFORM FOR SELECTING ELITE GENETICS AND SELECTIVE BREEDING

Micropropagation is most often used for the mass production of elite genetics that have been selected for various traits. However, in vitro technologies can also provide a valuable tool to improve the breeding process. In a standard cannabis breeding program focused on developing new genetics for cannabinoid content, large numbers of seeds are planted and grown to maturity to identify the elite plants with superior yield, chemical profile, disease resistance, etc. However, since the plants need to be evaluated as unfertilized plants at full maturity, cuttings are typically taken before floral induction and maintained throughout the process. Following the selection process, these cuttings provide propagation material to multiply the plants as elite clonal lines or for parents in a selective breeding program.

An alternate approach to this would be to first germinate the seeds in vitro. Once they have reached a sufficient size, an apical cutting from each seedling can be rooted and transferred to the production facility to be evaluated. During the selection process, the remaining basal portion of the seedling can be maintained in vitro. If the in vitro seedlings are maintained at a reduced temperature (10–15°C), no further subculturing is required. Once the elite genotypes are selected from the plants in the production facility, they can be rapidly multiplied for subsequent purposes using original source in vitro seedling. The advantage of this approach is that there is no need to take stem cuttings and maintain copies of each plant in the greenhouse. Instead, large numbers of genotypes can be maintained in a very small space and are available when needed. In general, introducing cannabis seeds into culture is relatively easy following the same protocol as shoot explants, and they grow well on MS or DKW basal media without the need for PGRs.

Another technique that can be adopted to enhance plant breeding programs is the use of floral regeneration to establish cultures from flowering plants. As previously described, florets from greenhouse plants can be used to establish plants in vitro, in some cases up to the day of harvest (Piunno et al. 2019). While further refinement is needed before this would be reliable enough to replace vegetative cuttings, it provides a viable approach to propagate plants from flowering plants when there are no other clones available. Together, these techniques provide valuable tools to improve the efficiency of selection/breeding programs for cannabis.

4.5.3 VIRUS REMEDIATION

As previously discussed, plant tissue culture is often used to produce plants free of insects, as well as microbial and viral diseases. However, standard micropropagation alone does not eliminate viral infections, and using these techniques to propagate an infected individual will result in infected propagules. However, specialized techniques using meristem culture in combination with thermal or chemical therapy can eliminate viruses from infected plants and can be further multiplied through standard micropropagation techniques to produce virus-free propagules.

Traditionally, it was believed that the underlying mechanism was that viral particles travel through plant vasculature, and since young meristematic cells are not yet connected, they are often virus free. By regenerating plants from these cells, it is therefore possible to produce virus-free shoots even from an infected plant. Although recent research has demonstrated that it is more complex than this, and that meristematic cells express genes that inhibit virus replication (Wu et al. 2020), the general principle remains the same in that we know these cells tend to be free from the virus and provide a source of clean plants.

In general, this is used in conjunction with heat, cold or chemical treatments to increase the percentage of virus-free plants that are produced. The optimal combination will depend on the plant species, as well as the specific virus. In the case of cannabis, it appears that a high temperature pre-treatment combined with meristem culture works to effectively eliminate hop latent viroid, and potentially others, but more work is needed in this area.

4.5.4 LOW TEMPERATURE CULTURE TO MAINTAIN GENETICS

Somatic mutations occur naturally during plant growth, and maintaining genetic fidelity in clonally propagated species can be a challenge. This phenomenon is well illustrated by the development of new apple cultivars from "sports," which are naturally occurring spontaneous mutations in a single branch of a tree. When this mutation leads to unique traits with commercial potential, they are propagated and established as a new clonal cultivar. It should be noted that the mutations that we see only represent a fraction of the mutations that are accumulating, and there is significant genetic diversity within clonal populations—and even within a single plant. While this has not been documented in cannabis, based on what we know from other species, it is almost guaranteed to occur during long-term maintenance of a clonal line.

The rate of somatic mutations is thought to be higher in vitro, but depends on the micropropagation methods that are employed, with de novo regeneration techniques being much greater than shoot proliferation. Genetic analysis of micropropagated cannabis plants have not identified any mutations, but these studies used molecular markers (i.e., ISSR) that could have missed many mutations if they occurred in other regions of the genome. Regardless, somatic mutations can occur in vitro, or in a traditional production setting, and other approaches should be considered for long-term genetic maintenance.

Since somatic mutations primarily occur during mitosis, the best way to prevent them is to slow down this process. The most effective way to do this is by reducing the temperature to slow plant growth, and thereby reduce the rate of mutations. One study accomplished this by culturing encapsulated cannabis explants at low temperature, whereby they could be stored for several months without the need for subculturing (Lata et al. 2012). While genetic fidelity was not quantified, this approach is expected to substantially reduce the chances for mutations and provides a good method for genetic maintenance.

This concept can be taken one step further and tissues can be stored at cryogenic temperatures (−196°C). At this temperature, cell division ceases and genetic mutations during the storage period are not possible and tissues can be preserved indefinitely. A recent study (Lata et al. 2019) provides a protocol to cryopreserve axillary buds of cannabis using a droplet-vitrification method. This provides an ideal method for long-term preservation of elite cannabis genetics. Despite the higher upfront costs of cryopreservation, an economic analysis (Li and Pritchard 2009) has found that this approach is more cost effective in the long term for many species.

4.6 CONCLUSION

Cannabis can be propagated using a wide array of techniques ranging from traditional approaches such as open pollinated seeds, stem cuttings, grafting or layering, to more advanced methods to produce feminized seed, F1 hybrids and micropropagation. While each of these methods has distinct advantages, limitations and challenges, there is no one approach that is ideal for all applications, and the propagation system needs to be tailored to the production system. Currently, seed-based propagation dominates the hemp industry, while production of drug-type cannabis is generally accomplished through clonal propagation. As the industry evolves and more stable, seed-based cultivars are developed, it is likely that more producers will adopt seed-based propagation, but for the foreseeable future, clonal propagation will play an important role in the industry and help facilitate breeding progress.

REFERENCES

Baxter, B. J., and G. Scheifele. 2000. *Growing industrial hemp in Ontario*. Ministry of Agriculture, Food & Rural Affairs. https://www.cesarnet.ca/biocap-archive/files/Position_Paper_OCBBE.pdf.

Beard, K. M., A. W. H. Boling, and B. O. R. Bargmann. 2021. Protoplast isolation, transient transformation, and flow-cytometric analysis of reporter-gene activation in cannabis sativa L. *Industrial Crops and Products* 164: 113360.

Cabezudo, B., M. Recio, J. M. Sánchez-Laulhé, M. Del Mar Trigo, F. Javier Toro, and F. Polvorinos. 1997. Atmospheric transportation of Marihuana Pollen from North Africa to the Southwest of Europe. *Atmospheric Environment* 31(20): 3323–3328.

Cameron, A. C., R. D. Heins, and H. N. Fonda. 1985. Influence of storage and mixing factors on the biological activity of silver thiosulfate. *Scientia Horticulturae* 26(2): 167–174.

Campbell, L. G., S. G. U. Naraine, and J. Dusfresne. 2019. Phenotypic plasticity influences the success of clonal propagation in industrial pharmaceutical cannabis sativa. *PloS One* 14(3): e0213434.

Campbell, S. M., S. L. Anderson, Z. Brym, and B. J. Pearson. 2021. Evaluation of substrate composition and exogenous hormone application on vegetative propagule rooting success of essential oil hemp (Cannabis Sativa L.). *BioRxiv*, January. https://doi.org/10.1101/2021.03.15.435449.

Caplan, D., J. Stemeroff, M. Dixon, and Y. Zheng. 2018. Vegetative propagation of Cannabis by stem cuttings: Effects of leaf number, cutting position, rooting hormone, and leaf tip removal. *Canadian Journal of Plant Science* 98(5): 1126–1132. https://doi.org/10.1139/cjps-2018-0038.

Cockson, P., G. Barajas, and B. Whipker. 2019. Enhancing rooting of vegetatively propagated Cannabis Sativa 'BaOx' cuttings. *Journal of Agricultural Hemp Research* 1(1): 2.

CSGA. 2019. *Canadian regulations and rocedures for pedigreed seed crop production, circular 6*. CSGA. https://seedgrowers.ca/wp-content/uploads/2019/01/Circ6-SECTION-00-ENGLISH_Rev02.01-2019_20190124.pdf.

DiMatteo, J., L. Kurtz, and J. D. Lubell-Brand. 2020. Pollen appearance and in vitro germination varies for five strains of female hemp masculinized using silver thiosulfate. *HortScience* 1(aop): 1–3.

Dolan, L. 1997. The role of ethylene in the development of plant form. *Journal of Experimental Botany* 48(2): 201–210.

Eccleston, A., N. DeWitt, C. Gunter, B. Marte, and D. Nath. 2007. Epigenetics. *Nature* 447(7143): 395.

Hesami, M., M. Pepe, M. Alizadeh, A. Rakei, A. Baiton, and A. M. P. Jones. 2020. Recent advances in cannabis biotechnology. *Industrial Crops and Products* 158: 113026.

Howe, P. D., S. Dobson, and World Health Organization. 2002. *Silver and silver compounds: Environmental aspects*. Geneva: World Health Organization.

Lata, H., S. Chandra, I. A. Khan, and M. A. ElSohly. 2009. Thidiazuron-induced high-frequency direct shoot organogenesis of *Cannabis sativa* L. *In Vitro Cellular & Development Biology – Plant* 45: 12–19. https://doi.org/10.1007/s11627-008-9167-5.

Lata, H., S. Chandra, I. A. Khan, and M. A. ElSohly. 2010. High frequency plant regeneration from leaf derived callus of high Δ9-tetrahydrocannabinol yielding Cannabis Sativa L. *Planta Medica* 76(14): 1629–1633.

Lata, H., S. Chandra, Z. Mehmedic, I. A. Khan, and M. A. ElSohly. 2012. In vitro germplasm conservation of high Δ 9-tetrahydrocannabinol yielding elite clones of Cannabis Sativa L. under slow growth conditions. *Acta Physiologiae Plantarum* 34(2): 743–750.

Lata, H., S. Chandra, N. Techen, I. A. Khan, and M. A. ElSohly. 2016. *In vitro* mass propagation of *Cannabis sativa* L.: A protocol refinement using novel aromatic cytokinin meta-topolin and the assessment of eco-physiological, biochemical and genetic fidelity of micropropagated plants. *Journal of Applied Research on Medicinal and Aromatic Plants* 3: 18–26. https://doi.org/10.1016/j.jarmap.2015.12.001.

Lata, H., E. Uchendu, S. Chandra, C. G. Majumdar, I. A. Khan, and M. A. ElSohly. 2019. Cryopreservation of axillary buds of Cannabis Sativa L. by V-cryoplate droplet-vitrification: The critical role of sucrose preculture. *Cryoletters* 40(5): 291–298.

Li, D. Z., and H. W. Pritchard. 2009. The science and economics of ex situ plant conservation. *Trends in Plant Science* 14(11): 614–621.

Lubell-Brand, J. D., and M. H. Brand. 2018. Foliar sprays of silver thiosulfate produce male flowers on female hemp plants. *HortTechnology* 28(6): 743–747.

Lubell-Brand, J. D., L. E. Kurtz, and M. H. Brand. 2021. An in vitro – ex vitro micropropagation system for hemp. *HortTechnology* 31(2): 199–207.

Mestinšek-Mubi, Š., S. Svetik, M. Flajšman, and J. Murovec. 2020. In vitro tissue culture and genetic analysis of two high-CBD medical cannabis (Cannabis Sativa L.) breeding lines. *Genetika* 52(3): 925–941.

Moher, M. 2020. *Lighting strategies for indoor cannabis propagation, vegetative growth, and flower initiation*. https://atrium.lib.uoguelph.ca/xmlui/handle/10214/23724.

Moher, M., M. Jones, and Y. Zheng. 2020. *Photoperiodic response of in vitro Cannabis Sativa plants*. https://journals.ashs.org/hortsci/view/journals/hortsci/56/1/article-p108.xml.

Monthony, A. S., S. Bagheri, Y. Zheng, and A. M. P. Jones. 2021a. Flower power: Floral reversion as a viable alternative to nodal micropropagation in Cannabis sativa. *In Vitro Cellular & Development Biology – Plant* 21: 1–13. https://doi.org/10.1007/s11627-021-10181-5.

Monthony, A. S., S. T. Kyne, C. M. Grainger, and A. M. P. Jones. 2021b. Recalcitrance of Cannabis sativa to de novo regeneration; a multi-genotype replication study. *PLoS One*. https://doi.org/10.1371/journal.pone.0235525.

Monthony, A. S., S. R. Page, M. Hesami, and A. M. P. Jones. 2021c. The past, present and future of Cannabis sativa tissue culture. *Plants* 10: 185. https://doi.org/10.3390/plants10010185.

Murphy, R., and J. Adelberg. 2021. Physical factors increased quantity and quality of micropropagated shoots of cannabis sativa l. in a repeated harvest system with ex vitro rooting. *In Vitro Cellular & Developmental Biology-Plant* 1–9.

Page, S. R. G., A. S. Monthony, and A. M. P. Jones. 2021. DKW basal salts improve micropropagation and callogenesis compared with ms basal salts in multiple commercial cultivars of Cannabis Sativa. *Botany* 99(5): 269–279. https://doi.org/10.1139/cjb-2020-0179.

Piunno, K. F., G. Golenia, E. A. Boudko, C. Downey, and A. M. P. Jones. 2019. Regeneration of shoots from immature and mature inflorescences of cannabis sativa. *Canadian Journal of Plant Science* 99(4): 556–559.

Ram, H. Y. M., and V. S. Jaiswal. 1972. Induction of male flowers on female plants of cannabis sativa by gibberellins and its inhibition by abscisic acid. *Planta* 105(3): 263–266.

Ram, H. Y. M., and R. Sett. 1979. Sex reversal in the female plants of cannabis sativa by cobalt ion. *Proceedings of the Indian Academy of Sciences-Section B. Part 2, Plant Sciences* 88(4): 303–308.

Ram, H. Y. M., and R. Sett. 1982a. Induction of fertile male flowers in genetically female cannabis sativa plants by silver nitrate and silver thiosulphate anionic complex. *Theoretical and Applied Genetics* 62(4): 369–375.

Ram, H. Y. M., and R. Sett. 1982b. Modification of growth and sex expression in cannabis sativa by aminoethoxyvinylglycine and ethephon. *Zeitschrift Fuer Pflanzenphysiologie* 105(2): 165–172.

Rein, W. H., R. D. Wright, and D. D. Wolf. 1991. Stock plant nutrition influences the adventitious rooting of 'rotundifolia' holly stem cuttings. *Journal of Environmental Horticulture* 9(2): 83–85.

Small, E. 2016. *Cannabis: A complete guide.* Boca Raton: CRC Press.

Small, E., and B. Brookes. 2012. Temperature and moisture content for storage maintenance of germination capacity of seeds of industrial hemp, marijuana, and ditchweed forms of cannabis sativa. *Journal of Natural Fibers* 9(4): 240–255.

Smýkalová, I., M. Vrbová, M. Cvečková, et al. 2019. The effects of novel synthetic cytokinin derivatives and endogenous cytokinins on the *in vitro* growth responses of hemp (*Cannabis sativa* L.) explants. *Plant Cell, Tissue and Organ Culture* 139: 381–394. https://doi.org/10.1007/s11240-019-01693-5.

Spitzer-Rimon, B., S. Duchin, N. Bernstein, and R. Kamenetsky. 2019. Architecture and florogenesis in female cannabis sativa plants. *Frontiers in Plant Science* 10: 350. https://doi.org/10.3389/fpls.2019.00350.

Wróbel, T., M. Dreger, K. Wielgus, and R. Słomski. 2020. Modified nodal cuttings and shoot tips protocol for rapid regeneration of cannabis sativa L. *Journal of Natural Fibers* 1–10.

Wu, H., X. Qu, Z. Dong, L. Luo, C. Shao, J. Forner, J. U. Lohmann, M. Su, M. Xu, and X. Liu. 2020. WUSCHEL triggers innate antiviral immunity in plant stem cells. *Science* 370(6513): 227–231.

5 Rootzone Management in Cannabis Production

Youbin Zheng

CONTENTS

DOI: 10.1201/9781003150442-5

5.1   INTRODUCTION

A healthy root system is essential to maximize plant performance. The roots of a cannabis plant function to: 1) take up and transport water and nutrients; and 2) anchor and provide physical support for the plant. Plant roots need water, nutrients, dissolved oxygen (DO), a suitable temperature and minimal biotic (e.g., pathogens) and abiotic (e.g., physical restriction) stress. As long as these conditions are met, cannabis can either be grown in soil or soilless media. Rootzone is the environment where roots reside; it is influenced by several factors (e.g., water content and temperature), and rootzone conditions affect root health and function. All the elements in the rootzone and the environment surrounding it are interconnected; changing one can affect the others. For example, when cannabis is grown in a container with a solid growing medium, too much water can cause an oxygen deficiency and create a rootzone conducive to pathogens, while insufficient water can cause drought stress. For a healthy root system, an integrated approach should be taken to ensures optimal water, DO and nutrients. This approach is called integrated rootzone management (IRM), which mirrors the term integrated pest management (IPM).

     Cannabis can be grown either in soil or soilless cultivation systems (SCS); however, the majority of modern controlled environment cannabis cultivators are using SCSs. An SCS refers to a plant cultivation system that does not use natural soil, or at least the majority of the growing media constituents are not natural soil; they include systems using soilless growing media (e.g., peat-based or coir-based) in containers (e.g., in pots or bags), deep water culture (DWC), nutrient film technique (NFT), aeroponics, aquaponics. Sometimes the term hydroponics is used synonymously with soilless culture, especially when nutrients and water are mainly supplied through nutrient solution. More commonly, hydroponics refers strictly to the cultivation of plants in liquid media. This book avoids the term hydroponics to avoid confusion.

This chapter will first discuss the needs of cannabis root systems, then outline various types of cultivation systems and how to use the IRM approach in cannabis cultivation.

5.2 WATER, WATER QUALITY AND WATER TREATMENT

Cannabis cultivation requires plenty of high-quality water, which may also be eventually discharged from the facility. When choosing the type and location of your cultivation facility, it is essential to consider water quality, availability, and its ability to discharge the nutrient solution.

5.2.1 WATER SOURCES AND REQUIRED QUANTITY

There are a few different water sources that can be used in cannabis cultivation, depending on where you are and what type of production system you are using. Different water sources have their advantages and disadvantages. It's important to take into consideration which water sources to use during the facility planning stage. The following are some common sources of water.

5.2.1.1 Municipal Water

Municipal water is normally reliable at an almost unlimited supply if your facility is within the vicinity of a city or town. However, there is normally a fee for both its use and discharge. Most municipalities use chlorine for water treatment; therefore, both the chlorine and chloride could be a concern if their levels are too high. For example, one should be cautious when using water that contains active chlorine higher than 2.5 mg/l—which may be toxic to cannabis, especially in solution culture.

5.2.1.2 Surface Water

Surface water includes water from rivers, lakes, reservoirs, and so on. If your facility is close to one of these, the water is normally free; however, you need to check with the authorities to see whether a permit is needed. Depending on where you are and the season, both water quality and quantity can vary dramatically. For example, if the river is running through farmlands, there may be pesticide and herbicide contamination.

5.2.1.3 Well Water

Well water can be a good water source, depending on the height of the water table and some other factors. Some well water can have high alkalinity and hardness (see Section 5.2.2 for more details). It may contain large amounts of certain elements, as well. For example, in certain areas of southwest Ontario, Canada, well water can have elevated levels of sulfur, which can result in too high of a nutrient solution SO_4 content, especially when nutrient solution is reused. A permit to take well water may be needed in certain regions, as well.

5.2.1.4 Rain and Snow

Rain and snow are normally of good quality for irrigation. These sources can be sustainable and are free. However, you need to either build a pond or a cistern to store the water. These can be inside or outside, aboveground or below ground. Below ground reservoirs can save space and keep the water temperature constant. The amount is season- and location-dependent. Snow must also be melted. For both rain and snow, beware of the collecting surface and other external sources of contaminations.

5.2.1.5 Reverse Osmosis and Deionized Water

The water quality of all the aforementioned water sources sometimes do not meet the requirements (see water quality guidelines in Table 5.1) at certain locations and/or times of the year, or the cultivator may prefer to reuse leachate from the cultivation system. In these cases, there may be a need to treat the water/solution to improve water quality. Reverse osmosis (RO) or deionization (DI) can be

TABLE 5.1

Guidelines for Irrigation Water Used in Controlled Environment Cannabis Cultivation

Parameter	Units	Upper Limit	Optimum Range
Alkalinity	mg/L CaCO$_3$	150	40–100
	meq/L CaCO$_3$	3	0.8–2
Bicarbonate (HCO$_3^-$)	mg/L	150	30–50
	meq/L	2.5	0.5–0.8
EC	µS/cm	1250	150–300
pH			5.5–6.5
Total Dissolved Solids (TDS)	mg/L	800	0–190
Hardness	mg/L CaCO$_3$	200	20–150
Na	mg/L	115	0–70
	meq/L	5	0–3
SAR		4	0–3
Ca	mg/L	120	40–100
	meq/L	6	2–5
Mg	mg/L	50	5–25
	meq/L	4.2	0.4–2.1
Nitrate-N	mg/L	50	0–10
	meq/L	3.6	0–0.7
SO$_4^{2-}$	mg/L	196	0–100
	meq/L	4	0–2
Cl$^-$	mg/L	140	0–50
	meq/L	3.9	0–1.4
Free chlorine	mg/L	2.0	
P	mg/L	5	0–1
K	mg/L	10	0–10
	meq/L	0.3	0–0.3
B	mg/L	0.5	0–0.5
Cu	mg/L	0.2	0–0.15
Fe	mg/L	4	0–3
Mn	mg/L	0.5	0–0.5
Mo	mg/L	0.01	0–0.01
Zn	mg/L	1	0–0.05
Microbial contamination[*]		0	
Pesticides[*]		0	
Herbicides[*]		0	
Heavy metals[*]		0	

Notes: [*]Although the upper limits for these contaminants are not listed here, it is important to minimize their levels; otherwise, the end-product (e.g., inflorescence) may not meet the acceptable standards set by regulatory bodies. Herbicides in irrigation water above certain levels can cause injury to cannabis plants.

effective solutions in such cases. RO and DI water can either be generated on-site or trucked in, but both RO and DI water can be costly. Also, because of their low alkalinity, it is important to make sure the rootzone pH does not get too low, which is quite common in operations using RO or DI water.

The volume of water needed for a facility depends on many factors. Water needed for cultivation depends on the size of the plants, the number of plants, the type of cultivation system, the environment conditions (especially radiation and vapor pressure deficit [VPD]) and whether the nutrient solution is reused. Plants transpire about 90% of the water taken up by roots and only about 10% remains in the plant. A flowering cannabis plant in a controlled environment may require anywhere from 1–5 liters of water per day. Other areas of a production facility, such as cleaning, also need water; estimates of the water requirements for a cultivation facility should take all these demands into consideration.

5.2.2 Water Quality

Water that meets the quality standards for human and animal consumption is not necessarily good enough for cannabis cultivation. For example, based on the *Guidelines for Canadian Drinking Water Quality* (Health Canada 2020), the maximum acceptable concentrations for both calcium and chloride are not required, and for copper it is 2 mg/L. You can see from Table 5.1 that there are limits for both calcium and chlorine for irrigation water, and copper content as low as 0.3 mg/L may cause root injury. Table 5.1 is a general guideline for source water quality for controlled environment cannabis cultivation. Explanations for some of the parameters are provided below.

5.2.2.1 Alkalinity and pH

Alkalinity is a measure of the acid-neutralizing capacity of a solution. The higher the alkalinity, the more acid is needed to bring down the solution's pH. Alkalinity is normally represented by the equivalent amount of $CaCO_3$ in the water. The units are expressed as mg/L or meq/L, with 1 meq/L $CaCO_3$ = 50 mg/L $CaCO_3$. It can be challenging to grasp the difference between pH and alkalinity. pH is a scale, from 0–14, used to specify how acidic or basic water or a solution is, with 7 being neutral. For most plant species, rootzone pH should be between 5.5 and 6.5, which will be discussed in more detail later in this chapter. For example, when two solutions both have a pH of 7.5 but with different alkalinities, it will take a different amount of added acid to bring them down to an acceptable pH (e.g., 6) for cannabis. The water with higher alkalinity will need more acid.

Acids that are commonly used for pH reduction include nitric acid, sulfuric acid, phosphoric acid and citric acid. Citric acid (among others) is typically used more in organic crop production. When alkalinity is very high, the acid used to neutralize the water can contribute high quantities of nitrate nitrogen, or sulfate sulfur or phosphate phosphorus. The method of calculating the amount of acid needed to neutralize water will be discussed in Section 5.2.3. It is important to bear in mind that when using pH-adjusted nutrient solution, the acid-derived nutrients (e.g., nitrate nitrogen) should be subtracted from the intended fertilizer recipe.

A solution's pH can easily be measured using a pH meter, but alkalinity measurements can be more complicated. The pH values of water, nutrient solution and growing media need to be tested regularly, while alkalinity normally requires testing less frequently. It is essential to at least have one good quality pH meter in each cultivation facility. There are commercially available kits to test alkalinity; however, water samples are more commonly sent to commercial laboratories to test for alkalinity, along with other water quality parameters.

5.2.2.2 EC and TDS

Electrical conductivity (EC) is the measurement of the total electrically charged ions dissolved in water. The higher the EC value, the higher soluble salt concentration is in the solution. The most common unit for EC is µS/cm, where S stands for siemens. The following are conversion factors for various units of EC: 1 S = 1 mho, 1 S/m = 1,000,000 µS/m, 1 S/m = 0.01 S/cm = 10,000 µS/cm, 1 S = 10 dS. TDS stands for total dissolved solids in water, and the unit is in mg/L. When water has too high of a TDS, it is not suitable for irrigation.

5.2.2.3   Sodium and SAR

Not only can high concentration of sodium be harmful to cannabis, but high ratio of sodium to the total Ca and Mg in water can be harmful, as well. The sodium absorption ratio (SAR) is a measure of the amount of Na relative to the amount of Ca and Mg, and is calculated using the following equation.

$$SAR = \frac{Na}{\sqrt{Ca + Mg}}$$

Where all the cation concentrations are in millimoles per liter (mmol/L). When the concentrations are expressed as mmeq/L, the following equation is used.

$$SAR = \frac{Na}{\sqrt{(Ca + Mg)/2}}$$

5.2.2.4   Microbial Contamination

Water can be contaminated by both human and plant pathogens; however, there are still no guidelines for irrigation water microbial contamination available for cannabis production. As a rule, it is risky to use pathogen-containing water in cultivating and processing cannabis. Water-borne pathogens commonly found in cannabis production include *Pythium* spp. and *Fusarium* spp.

5.2.2.5   Pesticides, Herbicides and Heavy Metals

High concentrations of certain herbicides can harm cannabis plants. Typically, regulatory bodies set strict limits for the concentrations of certain pesticides and heavy metals in finished cannabis products (e.g., dry inflorescence). There are still no guidelines for the concentration limits of pesticides, herbicides and heavy metals in irrigation water; however, as a precaution, it is advisable to maintain levels as close to zero as possible. If your facility is located in an agricultural area, and you are using surface water, then the pesticide and herbicide content of the irrigation water certainly needs to be monitored, and the pesticide and herbicide can be removed when necessary.

5.2.3   Water Treatment

If your source water does not meet the water quality guidelines, or if you are reusing your nutrient solution, you may need to treat your water. Water treatment options for cannabis production commonly includes: 1) pH adjustment; 2) particle and debris removal; 3) pathogen removal; and 4) nutrients and other contaminant removal. There are different technologies to choose from to address these needs. For example, for pathogen control, there are a few commonly used technologies such as ozonation, chlorination, hydrogen peroxide (H_2O_2; major component of ZeroTol), UV, RO, heat treatment (pasteurization) and slow sand filtration. Each technology has its advantages and disadvantages. To help cultivators choose the right technologies, our group has created a Greenhouse and Nursery Water Treatment Information System (www.ces.uoguelph.ca/water/index.shtml). The information system is free and regularly updated.

5.3   CANNABIS NUTRITION AND FERTILIZATION

5.3.1   Plant Nutrition

Cannabis needs nutrients to grow and flower. There are essential nutrients and beneficial nutrients. The following are three criteria for essential mineral nutrient elements.

1. A given plant cannot complete its lifecycle without the mineral element.
2. The element's function cannot be replaced by another mineral element.
3. The element must be directly involved in plant metabolism, or it must be required for a distinct metabolic step such as enzyme reaction.

The following are the known essential plant mineral nutrient elements:

- *Macronutrients*: N, P, K, S, Mg, Ca
- *Micronutrients*: Fe, Mn, Zn, Cu, B, Mo, Cl, Ni

Plants normally need higher quantities of macronutrient elements relative to micronutrients, which are mainly for physiological and biochemical processes, such as certain metabolic activities.

In addition to the essential nutrient elements, there are also beneficial elements. For example, the following elements are beneficial to certain plants: cobalt (Co), selenium (Se), silicon (Si) and sodium (Na). Si can help cannabis and some other plants in their resistance to pathogens such as powdery mildew.

Essential and beneficial nutrients play varying roles in plant growth and development, and are used by plants differently depending on their developmental stage and ratio to one another.

5.3.1.1 Nitrogen (N)

5.3.1.1.1 Functions and Root Uptake

Nitrogen is required by plants in quantities only surpassed by carbon. It is the constituent of proteins, nucleic acids, co-enzymes, chlorophyll, phytohormones and some secondary metabolites. Nitrogen promotes rapid vegetative growth.

The major forms of nitrogen that can be taken up by plant roots are: NO_3^-, NH_4^+, amino acids and urea. Plant roots can also take up NO_2^-; however, too much of NO_2^- is toxic. NO_3^- can suppress the uptake of other essential anions such as HPO_4^{2-}, $H_2PO_4^-$, SO_4^{2-} and Cl^-; NH_4^+ can suppress the uptake of other essential cations such as K^+, Ca^{2+} and Mg^{2+}.

5.3.1.1.2 Disorder Diagnosis

When there is too much N, plant leaves can become dark green and the plant can become overgrown, which can result in reduced air flow within the crop and the plant more suscetable to pathogens and insect pests. When there is an N-deficiency, the lower leaves start to become chlorotic first and the yellowing moves up the plant gradually (Figure 5.1). When there is limited supply of N in the rootzone, N can move from older leaves to younger leaves and is thus considered a mobile nutrient. Severe N-deficiency can cause leaf tip and edge necrosis, eventually stunted plant growth and even death. Among the nutrient elements, N-deficiency symptoms are normally first to appear. When a well-balanced nutrient recipe is used and N-deficiency symptoms are observed, it is usually an indication that an increased fertilizer application rate is needed. Applying N as NO_3^- or NH_4^+ can have varying impacts on plants and their rootzones (more discussions in Section 5.3.3.1); too much NH_4^+ and ammonia can be toxic to plants. The applied ratio of NO_3^- to NH_4^+ needs to be taken into consideration, especially in organic cultivation systems.

5.3.1.2 Phosphorus (P)

5.3.1.2.1 Functions and Root Uptake

Phosphorus is the constituent of many organic compounds such as sugar phosphates, nucleic acids, phospholipids and certain co-enzymes. It also functions as an energy carrier and can buffer cell pH. P stimulates root development, stimulates and enhances flower and seed formation, and increases stem strength. However, too much P not only can reduce cannabis growth and yield, but also can harm the environment.

FIGURE 5.1   Typical nitrogen deficiency symptoms of *Cannabis sativa*, with lower leaves showing chlorosis first.

Source: Photo by Scott Golem

Plant roots can take up P as HPO_4^{2-} and $H_2PO_4^-$. Too much P can decrease the uptake and translocation of Zn^{2+}, Fe^{2+} and Cu^{2+}.

5.3.1.2.2   *Disorder Diagnosis*

P deficiency is difficult to identify, especially at the early stage. Look for leaves with necrosis at their tips and edges, and also spots that look like water marks more or less as the symptoms are caused by waterlogging (Figure 5.2A). When P deficiency is not corrected, leaves become further necrotic and eventually dry out (Figure 5.2B).

5.3.1.3   **Potassium (K)**

5.3.1.3.1   *Functions and Root Uptake*

Potassium plays an essential role in water and nutrient movement within the plant and regulates cell turgor pressure and stomatal movement. It can increase plant vigor, stem strength, disease resistance

FIGURE 5.2 Phosphorus (P) deficiency symptoms of *Cannabis sativa*. A: showing leaf with P deficiency progressing from no symptom to severe injury (left to right); B: P deficiency symptom showing at a whole plant level.

Source: Photos by Scott Golem

and inflorescence quality and yield. Potassium is taken up by plants as K^+. K^+ suppresses Ca^{2+} and Mg^{2+} uptake.

5.3.1.3.2   Disorder Diagnosis

When cannabis is deficient in K, some of its leaves start to show interveinal chlorosis, and eventually develop into scorching or browning leaf margins (Figure 5.3). Also, the stems are normally weak.

FIGURE 5.3   Potassium (K) deficiency symptoms of a *Cannabis sativa*. A: showing leaf K deficiency progressing from no symptom to severe injury (from left to right); B: showing whole plant K deficiency symptoms.

Source: Photos by Scott Golem

5.3.1.4 Calcium (Ca)

5.3.1.4.1 Functions and Root Uptake

Calcium is important for the stability of cell wall and membrane. Ca supports cell division and plays an important role in cell elongation in both shoots and root growing tips. Ca is also involved in many developmental and physiological processes, such as stomatal opening and closing.

Ca is taken up by plants as Ca^{2+} and mainly at the tip of the young roots. That's why Ca uptake can be depressed if the root tip is damaged, commonly during transplanting, by insects, disease or water stress. Ca^{2+} uptake can also be suppressed by other cations such as NH_4^+ and K^+. It can be influenced by transpiration rate, and low transpiration rates can cause Ca deficiency. Ca can increase P uptake.

5.3.1.4.2 Disorder Diagnosis

When plant is deficient in Ca, its leaves start to show interveinal brown spots or necrotic tissues, especially toward the leaf tips and edges (Figure 5.4). Too much Ca can cause Mg or K deficiency, as well.

FIGURE 5.4 Calcium (Ca) deficiency symptoms of a *Cannabis sativa*. A: Ca deficiency symptoms at a whole plant level; B: showing Ca deficiency leaf progressing from no symptom to severe injury (from left to right).

Source: Photos by Scott Golem

5.3.1.5 Magnesium (Mg)

5.3.1.5.1 Functions and Root Uptake

Magnesium is a constituent of chlorophyll, vitamins and some enzymes. It also activates enzymes and is involved in biosynthesis of ATP and proteins in plants. Mg uptake competes with the uptake of other cations such as NH_4^+, K^+, Ca^{2+} and Na^+. Mg can subpress Mn^{2+} uptake.

5.3.1.5.2 Disorder Diagnosis

When a cannabis plant is deficient in Mg, the younger leaves start to show interveinal chlorosis (yellowing), with the veins remaining green (Figure 5.5). The chlorosis eventually leads to tissue death.

5.3.1.6 Sulfur (S)

5.3.1.6.1 Functions and Root Uptake

Sulfur is a component of amino acids, proteins, lipids, coenzyme A and some vitamins. S is essential for the formation of glucosides and some volatile compounds. S is mainly taken up by roots as SO_4^{2-}. SO_4^{2-} uptake is regulated by plant hormones, and S metabolism is regulated by cytokinins and auxins. S can reduce the uptake of boron and molybdenum.

5.3.1.6.2 Disorder Diagnosis

When a cannabis plant is deficient in S, the fan leaves start to show symptoms of chlorosis (Figure 5.6). The symptoms are similar to N-deficiency; however, unlike N-deficiency, it takes a longer time for the plant to display symptoms of S deficiency; and when they do, yellow leaves can be seen throughout the plant.

FIGURE 5.5 Magnesium (Mg) deficiency symptoms of a *Cannabis sativa*. A: Mg deficiency symptoms at the whole plant level; B: close look of the young leaves right below the inflorescences showing Mg deficiency symptoms.

Source: Photos by Scott Golem

FIGURE 5.6 Cannabis plant with Sulfur deficiency.

Source: Photo by Scott Golem

5.3.1.7 Chloride (Cl)

5.3.1.7.1 Functions and Root Uptake

Chloride plays a role in photosynthesis, stomatal opening and closing, stimulates the activity of some enzymes, increases tissue hydration and turgor pressure and contributes to the elongation of cells of roots and shoots. Chloride can readily be taken up by roots as Cl^-.

5.3.1.7.2 Disorder Diagnosis

Cannabis plants do not need much chloride; however, when a plant is deficient in Cl, it can cause chlorotic leaves and wilting. In soilless production systems, especially when the nutrient solution is reused, Cl can often accumulate to toxic levels.

5.3.1.8 Iron (Fe)

5.3.1.8.1 Functions and Root Uptake

Iron plays an important role in oxidation-reduction and electron transfer reactions. It activates certain enzymes and is part of ferredoxin. Fe is also involved in photosynthesis and respiration. Plant roots can take up both Fe^{2+} (ferrous form) and Fe^{3+} (ferric form) ions. Fe availability is greatly influenced by the rootzone environment, especially pH. Quite often cannabis Fe deficiency symptoms are observed when rootzone pH is > 6.5. To ensure Fe ions are soluble and available to plant roots, Fe is often supplied as Fe chelate, such as Fe-EDTA, Fe-DTPA and Fe-EDDHA.

5.3.1.8.2 Disorder Diagnosis

Fe is not mobile in plants, and when there is a Fe deficiency, the youngest leaves start to show interveinal chlorosis (Figure 5.7). Similar to S, it normally takes a relatively long period from when Fe deficiency begins until leaf symptoms are visible. High Cu and Zn concentrations in the rootzone can cause Fe deficiency, as well. A fast way of correcting Fe deficiency is through foliar application of chelated Fe. Foliar application can also be used for correcting other micronutrient deficiencies in cannabis.

5.3.1.9 Manganese (Mn)

5.3.1.9.1 Functions and Root Uptake

Manganese activates enzymes in fatty acid synthesis and in DNA and RNA formation. It is involved in chlorophyll synthesis and photosynthesis. Roots take up Mn as Mn^{2+}.

5.3.1.9.2 Disorder Diagnosis

Mn has relatively low mobility in plants. When a plant is deficient in Mn, the youngest leaves first show interveinal chlorosis, much like Fe deficiency. However, it seems cannabis does not need that much Mn and is very efficient at utilizing it, so Mn deficiency is uncommon. Mn toxicity in cannabis is more common. Mn toxicity symptom first appears on older leaves as brown spots surrounded by chlorotic tissue. When using low alkalinity water, such as reverse osmosis, distilled water or dionized water, and rootzone pH is not well controlled, cannabis plants often show Mn toxicity.

FIGURE 5.7 Young cannabis leaves showing iron deficiency.

Source: Photo by Scott Golem

5.3.1.10   Zinc (Zn)

5.3.1.10.1   *Functions and Root Uptake*

Zinc is a constituent of different enzymes essential for energy production, protein synthesis, the formation of some organic compounds and growth regulation. It is required for the formation of auxins. Zn also activates enzymes in biochemical reactions. Plant roots can take up Zn^{2+}, $(ZnCl)^+$ and Zn-chelates. Ammonium application can improve Zn uptake, and P application acts as an antagonist to Zn uptake.

5.3.1.10.2   *Disorder Diagnosis*

When a plant is deficient in Zn, the youngest leaves first show interveinal chlorosis, which is similar to deficiency symptoms of Fe and Mn. Leaf tips can be distorted and eventually become dry. Zn toxicity does not often appear. Too much Zn can reduce root growth and leaf expansion. High Zn concentration in the rootzone can induce Fe, Mn or P deficiency.

5.3.1.11   Copper (Cu)

5.3.1.11.1   *Functions and Root Uptake*

Copper is a constituent of some enzymes which are involved in photosynthesis, respiration and the synthesis of chlorophyll and lignin. Roots can take up Cu^{2+}, Cu^+ and Cu^{2+} chelates. Cu can stimulate Mn uptake. Elevated Cu can also suppress the uptake of Fe.

5.3.1.11.2   *Disorder Diagnosis*

When a plant is deficient in Cu, young leaves may show distortion and reduced or stunted growth. Cu has relatively low mobility in plants, so deficiency symptoms first show on the youngest leaves. Excess Cu can cause root tip damage (brown and stunted root tip growth) and reduced root elongation, but enhanced lateral root formation. Severe Cu toxicity can cause Fe deficiency and chlorosis on the young leaves. Cannabis seems to have favorable Cu uptake. Therefore, precautions should be taken to avoid Cu toxicity. For example, copper/brass components—such as copper pipes—should be avoided in a cannabis fertigation system.

5.3.1.12   Molybdenum (Mo)

5.3.1.12.1   *Functions and Root Uptake*

Molybdenum is essential for converting NO_3^- to NH_4^+ and in some other redox reactions inside plant tissues. Roots can uptake Mo as MoO_4^{2-} and in organic chelated forms.

5.3.1.12.2   *Disorder Diagnosis*

Mo deficiency normally appears first as chlorosis on lower and middle leaves. Young leaves may eventually show slow growth and chlorosis, with leaf tips rolling inward. Mo toxicity is rare. P and Mn can enhance the uptake of Mo; K and S can suppress the uptake of Mo.

5.3.1.13   Boron (B)

5.3.1.13.1   *Functions and Root Uptake*

Boron is involved in cell differentiation, maturation, division and elongation, and therefore meristematic growth. It can stimulate or inhibit certain metabolic pathways and is involved in lignin biosynthesis and in xylem vessel differentiation. B is important for flowering. Roots can take up B as boric acid, BO_3^{2-} and $B_4O_7^{2-}$.

5.3.1.13.2   *Disorder Diagnosis*

When there is a B deficiency, plant meristematic growth is affected, and young leaves may appear misshapen, wrinkled and/or thick. Leaves and stems can also become brittle. B is relatively immobile. Excess B can cause leaf tip and leaf marginal chlorosis and necrosis, appearing first on older leaves. N, P and K can all suppress B uptake.

5.3.1.14   Nickel (Ni)

Nickel was not identified as an essential plant nutrient until the 1980s. One of the primary reasons is that most plants only need a trace amount of Ni, and the amount of Ni existing in water and soil is normally sufficient. Therefore, there is normally no need to supply Ni in cannabis production system.

5.3.1.15   Silicon (Si)

Silicon can increase plant vigor and stem strength, and improve plant resistance to diseases such as powdery mildew. Roots can take up Si as SiO_3^{2-}, and it is commonly supplied as K_2SiO_3 or Na_2SiO_3. Si has very low solubility when pH < 8, and the solubility increases when pH > 8.

5.3.1.16   Selenium (Se)

Selenium can reduce UV-induced oxidative stress and increase tolerance to water stress.

5.3.1.17   Sodium (Na)

Sodium is not an essential nutrient element, though it can replace K in some biochemical reactions in plants. Too much Na in the cannabis rootzone can be harmful to plants. To evaluate the effects of Na, cannabis was grown in hydroponic solutions with NaCl concentrations of 1, 5, 10, 20 or 40 mM, and in aquaponic solutions with NaCl concentrations of 4, 8, 10, 20 or 40 mM. Cannabis inflorescence yield decreased linearly with increasing NaCl concentration in the hydroponic solutions, but not in the aquaponic solution (Yep et al. 2020b). Inflorescence total THC and CBD concentrations decreased linearly with NaCl from 5 mM to 40 mM in the hydroponic solutions. We don't know whether Cl or Na contributed more to these negative effects. Still, it remains important to monitor both Na and Cl concentrations in soilless cannabis cultivation, especially when the nutrient solution is reused.

As previously discussed, the presents and uptake of one element may affect the uptake of others. Nutrient elements interact with each other, and plant performance is affected by these interactions. Liebig's Law of the Minimum describes the relationship between plant performance (e.g., growth and yield) and plant nutrients. Plant performance is controlled not by the total available nutrients, but by the scarcest one (the limiting factor). For example, when there is a shortage of Mn in the rootzone, plant performance is dictated by the available Mn, regardless of the availability of other elements. Also, if one element (e.g., Na) is excessive, the plant performance is controlled by this excessive element. The Law of the Minimum also holds for some of the other growing conditions in cannabis cultivation. For example, one can provide everything (e.g., light, nutrients, water) the plant needs, but without adequate CO_2 in the air, the plant's growth and yield will be limited by the CO_2 concentration (see more detail in Chapter 6).

5.3.2   Tissue Sampling for Nutrient Analysis

As previously discussed, nutrient disorders (e.g., deficiency or toxicity) may cause visible symptoms; however, the visible symptoms may not be able to pinpoint the exact cause. For example, chlorosis on young leaves can be caused by an Fe deficiency, Mn deficiency or Cu toxicity. Also, it takes a long time for cannabis plants to display deficiency symptoms for some elements, such as S. When the symptoms appear, your crop's performance may have already been affected. A combination of visual inspection and nutrient analysis of plant tissue and the rootzone (e.g., nutrient solution and growing media) is a more reliable approach to avoid nutrient disorders.

There are two common reasons for leaf tissue nutrient analysis: 1) determining whether there is a nutrient disorder in the plants; and/or 2) determining the cause of a nutrient disorder symptom. To determine if there is nutrient disorder, the newest mature leaves should be sampled; to diagnose which element(s) has caused the observed symptom, the leaves with symptoms should be sampled.

Leaf samples can be analyzed in-house or sent to commercial laboratories. Most growers work with commercial laboratories. Depending on the analytical methods, the amount of tissue required varies, so it is best to first check with the laboratory before sampling. When sampling, take leaves from representative plants and combine them to make a composite sample. If you suspect that the leaves have been contaminated from spray or for other reasons, then gently wash the leaves with P-free detergent, rinse with deionized or distilled water, then blot dry with a paper towel. You may also dry in an oven at 80°C to a constant weight before sending out for analysis. It is important to record the stage of the crop, the time and location of sampling, as well. At different development stages the leaves can have different nutrient composition.

It is important to remember that sometimes when the rootzone is low in nutrients, plants can grow slower and smaller, but without obvious nutrient deficiency symptoms; and also that analysis can show adequate nutrient contents in the leaf tissue. Therefore, it is also important to test rootzone nutrient status from time to time. This will be discussed later in this chapter.

5.3.3 Fertilizer and Fertilization

5.3.3.1 Nutrient Recipes

The way in which plants respond to essential and beneficial nutrients can depend on how much and in what ratio they are supplied. Different plant species, cultivars and even the same plant at different development stages may have different nutrient demands. Having said that, plants are adaptable, so long as nutrients are supplied within certain ranges, both in terms of amount and ratio. Nutrient levels and ratios outside of these ranges can cause deficiency or toxicity, and subsequently, suboptimal yield and quality in cannabis. For example, when there is an extreme N-deficiency, cannabis inflorescence may have higher concentrations of certain cannabinoids and terpenes, while an oversupply of N can decrease both cannabinoid and terpene concentrations; of course, N-deficiency can cause yield reductions, as well. Therefore, an optimal N is required to achieve both cannabis yield (both dry weight and cannabinoid production per production area) and quality.

There is a long history of nutrient recipe development for soilless plant production; the most well-known one is the Hoagland solution (Table 5.2). Many other recipes are based on modifications of the Hoagland solution to suit a specific plant and its development stages.

TABLE 5.2
Hoagland Solution

Nutrient Element	Concentration	
	mg/L	mmol/L
N	210	15
K	235	6.03
Ca	200	5
P	31	1
S	64	2
Mg	48	2
B	0.5	0.05
Fe	1–5	0.018–0.09
Mn	0.5	0.009
Zn	0.05	0.0008
Cu	0.02	0.0003
Mo	0.01	0.0001

Based on some recent scientific studies (e.g., Bevan et al. 2021), the following are the recipes I would recommend for soilless cannabis production. As previously mentioned, ideal nutrient recipes may vary based on cultivars, development stage and by production system; however, as plants normally have a certain flexibility when it comes to nutrient ratios and amount, the recipes in Table 5.3 can be used as baselines for fine tuning to meet specific needs.

TABLE 5.3
Nutrient Recipes for Soilless Cannabis Production

	Concentration			
	Vegetative Stage		Flowering Stage	
Nutrient Element	mg/L	mmol/L	mg/L	mmol/L
N	200	14.3	195	13.9
P	30–40	0.97–1.3	60	1.9
K	180	4.6	190	4.9
Ca	180	4.5	190	4.75
Mg	36	1.5	36	1.5
S	48	1.5	60	1.88
Cl	2	0.06	2	0.06
B	0.4	0.037	0.4	0.037
Fe	1.5	0.027	1.5	0.027
Mn	0.6	0.011	0.6	0.011
Zn	0.4	0.006	0.4	0.006
Cu	0.02	0.0003	0.02	0.0003
Mo	0.02	0.0002	0.02	0.0002

The recipes can vary depending on many factors, such as the following.

- Cannabis cultivar
- Developmental stage
- Growing environment (e.g., higher concentration in winter months and lower concentration in summer months)
- Type of rootzones or production strategies
- Whether stress strategies (e.g., low N stress) are used
- Method of fertigation
- Nutrient solution reused or not
- Disinfection technology used

The provided recipes are for soluble chemical fertilizers. For organic fertilizers, not all the elements are immediately available for plant uptake. For example, nitrogen in most organic fertilizers must go through a mineralization process before becoming available to plants. Depending on the environmental conditions and the nature of the fertilizer, within ten weeks, only 20–70% of the total nitrogen may become available for plants. Therefore, when using the given recipe for organic fertilizers, nutrient release rate needs to be taken into consideration. For example, our recent studies used 4.0 N/1.3 P/1.7 K liquid organic fertilizer during vegetative stage and 2.00 N/0.87 P/3.32 K liquid organic fertilizer during flowering stage; results showed that highest yields were achieved when 389 mg N/L was applied at the vegetative stage (Caplan et al. 2017a), and 261 mg N/L was applied at the flowering stage (Caplan et al. 2017b). Both rates are much higher than the suggested N rates in Table 5.3.

When making nutrient solutions, one not only needs to meet the target concentration of each element, but it is also important to use the correct chemical form—and possibly different forms, based

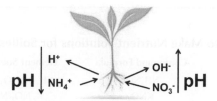

FIGURE 5.8 Illustration of how plant uptake of NH_4^+ and NO_3^- influences rootzone pH.

on your growing media and water quality. This is especially important because chemical form can affect rootzone pH. An acidic fertilizer increases rootzone H^+ concentration, while a basic fertilizer increases rootzone OH^- concentration. There are also fertilizers that do not change rootzone pH. For example, N can be supplied using ammoniacal nitrogen, nitrate nitrogen or both. When ammoniacal nitrogen is applied to the rootzone, pH may decrease; when nitrate nitrogen is applied, rootzone pH may increase (Figure 5.8). When roots take up cations, they release equivalent protons (H^+); and when they take up anions, they release equivalent amounts of OH^- or HCO_3^- to stay electrically neutral.

Bacteria in the rootzone can also convert NH_4^+ to NO_2^- and NO_3^-, and produce H^+ to decrease rootzone pH (see following equation).

$$NH_4^+ + 1.5\ O_2 = NO_2^- + 2H^+ + H_2O$$
$$NO_2^- + 1/2O_2 = NO_3^-$$

Generally, it is recommended to maintain NH_4^+/NO_3^- at between 1/5 and 1/3. If the rootzone pH is increasing, then this ratio can be increased.

Rootzone pH needs to be monitored regularly, and the pH of the feeding nutrient solution needs to be adjusted accordingly.

Table 5.4 outlines the commonly used chemical compounds, also called straight fertilizers, used to make nutrient solutions for soilless production systems.

5.3.3.2   How to Calculate Fertilizer Amount and Make Nutrient Solution

When making nutrient solution, we first need to have a nutrient recipe such as the ones listed in Table 5.3. For example, to make 100 liters of nutrient solution with a concentration of 30 ppm (mg/L) of P using the straight fertilizer diammonium phosphate ($[NH_4]_2HPO_4$), the following steps are followed to calculate the amount needed for 100 liters of water.

Step 1: Calculate how many grams of P for 100 liters of nutrient solution at 30 mg/L using the following equation.

$$C \times V = 100\ L \times 30\ mg/L = 3,000\ mg = 3\ g$$

Where C is the final concentration (mg/L) of an element and V (liter; L) is the final volume of the nutrient solution.

Step 2: Calculate the mass of straight fertilizer needed to provide the amount of P (3 grams, in this case).

$$3 \times \frac{(N+H \times 4) \times 2 + H + P + O \times 4}{P} = \frac{(14+1 \times 4) \times 2 + 1 + 31 + 16 \times 4}{31} = 12.77g$$

Where the N, H, P and O are molar mass (g/mol) of nitrogen, hydrogen, phosphorus and oxygen, respectively.

TABLE 5.4

Common Chemicals Used to Make Nutrient Solutions for Soilless Plant Production

Compound Name	Compound Formula	Element Source	Reaction in Rootzone[a]
Ammonium nitrate	NH_4NO_3	NH_4-N, NO_3-N	Acid
Potassium nitrate	KNO_3	K, NO_3-N	Basic
Calcium nitrate	$Ca(NO_3)_2$	Ca, NO_3-N	Basic
Magnesium nitrate	$Mg(NO_3)_2$	Mg, NO_3-N	Basic
Nitric acid	HNO_3	NO_3-N	
Urea	$CO(NH_2)_2$	NH_4-N, urea	Acid
Ammonium sulfate	$(NH_4)_2SO_4$	NH_4-N, S	Acid
Monoammonium phosphate	$(NH_4)H_2PO_4$	NH_4-N, P	Acid
Diammonium phosphate	$(NH_4)_2HPO_4$	NH_4-N, P	Acid
Phosphoric acid	H_3PO4	P	
Monopotassium phosphate	KH_2PO_4	K, P	Acid
Dipotassium phosphate	K_2HPO_4	K, P	Basic
Potassium sulfate	K_2SO_4	K, S	Neutral
Potassium chloride	KCl	K, Cl	Neutral
Magnesium sulfate	$MgSO_4$	Mg, S	Neutral
Sulfuric acid	H_2SO_4	S	
Calcium chloride	$CaCl_2$	Ca, Cl	Neutral
Iron chelate 11% (DTPA)		Fe	
Iron chelate 13% (EDTA)		Fe	
Iron chelate 6% (EDDHA)		Fe	
Iron sulfate 21%	$FeSO_4$	Fe, S	
Manganese chelate 6% (EDTA)		Mn	
Manganese chelate 13% (EDTA)		Mn	
Manganese sulfate (32%)	$MnSO_4$	Mn, S	
Borax (sodium borate)	$Na_2H_4B_4O_9 \cdot nH_2O$	B, Na	
Boric acid	H_3BO_3	B	
Zinc sulfate	$ZnSO_4$	Zn, S	
Zinc chelate 14% (EDTA)		Zn	
Copper chelate 14% (EDTA)		Cu	
Copper sulfate	$CuSO_4$	Cu, S	
Sodium molybdate	Na_2MoO_4	Mo, Na	
Potassium silicate	$K_2Si_2O_5$	Si, K	

Note: [a]Reaction in rootzone denotes whether the compound applied to the rootzone can increase H^+ (Acid), OH^- (Basic) or has no effect (neutral) on rootzone acidity.

If the straight fertilizer is of 100% purity, then you need to add 12.77 g of $(NH_4)_2HPO_4$ to 100 L of water to make a solution containing 30 mg/L P. Quite often, the straight fertilizers are not 100% pure. If the label indicates the $(NH_4)_2HPO_4$ is 95% pure, then you need to add $12.77 \div 95\% = 13.45$ g. If the fertilizer analysis (label) shows 18–46–0, it means this fertilizer contains 18% N, 46% P_2O_5 and 0% K_2O. Then use the following equation for calculating the amount of the diammonium phosphate to add to 100 L of water.

$$3 \div 46\% \times \frac{P \times 2 + 0 \times 5}{P \times 2} = 3 \div 46 \times \frac{31 \times 2 + 16 \times 5}{31 \times 2} = 14.94g$$

You may have noticed that when adding $(NH_4)_2HPO_4$ to make a solution containing 30 mg/L P, you are also adding ammoniacal N. To calculate the amount of N added, you can use the following equation.

$$\text{g straight fertilizer} \times \text{purity} \times \frac{N \times 2}{(N + H \times 4) \times 2 + H + P + O \times 4}$$

$$= 13.45 \times 95\% \times \frac{14 \times 2}{(14 + 1 \times 4) \times 2 + 1 + 31 + 16 \times 4} = 2.71g$$

Which means you added 2.71 g N to the 100 L solution and the N concentration is 2.71 g ÷ 100 L × 1,000 = 27.1 mg/L.

These example equations can be modified to calculate the additions of any fertilizer. Of course, many fertilizer companies make things much easier for growers. For example, one company that supplies monopotassium phosphate (KH_2PO_4) provides the following table with the product analysis on the label.

Nutrient	P_2O_5	P	K_2O	K
Content (%)	52	22.7	34	28.7

With this information, the calculation becomes simpler. Here is how to calculate the amount of fertilizer needed to make 30 mg/L P:

$$C \times V \div \text{content} \div 1,000 = 30 \text{ mg/L} \times 100 \text{ L} \div 22.7\% \div 1,000 = 13.22 \text{ g}$$

Be aware when reading fertilizer labels. When the label indicates three numbers, such as 20–8–20, it indicates that this fertilizer contains (by weight) 20% N, 8% P_2O_5 (3.5% P) and 20% K_2O (9.9% K).

There are many commercially formulated fertilizers in the market which can contain most or all the essential plant nutrient elements. Here is how to calculate the amount of fertilizer to use when making nutrient solutions. Say the label of the fertilizer is 20–8–20 and you would like to make a 100 L solution with 200 ppm (mg/L) N; the equation is then as follows.

$$C \times V \div \% \text{ N content} = 200 \text{ mg/L} \times 100 \text{ L} \div 20\% = 100,000 \text{ mg} = 100 \text{ g}$$

Adding 100 g of 20–8–20 fertilizer into 100 L of water will provide:

$$\text{P}: 100 \text{ g} \times \%P \div V = 100 \text{ g} \times 3.5\% \div 100 \text{ L} = 0.035 \text{ g/L} = 35 \text{ mg/L}$$
$$\text{K}: 100 \text{ g} \times \%K \div V = 100 \text{ g} \times 9.9\% \div 100 \text{ L} = 0.099 \text{ g/L} = 99 \text{ mg/L}$$

Since the source water often also contains a certain amount of nutrients, the target concentration equals the concentration from fertilizer plus the concentration from water.

When nitric acid or phosphoric acid is used to acidify water for making nutrient solution, the added N or P need to be taken into consideration; also, if KOH or K_2CO_3 are used to bring up solution pH, then the added K needs to be taken into consideration.

Straight fertilizers (e.g., KNO_3) can be added to water to make fertigation solutions with the targeted final concentration, but more commonly, concentrated stock solutions (100–200 times concentrated) are first made and then diluted to make a final solution. Most often, there are two stock solution tanks, A and B, to avoid Ca-containing fertilizers from mixing with P and S-containing fertilizers at high concentrations, which can cause precipitation. Some operations even have a third stock solution tank only containing micronutrients (See more details in Section 5.4.2).

5.4 IRRIGATION/FERTIGATION

When only water is supplied to plants, it is called irrigation, but when both water and nutrients are supplied together, it is called fertigation. For soilless production, fertigation is more common than irrigation; however, if controlled release fertilizer is used, or the growing media contain enough nutrients, then irrigation is used.

5.4.1 Types of Irrigation/Fertigation

There are several types of irrigation/fertigation depending on the size of the operation, the developmental stage of the plant, production system, and so on. Different types have their advantages and disadvantages.

5.4.1.1 Drip Irrigation

Drip irrigation normally relies on pressure-compensated emitters to deliver water/nutrient solution at specific flow rates to the plant rootzone. There are many types of emitters with different flow rates (e.g., 2, 4, 8 liters/hour) which can be chosen for different sizes of rootzone and canopy, and growing environment. Since drip irrigation can be precisely controlled to deliver the right amount of water at the right time, it is one of the most efficient irrigation systems. However, there are also the following disadvantages.

1. Each container needs one or more emitters; this may not be feasible if an operation grows many small plants.
2. Requires long lengths of piping.
3. Emitters can become blocked by debris or chemical deposits such as Ca and Mg.
4. If the quality of the emitters is not good, they can deliver an inconsistent amount of solution to different plants and cause non-homogeneous growth rates, and eventually induce disease in smaller and weaker plants.
5. Irrigation lines and emitters need to be cleaned and disinfected, at least between crops.
6. Can be time-consuming during harvesting, as the emitters are usually removed from containers.

5.4.1.2 Subirrigation

Subirrigation normally utilizes a gently sloped water-holding floor (e.g., troughs, benches or an actual concrete floor) which holds water/nutrient solution at a certain depth. Plants in containers with growing media are placed on the floor. When fertigating, water/nutrient solution reaches the rootzone from the bottom of the container, which is resting on the floor, and the water travels upward through growing media. Water/nutrient solution is delivered to the water-holding floor at the higher end, and drained at the lower end. Subirrigation is commonly used in operations with a large number of small plants.

The advantages can include the following.

1. Efficient and homogenous solution delivery.
2. Easy to operate.
3. Easy to move plants around; whole flood-benches can be easily moved from one area to another when needed.

The disadvantages can include the following

1. Fertilizer salts can accumulate at the upper section of the growing medium; therefore, it is important to monitor the EC of the growing medium and adjust the feed nutrient concentration accordingly.
2. If the floor is imperfectly sloped, nutrient solution can pool to promote green algae and pathogen proliferation.
3. Water-borne pathogens can spread from one plant to the others via nutrient solution.

5.4.1.3 Overhead Irrigation

Overhead irrigation involves watering plants from the top. This includes irrigation with a hose and water-breaker, or an overhead boom system. Overhead irrigation is more commonly used in the propagation stage when plants are small and need a high-humidity environment.

The advantages can include the following.

1. Homogeneous and easy-to-control solution delivery.
2. Plants can uptake water and nutrients through the leaves.
3. Can facilitate more frequent applications.

The disadvantages can include the following.

1. When leaves are wet, they are more susceptible to some pathogens; therefore, it is crucial to make sure the leaves are dry when there is low light or no light.
2. Manual irrigation is time-consuming and requires experienced personnel.
3. Can be wasteful if there are large gaps between pots.

5.4.2 Types of Fertigation Systems

There are different fertigation systems with varying degrees of sophistication that can be used in cannabis production. The basic fertigation system, as illustrated in Figure 5.9, consists of a single tank where all the fertilizers are mixed to their final concentrations. The nutrient solution is then delivered to the plants through the delivery system. The spent nutrient solution, including the leachate, is either directly drained out of the system or returned to the originating tank.

Since there is only one mixing tank, this system is practical for small operations in order to save space. If the solution is not reused, the system is easy to operate and manage. One only needs to mix a nutrient solution based on the recipe, the water and fertilizer analyses. Of course, it is essential to check the EC and pH before each fertigation to catch any mistake made during mixing. If the nutrient solution is reused, it is essential to regularly sample and analyze the nutrient solution and

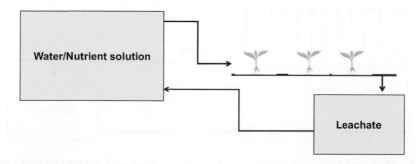

FIGURE 5.9 Single tank fertigation system.

adjust fertilizer components accordingly. Plants uptake different nutrient elements at different rates depending on many factors, including plant growth stage and the environmental conditions. Even if plants are fed a good balanced nutrient solution, the used solution may be depleted of or accumulate with certain elements more than the others. When using this system, growers should have a spread-sheet or software to automatically calculate the amount of each individual fertilizer to add based on the volume and chemical analysis of the used nutrient solution.

A slightly more advanced system is one with A and B tanks, and an acid/base tank (Figure 5.10). Concentrated stock solutions are mixed and stored in tanks A and B. The stock solutions and the acid or base are injected into the water line and mixed inline to reach the target nutrient concentration and pH. For this system, it is essential to install a backflow preventer at the water supply line.

When nutrient solution is reused, or the grower prefers to manage the feeding nutrient solution more precisely, then a multi-tank fertigation system (Figure 5.11) is necessary. This system has more flexibility but is also more complex to operate and maintain. For example, a multi-tank system offers the grower control over the nutrient ratios and rates based on plant performance, growing stage, and so on. A digital controller is a must for this type of system; there are several fertigation

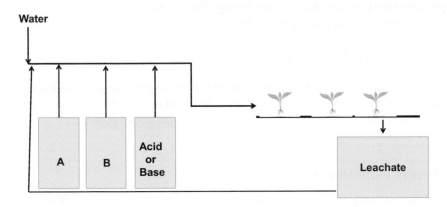

FIGURE 5.10 Fertigation system with A, B and acid or base tanks. The fertilizer injector is controlled by its flow rate. There are a few commercially available injection systems that are commonly used in the controlled environment plant production systems, such as those supplied by Dosatron. The fertilizer injector must be calibrated and verified from time to time by sampling and analyzing the downstream nutrient concentration. Verification can be done by measuring the solution EC in most of the cases. Again, if the spent nutrient solution is reused, a computer system should be used to calculate and add fertilizers accordingly, or this can be achieved manually.

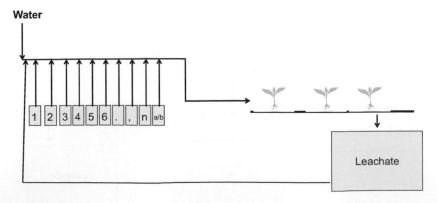

FIGURE 5.11 Multi-tank fertigation system. 1, 2, 3. . . n denote tanks with different straight fertilizers; a/b is acid/base solution.

and environmental control companies that can help growers set up these systems and provide training on how to operate them.

5.5 GROWING MEDIA

Growing media includes both soil and soilless media. For indoor cannabis production, soilless media is more commonly used. Growing cannabis in natural soil is easy and often more forgiving; however, there are many limitations of using soil in indoor production. For example, soil varies from location to location—even in a small area—due to the topography of the land and other factors. Growing one crop in the soil repeatedly can also lead to nutrient depletion, salt accumulation and pathogen issues. Soilless media can be prepared to the same quality (e.g., chemical, physical and biological properties) crop after crop, and is available for operations of any size.

Common soilless growing media used for cannabis production include those formulated with peat, coconut coir, rockwool and others. When selecting and evaluating growing media, their physical, chemical and biological properties must all be considered.

5.5.1 PHYSICAL PROPERTIES

Soilless media consist of solid particles or fibers, and the space between them. The proportion of a medium that consists of space is termed *total porosity* (%). When the medium is inside a container from which water can drain freely, the maximum amount of water the medium can hold is termed the *container capacity*, which is expressed as the percentage of the total media volume that contains water. When the medium is wetted to container capacity, the rest of the space is filled with air; this portion of the volume is termed the *air-filled porosity* (%). The dry weight of the medium divided by its volume is termed *bulk density*, which is expressed in g/cm³.

The physical properties of the same medium, except for bulk density, can differ depending on the size and shape of the container. Different regions have different standard container sizes (e.g., diameter and height) used for the measurement, and these may be very different from those used in cultivation. Therefore, these physical properties are best used to make direct comparisons between media to indicate, for example, which is more porous or able to hold more water.

5.5.2 CHEMICAL PROPERTIES

Commonly measured and reported growing media chemical properties include pH, EC, individual chemical element content, the cation exchange capacity (CEC) and the carbon/nitrogen (C/N) ratio. The pH is an extremely important parameter. Generally speaking, pH outside of 5.5–6.5 can cause nutrient disorders (e.g., Mn toxicity when pH is too low and Fe deficiency when pH is too high). Growing media manufacturers normally supply media within this range. If the pH is too low, for example when mixing media using sphagnum peat moss, dolomite lime is normally added to bring pH up.

EC is a measure of total soluble salt content. If the EC of a new growing medium is too high, it can be difficult to supply a balanced nutrient solution without causing salt stress to the plants. In the case of high medium EC, it is important to have the medium analyzed and to find out what is causing the high EC. For example, in the past, coconut coir normally had high Na content since coconut trees grow close to the ocean. More recently, coir used for growing plants is normally washed with fresh water, and some are also buffered with KCl to replace the Na; that is why most coir-based products have high K. If the EC of a new medium is too high, it may require a pre-treatment (e.g., washing with fresh water and then saturating with nutrient solution) before plants are transplanted.

CEC is the measure of the capacity of a medium to retain positively charged ions such as K⁺. This parameter is important if controlled or slow release fertilizer is used and the plants are irrigated with fresh water. When nutrient solution is used to fertigate plants, there are normally enough nutrients supplied to the rootzone; therefore, considering CEC is not critical. C/N is an indication whether the growing medium is biologically stable. This is important if the growing medium contains compost or other not biologically stable materials. When C/N is too low, the medium may further decompose.

Certain chemical properties of a growing medium or rootzone should be measured regularly to ensure the conditions are amenable for plants. At a minimum, the pH and EC should be checked before the medium is used for growing plants. During the cultivation period, the growing medium should be checked regularly, as well; how often depends on many factors. It can range from daily to weekly or even less frequently, but certainly should be checked whenever there is any sign of nutrient disorder.

To assess growing media chemical properties, there are in situ instruments—such as pH and EC probes—which can be inserted to rootzones for direct measurement, or a rootzone solution can be extracted and then analyzed. The most commonly used methods involve adding water to the growing media and analyzing the extracted solution. The commonly used extraction methods in North America include the following.

1. *The 1:2 Suspension Test:* Take one volume of growing medium, add two volumes of distilled or deionized water, mix thoroughly and let it to equilibrate for about one hour. Then measure pH and EC by inserting probes in the suspension.
2. *Pour-Through (PT) Method:* Pour deionized or distilled water to intact growing medium surface, just enough to displace about 50 ml leachate from the medium. This is normally conducted one hour after a normal fertigation event. The collected leachate can then be divided into two portions. One can be used for measuring pH and EC on-site, and another one can be sent to a laboratory to have the nutrient concentrations analyzed.
3. *Saturated Media Extract (SME):* Add distilled or deionized water to saturate the growing medium until the medium's surface is glistening. Let equilibrate for 30 minutes before vacuum filtering to collect the solution. Like the PT method, the collected solution can be analyzed on-site for pH and EC, and another portion can be sent to a laboratory for further analysis.

The on-site measurements are simple and can provide you with timely answers to questions such as whether your feeding solution is in the right concentration range and whether the rootzone pH and EC need corrective measures.

The pH values measured using these three methods may not be too different from each other; however, the EC values can vary quite a bit. For example, EC measured using 1:2 suspension test with a value from 0.76–1.25 mS/cm, would be equivalent to a value of 2.0–3.5 by SEM and 2.6–4.6 by PT. Therefore, it is important to choose your method and find the optimal EC levels for your crop. The optimal EC level can depend on cultivar, growing season, growing environment and other factors.

When submitting nutrient solution—either extracted from growing media, feeding solution or leachate—to a commercial lab for analysis, you can roughly assess whether the results are reliable by calculating whether the total cation concentration equals to the total anion concentration (mmeq/L), as follows.

1. Eq cations = $NH_4 + K + Na + Ca + Mg$
2. Eq anions = $NO_3 + Cl + SO_4 + HCO_3 + H_2PO_4$

5.5.3 How to Choose Growing Media

There are many varieties of growing media available for cannabis cultivation, but there is no "ideal" growing medium for all cultivation scenarios. The media you choose for your operation depends on many factors. These factors include—but are not limited to—the cultivation system, the cultivar and growth stage of your plants, the container size and shape, the fertigation method and the growing environment. Following are some examples.

5.5.3.1 Types of Rootzones

There are many types of rootzones in cannabis cultivation (See Section 5.8 for more details on different rootzones). For this purpose, we will focus on those with solid soilless growing media, including rockwool, peat-based and coir-based media. Using *rockwool*, cannabis can be grown in blocks, slabs, or block on slabs. There are different size of blocks (e.g., 4 in. or 6 in.) for different growth stages or plant sizes. Slabs have their fibers either vertically oriented, horizontally oriented or a combination of both, depending on their drainage requirements. For the block and slab combinations, normally smaller blocks (e.g., 4 in.) are used. Using peat-based or coir-based media, there are also many combinations. One can grow cannabis in pots, bags or slabs (with or without blocks on top). Different sizes and shapes of the pots, bags and slabs require their growing media to have different physical properties.

5.5.3.2 Container Size and Shape

Generally, when the same growing medium is used to grow plants in different-sized containers, the medium in the larger size container will hold more water, and therefore needs to be watered less frequently or needs to grow larger plant. When two containers have the same volume, but one is taller than the other (Figure 5.12), the shorter holds more water when the same growing medium is used. If one wants the two containers to hold the same amount of water, then the medium used in the taller container has to have higher water-holding capacity.

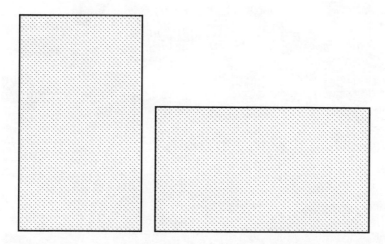

FIGURE 5.12   Two containers with the same volume but different shapes.

5.5.3.3 Plant Size

Smaller plants transpire less than larger ones; therefore, smaller plants are normally transplanted into finer growing media and smaller containers, while larger plants are transplanted to larger containers with coarser growing media. Mismatching can cause many issues, such as waterlogging when small plants are transplanted to large pots containing medium with high water-holding capacity (Figure 5.13). When the water-holding capacity of a medium is too high in a large pot, it can hold water for too long between irrigations, causing waterlogging stress. Typical symptoms of waterlogging stress on basil leaves are demonstrated in Figure 5.13 and on cannabis in Figure 5.14.

FIGURE 5.13 Basil (*Ocimum basilicum*) plant performs well when transplanted to the right size container, a smaller pot (left) compared with those transplanted to containers which are too large for the plants (right) (Photo A; the same peat and perlite mix medium was used). Photo B shows a close look of the typical water-logging injury symptoms observed on plants grown in the larger pots in Photo A. The plants were transplanted at the same size and on the same day.

Source: Adapted from Zheng (2020)

FIGURE 5.14 Cannabis transplants showing waterlogging stresses when watered and left in the dark for too long.

5.5.3.4 Growing Environment

When plants are growing in a drier environment, i.e., high vapor pressure deficit (VPD) and high wind velocity, the evapotranspiration rate is much higher than those in a more humid and lower air velocity environment. If everything is equal, then the crop growing in the drier environment can support a growing medium with a higher water-holding capacity.

If plants are grown outdoors in containers and the climate is hot with frequent wet weather, then a growing medium with higher drainage (i.e., high air-filled porosity) is a must. For example, in southwest Ontario, Canada, the temperature can be as high as 35°C with continuous rainy days in August. When the growing medium used does not have high enough air-filled porosity, it is common for the rootzone to experience low levels of oxygen which can induce fungal diseases such as *Pythium* root rot or *Fusarium* wilt. Therefore, media with higher air-filled porosity are essential. Porous media needs more frequent irrigation, but while irrigation is manageable, we cannot stop the rain.

5.6 BENEFICIAL MICROORGANISMS

Soil hosts a diverse and abundant community of microorganisms. Some microorganisms are beneficial to plants, while others are pathogenic. For indoor cannabis production, most cultivation systems start off clean or even sterile. However, as soon as plants enter the system, microorganisms quickly follow, though at lower levels than found in natural soil.

In recent decades, beneficial microorganisms are increasingly applied intentionally in plant cultivation systems. Some beneficial microorganisms improve plant nutrient uptake, some promote plant growth and others suppress plant pathogens. For soilless cannabis production, pathogen-suppressing microorganisms are more important. It is commonly believed that the pathogen-suppressing mechanisms of beneficial microorganisms are the following: 1) competing for nutrients and space; 2) increasing resistance to pathogen infection; and 3) directly interacting with pathogens via antibiosis or hyperparastism.

Based on the aforementioned modes of action, it is expected that beneficial microorganisms are applied to the production system at the early stage before other uninvited microorganisms colonize the system. There are commercially available beneficial microorganisms, but the vast majority of them are not tested on cannabis plants or in soilless production systems; therefore, it is recommended to first test their efficacy in one's system before incurring unnecessary costs.

5.7 OXYGEN AND OXYGENATION

Rootzone oxygen is essential for root formation, growth, respiration, health, water and nutrient uptake. Rootzone oxygen exists as gaseous oxygen and dissolved oxygen (DO). Rootzone oxygen levels can be assessed by measuring DO in situ; special oxygen sensors are needed for these type of measurements (Figure 5.15). Normal DO probes can only measure DO in solution which may not represent the DO at the actual rootzone. For example, during fertigation, leachate comes out from plant rootzone and can quickly equilibrate with the atmosphere air, which contains about 21% oxygen. When the leachate DO is measured, it may be higher than the DO at the rootzone.

Rootzone oxygen levels are mainly controlled by the speed at which atmospheric oxygen diffuses into the rootzone. It is estimated that the rate of oxygen diffusion through the air is more than 10,000 times faster than through water. That is why it is important to ensure that growing media have adequate air-filled porosity and are irrigated to allow dry-downs so that oxygen from the atmosphere can adequately supply to plant rootzone. Other factors can also affect rootzone oxygen, such as root and microbial respiration and temperature. Increased temperature reduces the DO concentration in water (Figure 5.16). Oxygen in the air remains constant at about 21% at sea level and at one atmospheric pressure; when pure water is well aerated and equilibrated with the atmosphere, the DO is at about 8.5 mg/L at 25°C. It drops to below 8 mg/L when the temperature increases to 30°C. Under the same conditions, rootzone DO can be much lower due to the respiration processes of both roots and microorganisms, and restricted air diffusion into rootzone. There are technologies that can supersaturate water with oxygen, which means DO in the water or nutrient solution can be higher than illustrated in Figure 5.16; however, it is hard to increase rootzone DO levels more than indicated in Figure 5.16 when the plant is grown in solid growing media (e.g., rockwool, peat-perlite

FIGURE 5.15 Inserting oxygen and temperature probes into a plant rootzone for in situ oxygen measurement. The probe marked red is the oxygen probe attached to a fiber optic cable; the one marked yellow is a temperature probe.

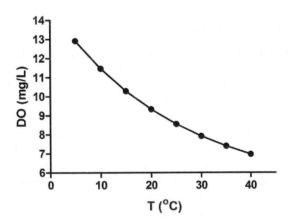

FIGURE 5.16 Dissolved oxygen (DO) concentration in pure water at different temperatures. The curve is created under one atmospheric pressure with 21% oxygen in the air.

mix). It is more effective and economical to manage rootzone DO by choosing growing media with adequate air-filled porosity and fertigate at appropriate intervals. This practice is also good for purging CO_2, ethylene and other gases from reaching harmful levels at the rootzone.

Plant rootzones require adequate oxygen; too little can slow plant growth and be inducive to diseases such as *Pythium* root rot (Figure 5.17), too much can also cause plant stress. Rootzone DO

FIGURE 5.17 *Pythium* root rot caused by rootzone oxygen deficiency.

Source: Photo by Youbin Zheng

should be maintained between 5–10 mg/L. Higher than 10 mg/L can be difficult to achieve and not economical; higher than 20 mg/L may cause root injury.

5.8 TYPES OF ROOTZONES IN CONTROLLED ENVIRONMENT CANNABIS CULTIVATION

Most controlled environment cannabis production uses soilless culture. Soilless culture is the practice of cultivating plants without using natural soil; however, the growing media may contain small amounts of natural soil in some cases. Plants may be grown in containers (e.g., pots) with solid media (e.g., peat-, or coir-based), in deep water culture (DWC) systems, using nutrient film technique (NFT), aeroponics, aquaponics or using other less common systems.

For indoor cannabis production, the following types of rootzone are most common: 1) container production; 2) rockwool; 3) aeroponics; 4) NFT and DWC; 5) aquaponics; and 6) in-ground culture (in no particular order). Of course, within each, there can be many variations. Different rootzones have their advantages and disadvantages, which are discussed in detail in the next section.

5.8.1 CONTAINER CULTIVATION

In container cultivation, plants spend most of their production period in containers (e.g., pots or bags). Depending on the system, the plants can be in one size of container throughout the vegetative and flowering stages or in a smaller container size at the vegetative stage and a larger one in the flowering stage.

Two types of growing media are mostly used. The first is a peat-based mix, as in Figure 5.18, and the other is coir-based, as in Figure 5.19. Of course, other growing media, such as shredded rockwool and wood fiber, can also be used. As discussed previously, it is important to choose the right media for your cultivation strategy. Each medium has unique characteristics. For example, coir-based media normally consist of different-sized coir chips or fibers. The proportions of different sizes can result in varying physical and chemical characteristics. Some advantages of using coir include: 1) there is no need to add any non-coir material to achieve desired physical properties (e.g., larger chips can be used to increase air-filled porosity), and the used media can be easily composted with or without cannabis plant residue to be reused; and 2) there is no need for a wetting agent since coir can be easily rewetted even when dried to an extremely low water content. Conversely, it is important to ensure coir components do not arrive with plant pathogens. Also, coir often contains high K when fresh, then K gradually drops due to the uptake by plant and leaching. It is important to keep an eye on the rootzone K concentration and supply nutrients accordingly.

FIGURE 5.18   Cannabis plants growing in a peat-perlite mix in pots placed on a subirrigation bench.

Source: Photo by Youbin Zheng

FIGURE 5.19 Cannabis plants growing in coir in upright bags placed in troughs and fertigated by a drip irrigation system.

Source: Photo by Youbin Zheng

Peat-based media normally come with added perlite or vermiculite for aeration. Recently, wood fiber has also been incorporated to improve the aeration and water conduction of peat-based media. In North America, most peat moss used in growing media is harvested in Canada from large peat bogs. Peat-based growing media can normally be purchased in large quantities, normally with consistent quality from batch to batch. This is vitally important for container cultivation. Each cultivation scenario should have all its horticultural management practices developed to suit each other. If the growing medium is changed, then other related practices, especially the fertigation schedule, must be changed accordingly. This will be discussed in greater detail in Section 5.9. Peat-based media has its disadvantages, as well, of course. One is that when peat moss is too dry, it is very hard to rewet; therefore, wetting agents are normally added when producing the media. Not all wetting agents meet organic criteria, so be aware if you are producing cannabis organically.

Containers are normally fertigated either using subirrigation (Figure 5.18) or drip irrigation (Figure 5.19). When growing plants in small containers at a high planting density, subirrigation is preferable, but when the containers are large and spaced out, then drip irrigation is preferable. Depending on the size of the plants and their containers, there can be one or several drippers per pot.

Container production can have the following advantages.

- Plants are physically supported to grow upright by the solid growing media.
- Flexibility in that containers can be small or large, plants can be up potted as needed and plant density can be changed according to plant growth stage and size.
- Drought stress can easily be applied to enhance cannabis potency.
- Diseased plants can be easily removed from the system to prevent pathogen spread.
- Easy harvesting. Plants in containers can be moved to a separate harvesting area.

Disadvantages of container production:

- Cost for containers.
- Not environmentally friendly if containers are not reused.
- Reusing containers is water- and labor-intensive.
- Subirrigation can cause salt accumulation to harmful levels at the upper section of the growing media if not managed correctly.
- If the container size is not optimized, it can cause issues such as root restriction (i.e., pot is too small for the plant to grow large) and waterlogging.

5.8.2 ROCKWOOL CULTIVATION

Rockwool, also called mineral wool or stone wool, is a fibrous material made by melting rock at a temperature of about 1,600°C. It was originally developed for insulation and noise absorption for buildings. With some modifications and fiber orientation arrangement, it has been successfully adapted for use as growing media in soilless plant cultivation for decades. As growing media, it is supplied as plugs, blocks, slabs and loose cubes or fibers. There are assorted sizes of each variant. As discussed before, the fiber orientations can vary to meet specific objectives. For example, some slabs have a top layer of horizontally arranged fibers and a bottom layer of vertically arranged fibers to improve drainage. Plugs are generally used for propagation, while blocks can be used for growing plants during the vegetative stage or from the vegetative stage to maturity if the block is large enough; some cultivators also use a smaller block on top of a larger block (Figure 5.20). Drip irrigation is most common in rockwool production, though subirrigation is possible using smaller blocks or cubes.

Rockwool is inert but some can arrive with a high pH. It is common practice to condition the new growing medium with nutrient solution, saturating it for 24 hours, then drained before use.

Using rockwool in cannabis production offers the following advantages.

- Plants are physically supported to grow upright by the solid growing media.
- Flexible. If using blocks, plants can be grown in smaller blocks at a high density to save space, then transplanted to larger blocks to finish in the flowering room. Even in the same-sized blocks, plants can be spaced according to their sizes and growing stages when necessary.
- Diseased plants can be easily removed from the system to prevent spread when using blocks.
- Rockwool is chemically inert, so it does not chemically interfere with the feeding nutrient solution.
- Rockwool is almost sterile when new; therefore, there is little chance of introducing pathogens into the cultivation system.
- The products are homogeneous and consistent, simplifying fertigation scheduling to achieve uniform plants.

FIGURE 5.20 Rockwool products used for cannabis cultivation. A: plugs for propagation; B: each cannabis plant grown in a block with drip irrigation; C: each cannabis plant grown in a smaller block placed on top of a larger block.

Photos A and C are courtesy of Grodan Inc.; Photo B by Youbin Zheng

Disadvantages of rockwool include the following.

- Not environmentally friendly if the spent product is sent to landfill. Rockwool, unlike organic growing media (e.g., coir), cannot be composted or degraded. Most rockwool from plant production ends up in landfills. Recycling efforts are increasing; for example, certain companies are collecting used rock to make bricks; however, if the cultivation operation site is too far from the recycling site, then opportunities for recycling can be limited.

- When growing in blocks, it may cause physical restrictions on the roots (i.e., limited space for roots) and limit the size of the plants.

5.8.3 AEROPONICS

Aeroponics is a plant cultivation technology in which water and nutrients are supplied to plant roots in a misting form. Aeroponics consists of a closed and dark misting chamber, holes for inserting plants on top of the misting chamber, small baskets containing inert solid materials for anchoring plants, misting nozzles to provide the nutrient solution as small particles and pumps to deliver a pressurized solution from a nutrient reservoir to the nozzles. The nutrient solution that has dripped off the roots can be reused or discharged.

Using aeroponics in cannabis production offers the following advantages.

- Little solid growing medium is needed, which not only reduces growing media cost but also reduces the opportunity of introducing pathogens to the cultivation system through media.
- The rootzone always has adequate oxygen, which can stimulate root growth and reduce root diseases such as root rot.
- Nutrient solution can be easily collected, disinfected and reused to reduce waste.
- If maintained well and without power outage, no water stress will occur.

Disadvantages of aeroponics:

- High initial capital cost.
- Possibly difficult to scale.
- Requires high-quality water and fertilizers to prevent nozzle blockage.
- Requires high-quality nozzles and adequate inline pressure to avoid nozzle blockages.
- Limited buffer during equipment malfunction or power outage; plants can experience water stress shortly after mist stops.
- Pathogens can spread quickly through nutrient solutions.
- Physical support is needed to keep plants upright.

5.8.4 NFT AND DWC

Nutrient film technique (NFT) and deep water culture (DWC) are two soilless plant cultivation systems that traditionally fall under the category of hydroponics. In both, plants are normally grown in small baskets containing inert solid materials such as leca (lightweight expanded clay aggregate). The baskets are supported so that roots have access to the nutrient solution below. NFT involves growing plants in troughs with a film of constantly moving nutrient solutions. The troughs normally have a 1% slope in order for the nutrient solution to flow from the higher end to the lower end. A portion of the root system is in the air, and the remainder is in the nutrient solution. Oxygen is supplied to the roots from both the air and the flowing nutrient solution. DWC normally uses containers filled with constantly aerated nutrient solutions. Traditionally, the roots are completely submerged in the nutrient solution, but some modern DWC designs leave an air gap between the basket and the nutrient solution so that some of the roots are exposed to air. This gap is beneficial for plant roots to access oxygen when there is not enough oxygen in the nutrient solution. Other names for DWC include deep flow technique and floating raft technology.

Using NFT and DWC in cannabis production offers the following advantages.

- Little solid growing medium is needed, which not only reduces growing media cost but also reduces the opportunity of introducing pathogens to the cultivation system through growing media.
- Rootzone is always adequately oxygenated if the NFT system is well designed and the aeration system is working well in the DWC system.

- Nutrient solution can be easily collected, disinfected and reused to reduce waste.
- If maintained well and without power outage, no water stress will occur.

Disadvantages of NFT and DWC include the following.

- Limited buffer during equipment malfunction or power outage; plants can experience water stress shortly when the nutrient solution in the NFT system dries out, especially when plants are large with high transpiration rates.
- Possibly difficult to scale.
- Pathogens can spread quickly through nutrient solutions.
- Physical support is needed to keep plants upright.

5.8.5 AQUAPONICS

Aquaponics is a technique in which plants are grown in the same system as aquatic organisms (e.g., fish). It is a combination of aquaculture and soilless plant production. The effluent from the aquaculture, after some microbial transformation, is used as the nutrient source for at least 50% of the total nutrient needs of the plants. The plants are able to remediate the water for the aquatic organisms. The original design involved growing fish in a DWC system, using a water container larger and deeper than normal. Most modern aquaponic operations grow aquatic organisms in separate tanks. The water and nutrients from the fish tanks go through microbial transformation, mainly to convert the ammonia/ammonium nitrogen into NO_3-N, which is delivered to plants growing in a separated NFT or DWC system. While some of the advantages and disadvantages are the same for aquaponics as for NFT and DWC systems, others are unique to aquaponics.

Using aquaponics in cannabis production offers the following advantages.

- It is perceived to be a sustainable practice, since more than 50% of the plant nutrients are supplied through the aquaculture "waste" and both the water and nutrients can be reused.
- The system is easily maintained as an organic production system.
- Abundant plant beneficial microbes exist in aquaponics.
- It also produces other aquatic organisms such as fish.

Disadvantages of aquaponics:

- Different pH requirements among plants, nitrifying bacteria and aquatic organisms. For example, the optimal nutrient solution pH for plants is between 5.5 and 6.5; for nitrifying bacteria, it is between 7.0 and 8.3 (depending on the species), and for Nile tilapia (a popular fish species used in aquaponics), it is between 7.0 and 9.0.
- Nutrient limitation. The nutrient requirements are different between fish and cannabis plants; therefore, in most aquaponics systems, the nutrient composition and concentration in the effluent from aquaculture are not ideal for cannabis plants. Quite often, there is a lack of nutrients such as K, Ca, Mg and Fe.
- Accumulation of organic materials such as uneaten feed, algae, fungi and fish sludge. These organic materials can increase the nutrient solution oxygen demand, therefore causing plant rootzone oxygen deficiency.
- High initial capital cost.

5.8.6 IN-GROUND SOIL

In North America, most of the controlled environment cannabis is grown using soilless production systems; however, some systems make use of natural soil, especially in greenhouses and high tunnels.

Growing plants using natural soil in-ground is the easiest and most natural cultivation strategy. The greenhouse or high tunnel floor can be divided into different rows with a walkway between rows for easy access during crop work. The rows can be raised beds or recessed, so they are slightly lower than the walkway to prevent the soil and water from getting on the walkway. Soil physical and chemical qualities need to be assessed frequently to provide cannabis roots with the optimal environment. For example, nutrient status needs to be measured to decide the nutrient recipe. When air porosity decreases to a certain level, organic matter such as good quality compost can be added to amend the soil. Drip irrigation is one of the best choices for in-ground cannabis fertigation. Alternatively, controlled release or slow release fertilizers can be used, so nutrient solution does not need to be used with irrigation.

Using in-ground soil in cannabis production offers the following advantages:

- Naturally available at no cost.
- More forgiving. Most natural soils have good water and nutrient holding capacities; therefore, plants grown in natural soils do not need water or fertilizer as frequently as in soilless media.
- The rootzone environment, such as temperature, is more constant when the air temperature is fluctuating.
- Soil does not need to be changed each crop cycle as with most soilless media.

Disadvantages of in-ground soil include the following.

- Soil needs to be disinfected for both pathogens and insect pests; steam disinfection is most commonly used.
- Soil needs to be amended often.
- Soil needs to be analyzed almost before each crop cycle to decide what and how much fertilizer to apply; otherwise, salt accumulation and nutrient imbalance can occur.
- When high tunnels or greenhouses are built on infertile soil, good soil needs to be brought in or the existing soil needs to be amended, which can be costly.
- Needs some initial soil amendments and mixing, so the soil is as homogenous as possible in the entire growing area for easy fertigation management; this can be costly.

5.9 INTEGRATED ROOTZONE MANAGEMENT (IRM)

As discussed above, there are different types of rootzones in cannabis production, and within each type, there are many variations. For example, for container production systems, there are options for container sizes, shapes and even colors; there are also several options for growing media and fertigation systems (e.g., drip irrigation or subirrigation). The ultimate goal is to provide cannabis roots with adequate oxygen, water, nutrients and temperature while avoiding pests, both pathogens and insects, and contaminants (e.g., heavy metals). All the factors within the rootzone and the environmental factors aboveground are interconnected, so changing one can affect the others. For example, too much water can cause an oxygen deficiency, which can encourage root disease such as *Pythium* root rot. The choice of rootzone system and cannabis genotype will always influence the plant size, morphology, physiology and biochemistry. To meet the plants' needs and maximize crop yield and quality, integrated rootzone management (IRM) is essential. IRM is an approach used to manage the rootzone by taking into consideration all the factors which can affect rootzone oxygen, water and nutrients, and so on.

One of our recent cannabis rootzone management studies demonstrates how different rootzone factors are connected and how they can affect cannabis quality and yield. We grew drug-type cannabis plants in 11-liter containers, then applied water stress once during the flowering stage. The inflorescence concentrations of tetrahydrocannabinolic acid (THCA) and cannabidiolic acid

(CBDA) increased by 12% and 13%, respectively, compared with the control, and the yield per unit growing area of THCA, CBDA, THC and CBD was 43%, 47%, 50% and 67% higher than the control, respectively (Caplan et al. 2019). When the plants were grown in smaller (6-liter) containers and drought stress was applied at different growing stages once or several times, there was no drought treatment effect on THCA or CBDA concentrations or yields. However, regardless of whether drought was applied or not, plants grown in the smaller containers had comparable or slightly higher inflorescence THCA and CBDA concentrations and total yields of THCA and CBDA per growing area than their counterparts in the drought-stressed plants grown in the larger pots (Caplan 2018). It is suspected that the plants grown in the smaller pots might be already under stress due to the root volume restriction. This example also demonstrates that more research is needed to explore the potential of stress treatments to enhance cannabis potency.

For controlled environment cannabis production, IRM should be applied at minimum to make the following decisions.

- When choosing the type of rootzone. The type of rootzone should depend on many factors, such as the growth rates of your genotype and the desired final size of your plants. For example, if you intend to grow plants at high density and harvest them small, rockwool blocks may be a viable choice.
- When choosing growing media and container sizes. Smaller containers are normally best paired with finer growing media, while larger ones are better with coarser growing media. Finer growing media holds more water; when used in large containers, it can cause waterlogging and oxygen deficiency, especially when the plants are at their earlier stage.
- When choosing a fertigation system. The fertigation system can depend on the rootzone type and several other factors. For example, if you are using large pots, then a drip irrigation system is a better option than a subirrigation system.
- When irrigating/fertigating plants. Both the timing and volume of irrigation events influence rootzone oxygen levels, plant transpiration rates, nutrient availability and more. A thoughtful irrigation strategy can increase plant growth rates, increase inflorescence yield or induce drought or nutrient stresses to enhance concentrations of secondary metabolites (e.g., cannabinoids and terpenes). When and how much to irrigate/fertigate can depend on many factors, such as the size and stage of the plants, the physical and chemical properties of the growing media, environmental conditions (e.g., light intensity, temperature and VPD) and the purposes (e.g., to induce stress).

These examples demonstrate the importance of using the IRM approach, but its adoption may be difficult based on traditional thinking and research approaches. The rapid development in sensor technologies (e.g., soil moisture sensor and imaging systems) and artificial intelligence may become key components in IRM in the near future. In the meantime, it is important for those setting the environmental conditions and making day-to-day fertigation decisions to communicate and think holistically as their decisions often dictate the success of an operation.

5.10 ACKNOWLEDGMENT

The author would like to thank Dr. Deron Caplan for proofreading and providing suggestions to this chapter.

BIBLIOGRAPHY

Bevan, L., M. Jones, and Y. Zheng. 2021. Optimisation of nitrogen, phosphorus, and potassium for soilless production of *Cannabis sativa* in the flowering stage using response surface analysis. *Frontiers in Plant Science*. https://doi.org/10.3389/fpls.2021.764103.

Caplan, D. 2018. *Propagation and rootzone management for controlled environment cannabis production*. PhD diss. Guelph: University of Guelph.

Caplan, D., M. Dixon, and Y. Zheng. 2017a. Optimal rate of organic fertilizer during the vegetative-stage for cannabis grown in two coir-based substrates. *HortScience* 52: 1307–1312.

Caplan, D., M. Dixon, and Y. Zheng. 2017b. Optimal rate of organic fertilizer during the flowering-stage for cannabis grown in two coir-based substrates. *HortScience* 52: 1796–1803.

Caplan, D., M. Dixon, and Y. Zheng. 2019. Increasing inflorescence dry weight and cannabinoid content in medical cannabis using controlled drought stress. *HortScience* 54: 964–969.

Health Canada. 2020. *Guidelines for Canadian drinking water quality – summary table*. www.canada.ca/content/dam/hc-sc/migration/hc-sc/ewh-semt/alt_formats/pdf/pubs/water-eau/sum_guide-res_recom/summary-table-EN-2020-02-11.pdf.

Marschner, H. 2002. *Mineral nutrition of higher plants*, 2nd ed. New York: Academic Press.

Nelson, P. V. 2012. *Greenhouse operation and management*, 7th ed. Upper Saddle River: Prentice Hall.

Saloner, A., and N. Bernstein. 2020. Response of medical cannabis (*Cannabis sativa* L.) to nitrogen supply under long photoperiod. *Frontiers in Plant Science* 11: 572293. https://doi.org/10.3389/fpls.2020.572293.

Saloner, A., and N. Bernstein. 2021. Nitrogen supply affects cannabinoid and terpenoid profile in medical cannabis (*Cannabis sativa* L.). *Industrial Crops and Products*. https://doi.org/10.1016/j.indcrop.2021.113516.

Veazie, P., P. Cockson, D. Kidd, and B. Whipker. 2021. Elevated phosphorus fertility impact on *Cannabis sativa* 'BaOx' growth and nutrient accumulation. *International Journal of Innovative Research in Science, Engineering and Technology* 8: 346–351.

Yep, B., N. V. Gale, and Y. Zheng. 2020a. Comparing hydroponic and aquaponic rootzones on the growth of two drug-type *Cannabis sativa* L. cultivars during the flowering stage. *Industrial Crops and Products* 157: 112881. https://doi.org/10.1016/j.indcrop.2020.112881.

Yep, B., N. V. Gale, and Y. Zheng. 2020b. Aquaponic and hydroponic solutions modulate NaCl-induced stress in drug-type *Cannabis sativa* L. *Frontiers in Plant Science*. https://doi.org/10.3389/fpls.2020.01169.

Yep, B., and Y. Zheng. 2019. Aquaponic trends and challenges – a review. *Journal of Cleaner Production* 228: 1586–1599.

Yep, B., and Y. Zheng. 2020. Potassium and micronutrients fertilizer addition in aquaponic solution for drug-type *Cannabis sativa* L. cultivation. *Canadian Journal of Plant Science*. https://doi.org/10.1139/CJPS-2020-0107.

Zheng, Y. 2018. Current nutrient management practices and technologies used in North American greenhouse and nursery industries. *Acta Horticulturae* 1227: 435–442.

Zheng, Y. 2019. Development in growing substrates for soilless cultivation. In *Achieving sustainable greenhouse cultivation*, ed. L. M. Marcelis and E. Heuvelink, 225–240. Cambridge: Burleigh Dodds Science Publishing.

Zheng, Y. 2020. Integrated root-zone management for successful soilless culture. *Acta Horticulturae* 1273: 1–8. https://doi.org/10.17660/ActaHortic.2020.1273.1.

Zheng, Y. 2021a. Advances in understanding plant root behaviour and rootzone management in soilless culture system. In *Advances in horticultural soilless culture*, ed. N. S. Gruda. Cambridge: Burleigh Dodds Science Publishing, 23–44.

Zheng, Y. 2021b. Soilless production of drug-type *Cannabis sativa*. *Acta Horticulturae* 1305: 376–382. https://doi.org/10.17660/ActaHortic.2021.1305.49.

Zheng, Y., D. F. Cayanan, and M. Dixon. 2010. Optimum feeding nutrient solution concentration for greenhouse potted miniature rose production in a recirculating subirrigation system. *HortScience* 45: 1378–1383.

Zheng, Y., M. Dixon, and P. Saxena. 2006. Greenhouse production of *Echinacea purpurea* (L.) and *E. angustifolia* using different growing media, NO_3^-/NH_4^+ ratios and watering regimes. *Canadian Journal of Plant Science* 86: 809–815.

Zheng, Y., T. Graham, S. Richard, and M. Dixon. 2004. Potted gerbera production in a subirrigation system using low-concentration nutrient solutions. *HortScience* 39: 1283–1286.

Zheng, Y., L. Wang, and M. Dixon. 2004. Response to copper toxicity for three ornamental crops in solution culture. *HortScience* 39: 1116–1120.

Zheng, Y., L. Wang, and M. Dixon. 2007. An upper limit for elevated root zone dissolved oxygen concentration for tomato. *Scientia Horticulturae* 113: 162–165.

6 Lighting and CO_2 in Cannabis Production

Youbin Zheng and David Llewellyn

CONTENTS

Light and carbon dioxide (CO_2) are two of the most import and controllable environmental factors which can affect cannabis growth, development and ultimately the yield and quality. This chapter discusses some of the essential and basic knowledge about light and CO_2, how cannabis responds to light and CO_2, and ultimately how to apply light and CO_2 in cannabis cultivation.

6.1 LIGHT IN CANNABIS PRODUCTION

6.1.1 LIGHTING BASICS

Like the growth of any photosynthetic organism, the art of cannabis production is essentially an optimization of the process of converting energy from the electromagnetic spectrum (i.e., "light") into chemical energy (i.e., biomass) through the process of photosynthesis. Given the importance of lighting in cannabis production, it is critical for the cultivator to have the knowledge and skills to objectively evaluate different electric lighting technologies and then characterize their performance in the production environment. Before delving into the various electric lighting technologies that are available to cannabis cultivators, it is important to set the stage with a background understanding of what is light, its important properties and how we can characterize it to optimize production.

"Light" is a colloquial term that is commonly used to refer to the portion of the electromagnetic spectrum that is visible to humans (i.e., approximately 380–780 nm). Consequently, the colors of the rainbow span this range with blue and red wavelengths appearing near the shorter- and longer-wavelength ends of the visible spectrum, respectively. Conveniently, the visible spectrum also roughly matches the waveband of light that plants can effectively utilize for photosynthesis. Conventionally,

DOI: 10.1201/9781003150442-6

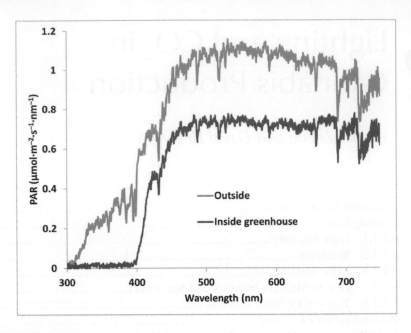

FIGURE 6.1 Relative photon flux distribution of natural sunlight and sunlight filtered by a polyethylene-covered greenhouse covering material. Spectrum scans were collected concurrently at noontime during a bright cloudy summer day in Ontario, Canada.

photosynthetically active radiation (i.e., PAR) has been defined as all wavelengths of between 400 nm and 700 nm since photosynthetic efficiency drops sharply outside of this waveband.

The natural sunlight that reaches the earth's surface contains a broad range of electromagnetic radiation, from < 300 nm to > 2000 nm, with about 45% of it comprising visible/PAR light. The earth's atmosphere (primarily the ozone layer) absorbs the majority of the sun's ultraviolet (UV) radiation, particularly wavelengths less than 300 nm. The UV light that does reach the earth's surface is predominantly longer-wave UVA (315–400 nm), with a small amount of UVB (280–315 nm). Wavelengths shorter than 280 nm, defined as UVC and strongly biocidal, are virtually completely removed by the ozone layer. Wavelengths of far-red (700–800 nm) are also important in mediating plant responses, especially morphology and flowering.

Greenhouse covering materials such as polyethylene, acrylic and glass can modify the sunlight spectrum—particularly in reducing the UV waveband—and also reduce the intensity, normally by 20–50%. For example, Figure 6.1 shows an example of sunlight spectrum measured at noontime on a cloudy summer day in Ontario, Canada, both outside and inside a greenhouse (covered with a double layer of polyethylene). In this case, the greenhouse covering reduced the PAR intensity by about 35% (as denoted by the reduction in the area under the respective curves). Also note the substantial reduction in UV wavelengths under the greenhouse covering; this may have important ramifications for cannabis production in greenhouse environments, as UV exposure has been linked with increases in secondary metabolite production, particularly THC (more details in what follows).

6.1.1.1 Light Intensity

The energy level of all electromagnetic radiation is inversely proportional to its wavelength, with shorter wavelength radiation thus containing higher energy per photon. Coincidentally, this is why UV radiation can be damaging to biological organisms (even biocidal) while visible light normally is not. The relationship between wavelength and energy is important in horticulture production because light is frequently quantified both in radiometric (i.e., energy) and quantum (i.e., photon) units. There are no easy conversions between these units, or with other common photopic units (e.g., foot candles, lumens and lux), without knowing the spectrum distribution. Further, while all photons in the PAR spectrum have sufficient energy to be used in photosynthesis, the per-photon energy reduces by approximately 40% as wavelength increases from 400 – 700 nm. As we will see later, this physical property of light has important implications when considering sources of electric lighting for horticulture because, all other things being equal, less energy is required to produce red vs. blue photons.

Historically, electric lighting technologies were most commonly characterized according to their photopic performance, but it has been well established that plants perceive and respond to their lighting environments very differently than do humans. Consequently, this is why human vision is relatively poor at evaluating intensity and quality of plant lighting environments. Specialized tools and procedures are required to get a clear picture of your crops' lighting environments so you can predict how they will respond and estimate their productivity, yield and quality.

The quantum metric of photosynthetic photon flux density (PPFD) is the overwhelmingly preferred unit for measuring and reporting PAR intensity in horticultural production environments. Briefly, PPFD is simply a count of the number of PAR photons (i.e., within the 400–700 nm waveband, regardless of actual wavelength distribution) that affect a horizontal surface in a given time. The most common units are micromoles per square meter per second (i.e., $\mu mol \cdot m^{-2} \cdot s^{-1}$). Note that a "mole" is a physical constant that is frequently used by scientists to conveniently abbreviate very large numbers; one mole is equal to the number of carbon atoms in 12 g, approximately 6.02×10^{23}. Since, by definition, there are 1 million micromoles in a mole, then 1 μmol of anything equals 6.02×10^{17}. Therefore, 1 μmol of PAR is equal to 6.02×10^{17} individual photons within the 400–700 nm waveband. At most inhabited latitudes on earth, the highest natural PPFDs attained (i.e., at noon on a sunny day during the summer equinox) are from 2,000–2,200 $\mu mol \cdot m^{-2} \cdot s^{-1}$. Conversely, the PPFD in most indoor office environments is < 10 $\mu mol \cdot m^{-2} \cdot s^{-1}$, which is too low for most plants to survive for very long.

While the PPFD metric is commonly used to characterize indoor crop production where "daytime" light intensity is relatively constant, this instantaneous value does not necessarily relate to how much light a crop receives in a day, particularly in outdoor or greenhouse environments. This is because of the immense variability that geographic location, time of day, time of year and weather can have on natural light intensity. For example, Figure 6.2 shows the PPFD measured over entire

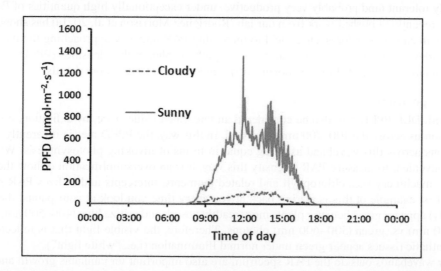

FIGURE 6.2 Example of logged outdoor instantaneous photosynthetic photon flux density (PPFD) measurements taken in Guelph, Ontario, Canada on cloudy and sunny days in January 2017. Integrating the PPFD data (i.e., area under the curves) results in daily light integrals (DLI) of 1.9 mol·m^{-2}·d^{-1} and 15 mol·m^{-2}·d^{-1} on the cloudy and sunny days, respectively.

days during both cloudy and sunny wintertime days in Ontario, Canada. Therefore, a more useful metric to describe a crop's exposure to PAR is to integrate each entire day's PPFD values into a single light sum called the daily light integral (DLI, $mol \cdot m^{-2} \cdot d^{-1}$). The DLI is essentially the area under the daily PPFD curves, such as in Figure 6.2 which equate to 1.9 $mol \cdot m^{-2} \cdot d^{-1}$ and 15 $mol \cdot m^{-2} \cdot d^{-1}$ on the cloudy and sunny days, respectively.

For outdoor and greenhouse growers, there are several tools available to help you estimate the range of DLIs you can expect over the entire year and across many regions of the globe. Faust and Logan (2018) have published monthly average DLI maps for the United States, by state. They have also developed an interactive online tool that has much higher (10 km^2) geographical resolution (https://myutk.maps.arcgis.com/apps/MapSeries/index.html?appid=d91ba9eb487d43f3a82161a12 47853b6). Further, Sun Tracker Technologies have developed an interactive online tool for estimating monthly average natural DLIs across the entire globe (https://dli.suntrackertech.com/). For greenhouse growers, please keep in mind how greenhouse transmission losses reduce the amount of light reaching the canopy.

For indoor growers, where the light intensity is kept constant, it is relatively straightforward to convert between PPFD and DLI by simply accounting for the number of seconds in the photoperiod. For example, for a flowering room (12-hour days) with an average canopy PPFD of 500 $\mu mol \cdot m^{-2} \cdot s^{-1}$, multiply PPFD by the number of seconds in the photoperiod (e.g., 60 s/ min × 60 min/hr × 12 h/d = 43,200) and then divide by 1,000,000 (to convert μmol to mol): [(500 × 43,200) / 1,000,000] = 21.6 $mol \cdot m^{-2} \cdot d^{-1}$. Note that the DLI associated with a PPFD of 500 and 12-h photoperiod is about 44% greater than the sunny January day in Figure 6.2, even though the PPFD during that day was frequently above 600 $\mu mol \cdot m^{-2} \cdot s^{-1}$ and briefly reached 1,400 $\mu mol \cdot m^{-2} \cdot s^{-1}$ (at around noontime). It should be noted that, while PPFD and DLI are very commonly used to relate the PAR intensity in horticultural production environments, they are somewhat limited in that they only quantify the PAR affecting a horizontal surface (normally "canopy level"). Since the cannabis canopy is complex, with widely varying plant architecture, cropping density and leaf shape, these area-based metrics do not directly relate to foliar interception and canopy penetration of PAR. Despite these limitations, PPFD and DLI are the most common and easily relatable metrics in use for evaluating PAR intensity in crop production environments.

Compared to most flowering crops grown in controlled environments, cannabis is known to be especially tolerant (and probably very productive) under exceptionally high quantities of PAR. For example, recently published work from our lab (Rodriguez-Morrison et al. 2021a) has shown linear responses in cannabis inflorescence yield to increasing PAR intensity approaching the intensity of full noon-day sun for the entire day; up to photosynthetic photon flux densities (PPFD) of at least 1,800 $\mu mol \cdot m^{-2} \cdot s^{-1}$ (i.e., DLI of ~78 $mol \cdot m^{-2} \cdot d^{-1}$) (see Section 6.1.4 for more details).

6.1.1.2 Spectrum

Along with DLI, PPFD can also be considered an integrated value since, by definition, it sums all of the photons across the 400–700 nm waveband. In this way, the PPFD metric inherently ascribes all photons across this waveband as being equal in terms of invoking photosynthesis. While certainly convenient to measure PAR intensity this way, it is an oversimplification of how the photosynthetic machinery (i.e., chlorophyll and related pigments) intercepts and utilizes PAR photons. The starkest example of this is staring back at you every time you look at your plants: the foliage (normally) appears green because plants more readily absorb and utilize blue (400–500 nm) and red (600–700 nm) vs. green (500–600 nm) photons. Therefore, the visible light that is reflected from photosynthetic tissues appear green under normal illumination (i.e., "white light").

Other wavebands outside the PAR spectrum are also important for cannabis growth and flower development. For example, far-red (700–800 nm) exposure is closely tied to mediating changes in cannabis morphology, as well as transitioning to flowering. Further, there is some (predominantly) ecological evidence that variations in natural UV exposure (~280–400 nm), related to geographic location (i.e., higher exposures at high altitudes and low latitudes), may have some influence on cannabis secondary metabolite composition. The importance of the distinctions of wavelengths both within (and beyond) the PAR waveband will become more apparent in the following discussion of spectrum and different electric light sources.

6.1.1.3 Light Sensors

There are two main types of sensors that are used to characterize horticultural lighting: quantum (i.e., PAR) sensors and spectroradiometers (i.e., a radiometrically calibrated spectrometer). Both types can provide PPFD information; spectroradiometers can also provide a precise characterization of the spectrum distribution, both within and beyond the PAR spectrum, as well as some other calculated spectrum characteristics such as light ratios.

Quantum sensors are an integrating style of sensor that are designed to estimate the photon flux density from 400–700 nm (i.e., PAR) regardless of spectrum. To do this accurately, the spectral response of quantum sensors must give equal weight to all photons within the PAR waveband while excluding photons outside of this range. This is easier said than done, and some quantum sensors are not well matched to narrow-band spectra generated from solid-state lighting technologies such as light-emitting diodes (LEDs). Therefore, if you are shopping for (or already have) this type of sensor, it is important to understand the spectral response of the sensor relative to the light spectrum or spectra in your facility. This is especially important when utilizing multiple lighting technologies, or technologies that emit with narrow-band spectra, particularly at wavelengths near the upper and lower extremes of the PAR spectrum. Normally the relative response spectrum is provided in the technical documentation for good quality quantum sensors, and it is well worth comparing this to your actual spectrum (if known). Some quantum sensors also have multiple settings for different light sources (e.g., "sunlight" vs. "electric light"); use caution when interpreting measurements from these different settings, as they may not match your spectrum at all. We find that quantum sensors are most useful for evaluating spatial variability in intensity in production areas that utilize single sources of PAR. Quantum sensors are not adequate for characterizing the relative intensity under different spectra (e.g., comparing different lighting technologies).

A spectroradiometer can provide absolute intensity data for very narrow wavelength intervals, providing a good measure of the spectral distribution from a light source. The underlying principle of operation is that the incoming light is physically divided into a continuous set of very narrow (e.g., 0.1–10 nm wide) individual spectrum bands (like a rainbow) which are then sensed on a corresponding detector array, usually a linear CCD (charge-coupled device). Spectroradiometers are normally factory calibrated for both wavelength and intensity (in radiometric units). Many modern spectroradiometers marketed for horticulture use software to convert radiometric measurements to quantum units and provide integrated photon flux density values over relevant wavebands [e.g., PPFD (400–700 nm), blue (B, 400–500 nm), green (G, 500–600 nm), and red (R, 600–700 nm)] in a simple user interface either built in to the device or accessible through PC software or a smartphone/tablet application). Some devices may also provide integrated data over user-selected wavebands, as well as other horticulturally relevant metrics such as the red to far-red ratio (R:FR) and phytochrome photoequilibrium.

Spectroradiometers are generally more expensive and more difficult to operate and interpret than quantum sensors, but the added information they can provide is often well worth the investment. There are now many commercially available spectroradiometers that are robust, (relatively) low cost and portable enough for use in production environments. Several of these (e.g., LI-1800 Portable Spectroradiometer, LI-COR Biosciences) are marketed specifically for use in horticultural research and production. However, there are some key design and functional parameters to keep in mind when shopping for a spectroradiometer. For example, while all of the systems marketed for horticultural use cover the PAR waveband, some devices do not entirely cover the entire UVB (280–315 nm), UVA (315–400 nm) and far-red (700–800 nm) wavebands, making the data from these wavebands difficult to interpret. So, if you are interested in more than the PAR waveband, take care in evaluating the wavelength capabilities in the technical specifications.

We strongly advise that a spectroradiometer be used in all horticultural lighting applications where the light spectrum is an important factor, especially when multiple light sources are used. See Llewellyn et al. 2021b for further info on proper setup and use of SR for characterizing light intensity and spectrum.

6.1.1.4   How to Measure Intensity and Spectrum

This subsection will cover strategies for evaluating the canopy lighting environment from the use of electric lighting systems to provide PAR. This can be done at any time (often easier when plants are not present) in indoor environments, or at night in greenhouses. The end of this subsection provides some more information on evaluating natural lighting in greenhouse environments. For the purposes of this subsection, we will focus on measuring PPFD, although the same principles normally apply to measuring spectrum, as well. One of the underlying principles of the PPFD metric is that the measurement surface is horizontal (pointing upwards). For this reason, sensors that measure PPFD are normally cosine corrected, meaning that the sensor has a 180° field of view with the sensor's angular sensitivity being proportional to the cosine of the angle of incidence. Changing the angle of the sensor from horizontal will probably give different results, depending on the characteristics of the light source(s) and their positioning relative to the sensor. Some sensors have built-in "fisheye" levels to help you keep the sensor in the horizontal plane.

Possibly the single largest factor that affects PPFD measurements (intensity and uniformity) from electrical lighting sources is the vertical distance between the light source and the measurement plane. This distance is normally called the "hang height" and typically reflects the distance from the emitting surface of the light fixtures to the top of the crop canopy. For a single light fixture, the relationship between hang height and intensity normally follows the inverse-squared law, which states that the intensity is proportional to the inverse of the square of the distance from the source. Figure 6.3 illustrates this relationship with PPFD measured at heights ranging from 0.2–1.4 m directly below a single light fixture. Note that when the hang heights on the left y-axis (blue trace, solid circles) are transformed using this inverse-squared relationship, the result is a straight line (red trace, open triangles). At larger hang heights, differences in height result in small differences in PPFD, while very large differences in PPFD can occur at smaller hang heights. For example, in Figure 6.3, the difference in PPFD between 70 cm and 80 cm is only 50 $\mu mol \cdot m^{-2} \cdot s^{-1}$,

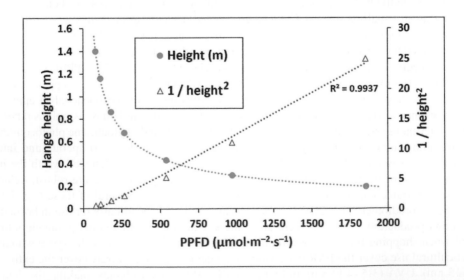

FIGURE 6.3   An example of the inverse-squared law that governs the relationship between the distance from a light source and the measured intensity level at that distance.

while the difference between 30 cm and 20 cm is 1,000 $\mu mol \cdot m^{-2} \cdot s^{-1}$. Note that the relationship between light intensity and hang height may be different under an array of light fixtures than it is below a single fixture.

In controlled environment production systems, average light intensity and uniformity are generally interdependent on hang height such that as intensity increases (e.g., by decreasing hang height), uniformity decreases. The delicate balance of light fixture vendors is to devise a light plan (comprised of hang height and lateral spacing of fixtures) that maximizes average PPFD over the entire crop while maintaining uniformity within acceptable limits, as economically as possible. A common uniformity target is to have the minimum PPFD be no less than 20% lower than the average PPFD (i.e., min ≥ average × 0.80). Some vendors would refer to this metric as the lighting "uniformity". Note that in some cases, heat output can also impose practical limits on the minimum hang height.

With the exception of cultivators who raise their light fixtures in accordance with the growth of their crop, all cannabis cultivators are encouraged to evaluate their lighting environments in the three-dimensional space that their plants normally occupy throughout their growth cycle. This requires selecting a grid of sufficient granularity as to be representative of the entire plot, including capturing the maximum and minimum PPFD areas (normally directly below and centered between adjacent fixtures, respectively). Note that proximity to reflective (or white) side walls will normally increase measured light levels at these locations. Measure at the location where each plant will be placed (at a minimum) in an area representative of at least the area covered by four adjacent fixtures. Measure and record the PPFD at each of these locations at a minimum of five different (evenly spaced) heights ranging from the minimum (e.g., at the top of the plants as they transition from vegetative to flowering photoperiod) to the maximum anticipated canopy height. From these data, you have built a 3-D model of how hang height affects light intensity and uniformity, which you can then relate to how your crops grow in subsequent production cycles. For example, Figure 6.4 shows the PPFD measured at each plant location in sea of green cannabis cultivation system (density 10/m^2) below the area covered by four fixtures. Measurements were done on an 8 × 8 grid at both 1.0 m

1.0 m hang height									0.5 m hang height							
Row	**Column**								**Row**	**Column**						
	1	**2**	**3**	**4**	**5**	**6**	**7**	**8**		**1**	**2**	**3**	**4**	**5**	**6**	**7**	**8**
1	418	515	495	461	481	521	446	315	1	291	439	311	194	206	352	300	158
2	359	500	558	516	498	538	541	425	2	302	843	1138	464	230	562	1082	580
3	451	552	525	493	516	544	459	323	3	684	1373	803	312	473	1195	935	313
4	333	465	514	474	458	497	500	400	4	253	507	635	341	200	390	637	398
5	327	462	511	472	457	496	495	392	5	296	455	331	217	248	409	371	194
6	423	515	493	461	481	520	460	328	6	347	917	1234	528	294	636	1106	608
7	449	542	517	488	514	548	476	347	7	731	1456	900	389	520	1230	1043	369
8	334	461	514	478	460	498	507	405	8	289	563	700	413	313	475	732	469

max	min	avg	min / avg		max	min	avg	min / avg
558	315	468	0.67		1456	158	558	0.28

FIGURE 6.4 Example of how fixture hang height (1.0 m and 0.5 m) can affect canopy-level PPFD ($\mu mol \cdot m^{-2} \cdot s^{-1}$) and uniformity in a high-density cannabis cultivation system. The lightbulb symbols represent the position of fixtures relative to the plants. Planting density was 10/m^2.

and 0.5 m hang heights. Note how the average PPFD increased at the lower hang height, but the uniformity (min/avg) was much lower, and in fact, the minimum PPFD was also lower. While the maximum and average PPFDs at the lower hang height may be impressive values, the high spatial variability could have unsatisfactory impacts on the uniformity of crop growth.

A cultivator can also measure PPFD centered directly below multiple fixtures (at the same height) in a production environment to quickly evaluate the relative uniformity of fixture output across a larger area—this information can be used to determine if and when fixture maintenance or replacement is needed.

When measuring light intensity and/or spectrum, one must also be mindful of how oneself and nearby objects (e.g., reflective walls) may impact light readings, either by blocking light from a nearby source or by reflecting light onto the sensor. It can be rather challenging to characterize a lighting environment without one's body affecting those measurements.

Keep in mind that while the light intensity at the top of a plant canopy is a good proxy for a crop's light exposure, it is not necessarily indicative of the PPFD that each leaf within that canopy is exposed to. Aside from the described effects of the distance between the fixture and a given leaf, vegetated shade (i.e., partial coverage by upper leaves) can drastically affect both the intensity and the spectrum of a leaf's localized lighting environment. For example, Table 6.1 summarizes a spectrum and intensity survey taken within a dense cannabis canopy (~1.0 m tall including 15 cm of substrate, untrimmed plants) that was growing in a greenhouse under natural lighting on a cloudy day at noontime. The measurement plane was horizontal and up-facing; eight replicate measurements were taken at each elevation (i.e., every 10 cm from the bottom to the top of the canopy) and averaged. The PPFD and red to far-red ratio (R:FR) dropped rapidly, but the percentage of PAR that was green (500–600 nm) remained relatively constant. The PPFD dropped to less than 10% of the "canopy-level" intensity within about the top 40 cm of the canopy, and the PPFD at the bottom of the canopy was only about 1.5% of the incident PPFD.

Leaf-level photosynthetic activity is very closely tied to leaf age, with recently expanded leaves being the most photosynthetically active (Bauerle et al. 2020; Rodriguez-Morrison et al. 2021a). Therefore, the light intensity at the top of the canopy can have relatively greater impact on whole-plant photosynthesis than the lighting environment within the inner canopy. However, the middle

---

TABLE 6.1

Survey of the Intra-Canopy Dynamic of Light Intensity and Spectrum within a Dense Cannabis Canopy Growing under Natural Lighting in a Greenhouse on a Cloudy Day (i.e., Diffuse Lighting Conditions)

Parameter	Distance below Top of Canopy (cm)								
	0	**10**	**20**	**30**	**40**	**50**	**60**	**70**	**80**
PPFD (μmol·m^{-2}·s^{-1}, 400–700 nm)	261	194	138	76.5	30.9	13.6	80.9	7.2	4.2
UVA (μmol·m^{-2}·s^{-1}, 315–400 nm)	8.7	5.7	4.1	2.4	1.1	0.6	0.5	0.4	0.3
Blue (μmol·m^{-2}·s^{-1}, 400–500 nm)	68.8	49.0	34.4	18.7	7.5	3.3	2.2	1.7	1.0
Green (μmol·m^{-2}·s^{-1}, 500–600 nm)	95.8	70.4	51.5	29.0	11.8	5.3	3.4	2.8	1.6
Red (μmol·m^{-2}·s^{-1}, 600–700 nm)	96.5	74.6	52.5	28.8	11.6	5.1	3.4	2.7	1.6
Far-red (μmol·m^{-2}·s^{-1}, 700–800 nm)	92.3	74.9	63.4	48.5	32.8	23.3	18.9	16.1	11.2
Red to far-red ratio (R:FR)	1.05	1.00	0.83	0.59	0.35	0.22	0.18	0.17	0.14
Phytochrome photoequilibrium	0.76	0.75	0.74	0.72	0.68	0.61	0.57	0.56	0.51
% of PAR that is "green"	37	36	37	38	38	39	38	39	38

Notes: The canopy was about 100 cm tall, including 15 cm of substrate and measurements were taken every 10 cm from the top of the canopy to just above the substrate (i.e., below the bottommost branches)

and lower leaves may still play important roles in light interception and biomass accumulation. Some cultivators may remove some of the lower leaves at certain points in the production, either to enhance canopy airflow (e.g., to reduce foliar diseases) and/or because (it is believed that) older leaves can transition from being a net source to net sink of photosynthate, potentially reducing yield. However, some research has also shown the potential for inner-canopy lighting to increase biomass production and modify secondary metabolite composition (Hawley et al. 2018). While the additional operational complexity of stringing light fixtures through densely-planted crops may limit the widespread adoption of inner-canopy lighting in commercial cannabis production, the average light intensity at the bottom of the canopy (from toplighting) may serve as a guideline for managing planting density and de-leafing practices. Obviously, cannabis cultivators don't want to waste any of the electric light they provide to their plants, so the average intensity at the bottom of a canopy should be only a fraction of the PPFD measured at the top. For example, if the average canopy level PPFD was 1,000 $\mu mol \cdot m^{-2} \cdot s^{-1}$, and the average lower canopy PPFD was 50 $\mu mol \cdot m^{-2} \cdot s^{-1}$, then ~95% of the incident light was intercepted by the crop. More research is still needed to evaluate how canopy architecture and light penetration affect cannabis inflorescence yield and quality from the entire plant.

Determining greenhouse transmission losses is an important component to evaluating the natural lighting in greenhouse cultivation systems. Transmission losses are relatively easy to estimate with various types of light meters. Even the camera on a modern mobile phone could be sufficient for this task (there are numerous "light meter" apps available to download). In simple terms, all you need to do is measure—and record!—light intensity at canopy level across various representative locations in your greenhouse and also measure outdoors. Keep the sensor horizontal and take care to avoid shading it with your body. You also need to be aware of changing weather conditions, particularly on variably cloudy days (e.g., the "sunny" curve in Figure 6.2). Also note that the direction of light hitting the sensor may have very large impacts on the measurements, particularly with sensors not specifically designed for quantifying incident PAR. Since the majority of the light hitting the earth's surface on sunny days is direct light (angle depending on the geographic location, time of year and time of day), but greenhouse coverings can scatter (i.e., diffuse) the light, some measurement error may be introduced when taking relative measurements (i.e., inside the greenhouse vs. outdoors) on sunny days. Therefore, it is recommended that transmission assessments be done near noon-hour on bright, cloudy days. Note that the actual values recorded by the light sensor are fairly unimportant (though if you use a PAR sensor, all the better!); rather, it is the relative values that are most relevant for evaluating transmission percentages (e.g., inside vs. outside) or spatial uniformity within a greenhouse facility (e.g., lowest vs. average). Also note that the angle of the sun relative to the shape of the structure can have substantial effects on greenhouse light transmission. Therefore, while noontime measurements provide a good benchmark, there will be some temporal variability in transmission percentages, both throughout the day and over the seasons.

In greenhouse and high tunnel production systems, global radiation measurements can be important, as well, since some cultivation systems use accumulated solar radiation to manage the plant growing environments, such as to trigger irrigation events. Global radiation is commonly measured using pyranometer in units of radiant flux density ($W \cdot m^{-2}$; where 1 W = 1 Joule s^{-1}) and comprising wavelengths from approximately 300–3,000 nm, depending on the pyranometer used. Since approximately 45% of the total global solar radiation is in the PAR waveband (400–700 nm), and the conversion between radiant flux density and photon flux density for the sunlight spectrum within the PAR waveband is $W \cdot m^{-2}$ = 4.6 $\mu mol \cdot m^{-2} \cdot s^{-1}$, then PPFD can be estimated from global solar radiation by multiplying these two factors together (i.e., 0.45 × 4.6 = 2.07). For example, if your global radiation sensor reads 100 $W \cdot m^{-2}$, then the corresponding PPFD is approximately 100 × 2.07 = 207 $\mu mol \cdot m^{-2} \cdot s^{-1}$. Note that this conversion factor is only appropriate for the sunlight spectrum, either outside or inside (under most greenhouse covering materials that do not alter the spectrum in the PAR waveband), and may not be appropriate for use with electric light spectra. Further, your radiation sensor is located outside then you must also account for the greenhouse transmission losses (discussed previously) when calculating canopy-level PPFD from outdoor global radiation.

6.1.1.5 Types of Lighting Technologies

In indoor production, cultivators rely totally on electric sources of PAR. Given the high PPFD requirements of cannabis, electric lighting (both infrastructure and energy) represents one of the most costly production inputs. Further, in some circumstances, additional energy is needed to manage the waste heat produced by electric lighting systems.

Like plants that convert the energy from sunlight into chemical energy (i.e., biomass), electric lighting systems are also fundamentally energy converters, designed to convert electrical energy into visible radiation (also some other wavelengths such as ultraviolet and far-red). None of these technologies are perfectly efficient in converting electricity into light, with the overwhelming majority of the waste energy being converted into heat. For example, legacy technologies such as incandescent lighting (i.e., the "Edison" tungsten filament lightbulb) can be terribly inefficient at producing visible radiation—with upwards of 95% of the electrical energy input being converted into (waste) heat. Therefore, the selection of lighting technologies to use in cannabis production represents a major decision. There are several standard electrical lighting technologies that are commonly used in cannabis production, each with its own optimum use scenario(s), benefits and drawbacks. The electric lighting technologies for cannabis production can be divided into three major classes: fluorescent, high-intensity discharge and solid state lighting. These technologies will be discussed in detail in the following subsections.

6.1.1.5.1 Fluorescent Lighting

Fluorescent lighting technologies have been in use for many decades for occupant lighting, as well as for low-intensity horticultural lighting applications, such as the production of young plants (e.g., seedlings, microgreens, clonal propagation). Fundamentally, fluorescent lighting is a low-pressure gas discharge lamp that passes an electric current through a mercury-containing vapor, generating ultraviolet (UV, 100–400 nm) light that then interacts with the fluorescent coating on the inside surface of the lamp, causing it to glow—i.e., emitting mostly visible wavelengths of radiation. Fluorescent lamps are relatively low efficacy, lose intensity output quickly and are hard to control (e.g., difficult to dim and emit light in all directions), and because they contain mercury, fluorescent lamps are considered hazardous waste when broken or at end of life. Fluorescent technologies also emit a small amount of UV radiation, primarily in the longer-wavelength UVA waveband (i.e., 315–400 nm). Due to fluorescent tube architecture (i.e., long and skinny), one advantage of fluorescent is relatively even heat distribution relative to photon flux distribution. From a practical standpoint, this allows fluorescent tubes to be placed relatively close to the plant canopy while still providing even photon flux distribution. This makes fluorescent lighting ideal for lower-intensity applications such as during clonal propagation, and even allows the setup of multi-layer production systems.

6.1.1.5.2 High-Intensity Discharge Lighting

High-intensity discharge (HID) lighting technologies produce broad wavelengths, ranging from UV through infrared, with the main target being in the visible wavelengths (380–750 nm). There are several types of HID technologies that are used in cannabis cultivation. These were also all developed primarily for human vision applications, and then adapted for uses in horticultural production. The most common types of HID used in cannabis are high-pressure sodium (HPS), metal halide (MH) and ceramic metal halide (CMH). Each of these technologies generates substantial waste heat due to very high operating temperatures of the luminaire. Much of this heat is emitted in the same direction as the emitted light (i.e., downwards, toward the crop canopy), meaning the heat generated by HIDs can increase the temperature of plant canopy and inflorescence, thus increasing rates of transpiration. Consequently, the hang height of HID fixtures must be kept relatively high in order to minimize negative impacts on the canopy. This requires taller grow spaces (eliminating the opportunity to grow in multi-layered systems in most circumstances) and generally reduces canopy PPFD but increases uniformity.

Due to their relatively long-term use for occupant and outdoor lighting, the development of HID technologies is relatively mature and are probably close to their physical maxima in terms of PAR efficacy levels. This means that HID technologies have begun to fall out of favor relative to more modern solid state lighting technologies (described in what follows), especially for indoor growing.

HPS produces a visible spectrum that is heavily weighted in the yellow and red wavebands, plus some FR (Figure 6.5). HPS fixtures and bulbs are typically the cheapest types of HID fixtures and also reportedly have the highest potential efficacy. In comparison, MH and CMH both produce "whiter" light, with relatively (to HPS) higher photon flux in the blue waveband. CMH are similar to MH, but use a ceramic material for the arc tube, which enables operation at higher temperature and promote much longer luminaire lifetimes. While HPS and MH are typically replaced after 10,000 hours of use (i.e., about 28 months of 12-hour days), CHM luminaires are reported to last up to twice as along. Among disadvantages of CMH are higher luminaire costs, plus the fixtures normally have lower wattage ratings, meaning more fixtures are needed to provide a given PPFD over a growing area.

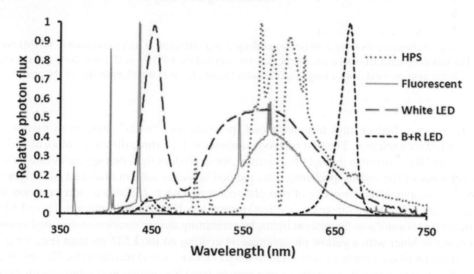

FIGURE 6.5 Comparative photon flux distributions of high-pressure sodium (HPS), fluorescent, phosphor-converted white light-emitting diode (LED) and blue plus red LED spectra emitted by different horticultural LED technologies. Note that units are in relative flux; the respective graphics do not represent the same intensity (i.e., dissimilar integrated areas under the curves).

6.1.1.5.3 Solid-State Lighting

Solid-state lighting technologies utilize a process called electroluminescence to convert electricity into visible radiation. The most common types of solid-state lighting technology are light-emitting diodes (LEDs). Fundamentally, LEDs utilize semiconducting materials that emit relatively narrow wavebands of radiation. There are many types of LEDs, with the chemical makeup of the semiconductor material being the major factor that defines the spectrum that is output.

The spectral output of LEDs is typically a narrow band of wavelengths, with the relative spectral intensity normally taking on the shape of a narrow bell curve (see Figure 6.6). LED spectral distributions are commonly defined by the wavelength at the highest intensity (i.e., peak wavelength) and the width of the peak at half of the maximum intensity level (called full width at half maximum, FWHM). Typical monochromatic horticultural LEDs have peaks in the blue (B, 400–500 nm) and red (R, 600–700 nm) wavebands (Figure 6.5), normally with FWHM between 10 nm and 25 nm (Figure 6.6).

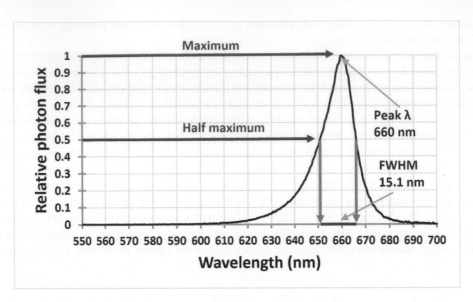

FIGURE 6.6   Schematic showing how peak wavelength and full width at half maximum (FWHM) are measured. For this deep red LED spectrum, the peak is 660 nm and the FWHM is 15.1 nm. Note that the shape of the peak is not symmetrical, with a longer and broader tail on the short wavelength side of the peak.

There are two distinct ways to make broad spectrum (i.e., "white") light with narrow band–emitting LED technologies. The first method combines at least three different colors of LED light (e.g., B, G and R)—mixed in the right proportions, these produce light that appears white (or almost any other color of the visible spectrum). This method is widely used in modern LED screens such as TVs and computer monitors. Use of phosphor-converted white LEDs is a more common way to generate white light for high-intensity applications (such as horticultural lighting). These LEDs use technology that is similar to fluorescent lights, by combining shorter wavelength (i.e., higher-energy) LEDs (usually blue) with a yellow phosphorescent coating on the LED package (i.e., lens)—this causes a portion of the light to be converted to longer (lower-energy) wavelengths. The spectral output of white LEDs normally comprises some narrow-band blue emission combined with the much broader, phosphor-converted yellow spectrum to produce light that appears "white" (Figure 6.5). The relative size and shape of these two peaks is largely dependent on the thickness and composition of the phosphorescent coating. Less dense coatings result in more B photons exiting the LED chip without interacting with the phosphor. This results in "bluer" spectra that are typical of spectra with higher correlated color temperatures (CCT). Ironically, luminaires with higher CCT are often referred to as appearing "cool," while lower CCT are associated with "warmer" spectra, which causes untold confusion when relating the appearance of a spectrum to its color temperature. For comparison, sunlight has a CCT of about 6,000 K (note that the Kelvin scale has the same magnitude as Celsius but 0 K [i.e., absolute zero] is equivalent to −273°C).

Since individual diodes are relatively small light emitters, horticultural LED fixtures are typically comprised of an array of multiple individual diodes, often as a mixture of different colors. Aside from the desired plant responses, the major considerations when developing a spectrum recipe relate to the cost of chips and their efficiency in converting electricity into usable light (i.e., efficacy).

Within the PAR spectrum, the vast majority of horticultural LEDs are either blue or red. If one were to consider the global uses of blue and red LEDs, it might seem that there are many more applications for red vs. blue (e.g., automobile indicator lights). However, remembering that virtually all phosphor-converted white LEDS (including occupant lighting) are actually blue, the economic driver for development of blue LED technology is considerably greater than red. Although blue LEDs are probably more technically advanced (i.e., closer to their projected technical limit of efficacy), the energy cost associated with making blue photons is substantially higher (~30%, depending on wavelength) than red photons. While blue LEDs currently have higher *efficiency* (0.93 W/W vs. 0.81 W/W), red LEDs have higher *efficacy* (4.7 µmol/J vs. 3.5 µmol/J) because of the reduced energy per photon (Kusuma et al. 2020). Conversely, the efficacy of phosphor-converted white LEDs is maximized when the percent blue is highest (i.e., less interaction between blue light and the phosphor). This is because of the energy lost in the phosphor conversion process which inherently result in lower efficacy values for white LEDs vs. their blue origins.

Horticultural LED manufacturers must balance the cost of chips, their efficacy and the resulting spectrum when developing a spectrum recipe to produce in a fixture. While plant response is obviously a major consideration, there are other factors—such as color rendering, human comfort and LED chip cost, availability and quality—that may guide the decisions.

Recently, an independent organization called the Design Lights Consortium has undertaken to standardize the methods of evaluating and reporting the characteristics and performance specifications of horticultural LED fixtures. Manufacturers are required to submit fixtures to third-party testing and reporting and qualified fixtures are listed with their characterizations, including spectrum and efficacy metrics. Cultivators are encouraged to seek out this information for any LED technology that they are considering purchasing, and proceed with caution when considering LED fixtures that are not listed on the DLC website (https://www.designlights.org/).

One persistent strategy has been to combine predominantly red LEDs with sufficient light from blue LEDs (normally 5–25% of total photon flux) to grow normal-looking plants under greenhouse or sole-source conditions (Ying et al. 2020). These technologies, which produce various shades of magenta-hued light (sometimes called "blurple"), have the potential to have very high efficacy. The spectra of technologies with the highest efficacy are normally heavily weighted in the red, often > 90% red. A major reason for this (as described previously) is the simple physics of the relationship between wavelength and energy: red photons require the least amount of electricity to produce. To illustrate how this phenomenon translates to commercial practice, Figure 6.7 illustrates the relationship between the percentage of red light of total PAR vs. efficacy or a range of DLC-listed fixtures, all from one horticultural LED company. Inset into this figure are sample spectra distributions that roughly illustrate the differences in spectrum distribution in groupings of LED fixtures that appear on different parts of the efficacy curve. The efficacy of red-heavy fixtures represented in the pink-hued oval in the upper right of this graphic are about 1.4 times higher than the very white-appearing light emitted from the fixtures represented in the blue oval at the bottom left. The fixtures in the green oval are representative of an intermediary spectrum, which will definitely have some pinkish undertones but still be reasonably pleasant to work under and provide workers with the ability to visually evaluate differences in plant color (e.g., nutrient deficiencies or pathogen caused diseases). In the grouping in the upper right, the small blue peak would come from blue LEDs. In the middle and lowest groupings, the majority of the blue light probably comes from phosphor-converted white LEDs (such as illustrated in Figure 6.7).

There are many different form factors of horticultural LED fixtures. A major consideration is heat management—even though much less heat is directed toward the crop canopy in LEDs vs. other electric lighting technologies, they still produce a considerable amount of heat (~20–40% of the electricity input is converted to heat in modern horticultural LED fixtures). Heat management systems are divided into active (fans, liquid cooling) and passive (heat sinks, cooling fins) systems. While active cooling adds complexity, it can also provide greater flexibility in fixture layout, whereas the LED chip density is somewhat more limited in passively cooled systems. Further, form factors can be broadly divided into greenhouse and indoor designs. A major consideration in greenhouse applications is the minimization of the fixture footprint to reduce the blocking of sunlight. Accordingly, many greenhouse LED fixtures have long and narrow shapes, designed to hang below already present structural members (e.g., below gutters and collar ties). Conversely, LED fixtures that are designed for indoor use (where sunlight-blocking shadows are not an issue) often have the

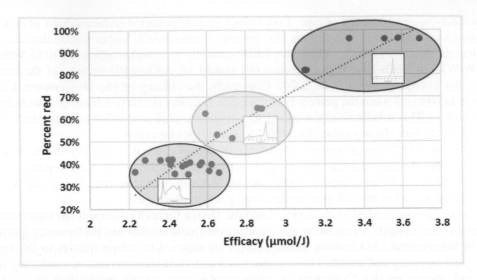

FIGURE 6.7 The relationship between LED fixture efficacy and proportion of the emission PAR spectrum that is comprised of red (600–700 nm) photons. These data were taken from the specifications of various LED fixtures manufactured by a horticultural lighting company as listed on the Design Lights Consortium (DLC) list of DLC-qualified products. The spectrum scans inset in each grouping of fixtures is representative of the spectral distributions of all fixtures in the respective grouping.

LED density spread out over a larger area (frequently in the form of an array of multiple parallel LED bars). This form factor helps with increasing PPFD uniformity, while also reducing the need for heavy and complex heat sinks.

When making a decision on what type of lighting to purchase, the photon flux (µmol/s), efficacy (µmol/s) and spectral distribution are certainly important descriptive parameters for any horticultural light fixture. However, while they may characterize the light produced, they do not describe where the produced light actually goes (i.e., direction). The direction of light emission is much more difficult to capture and conceptualize in a summary value. Many factors, such as primary (e.g., the physical shape of the light emitter) and secondary (e.g., HID diffuser geometry, reflectors and lenses in LED fixtures) optics influence the spatial distribution of the emitted light. The interplay between these factors and the lateral positioning of the fixtures in the growing environment will have a major impact on the lighting distribution. Things get even more complicated when multiple different fixture types are combined in a single production area, such as is the case with hybrid HPS-LED installations (where both spectrum and intensity can vary substantially). Most reputable horticultural manufacturers utilize light planning software to model and optimize the intensity and uniformity distribution in specific scenarios. In our experience, the models derived from careful light planning by the manufacturer have generally been well corroborated by actual measurements taken post-installation. One factor to keep in mind is the reflectivity of nearby surfaces, including the floor and (especially) walls—these structures can have profound effects on light distribution within your production environment.

Only a few publications have reported reliable information on the efficacy of different legacy horticultural fixture technologies, including Nelson and Bugbee (2014), Radetsky (2018), Wallace and Both (2016), Shelford et al. (2020) and Shelford and Both (2021). From these publications, the average reported efficacy values are presented in Table 6.2, along with the efficacy values for measured LED fixtures from that time period, plus the average efficacy of the fixtures presented in Figure 6.7, taken from the DLC in 2021. Note that while the efficacy of legacy technologies has remained relatively stagnant, LED technologies are still developing rapidly. For example, the maximum efficacy of the

TABLE 6.2

Average and Maximum Reported Efficacy Values (μmol·J^{-1}) for Different Types of Horticultural Lighting Technologies

Lighting Technology	Efficacy (μmol·J^{-1})	
	Average	Maximum
Incandescent	0.31	0.30
Metal halide	0.87	0.91
Fluorescent	0.92	0.97
Ceramic metal halide	1.42	1.58
High-pressure sodium	1.51	1.72
Light-emitting diode (LED) (2014–2018)	1.54	2.64
LED (2021)	2.72	3.69

Note: The efficacy values for the LEDs in 2021 were taken from the DLC-listed fixtures that are presented in the curve in Figure 6.7

LEDs listed in between 2014 and 2018 is less than the average efficacy of the fixtures listed in 2021. Overall, present-day efficacy of horticultural LEDs can be more than double the efficacy of legacy technologies, with perhaps 25% additional increases in LED efficacy projected in the next 20 years (Kusuma et al. 2020) as LED technology continues to improve at a rapid rate.

Currently, the purchase price of LED technologies is considerably more than HID systems of similar light output. The additional outlay for LEDs can be recuperated over time through reduced energy costs. Photon maintenance is an often-overlooked characteristic of light fixtures that should be factored into the selection of a lighting technology. For example, HPS bulbs are often rated to have a 90% photon maintenance levels between 10,000 hours and 20,000 hours (i.e., 2.25–4.5 years, if operating 12 hours/day year-round). Therefore, intermittent blub replacement costs should be considered with these technologies. In comparison, LED fixtures don't normally have replaceable LEDs, but their projected photon maintenance levels are typically higher than other technologies, often maintaining 90% photon maintenance levels from 30,000–50,000 hours.

6.1.2 PHOTOPERIOD FOR FLOWERING

Plant flowering responses to daylength are called photoperiodism. In fact, it is the length of the uninterrupted dark period between light periods that are responsible for the flowering responses. Plants are generally divided into three different photoperiod response groups: short-day, day-neutral, and long-day plants. Short-day (i.e., long-night) plants only flower when the uninterrupted dark period is increased to a certain length. Day-neutral plants can flower regardless of photoperiod. Long-day plants only flower when the dark period is reduced to a certain length, which can be accomplished either with a single dark period or when a longer dark period is interrupted with even short lit periods. Conversely, short-day plants will not flower when the required dark period gets interrupted with even a short period of light.

As discussed in Chapter 1, the majority of cannabis genotypes are short-day plants, although there are also some day-neutral genotypes. Day-neutral cannabis genotypes are also called autoflowering types. Autoflowering genotypes are generally seed-propagated and will initiate flowering in a certain number of days after germination, regardless of daylength. The vast majority of commercial controlled environment cannabis cultivations currently use a 12-hour photoperiod during the flowering stage, since almost all of the commonly used genotypes produce a strong flowering response under this photoperiod. In a light hungry crop such as cannabis (more information in what follows), limiting the daily photosynthetic activity during the flowering stage to only half the available time is a significant production constraint. Extending the photoperiod without negatively affecting the flowering response has the potential to shorten the production cycle or increase yield (or both).

There are reports that certain cannabis genotypes can even flower under a 15-hour photoperiod (i.e., as short as nine hours of uninterrupted darkness). For example, our research demonstrated that a dark period as short as 10.2 hours was sufficient to promote flowering in explants of one genotype used in commercial cultivation (Moher et al. 2020). With genotypes that still exhibit strong flowering responses under longer photoperiods, there is the potential to add between 5% and 20% more lighting time each day. Cultivators could therefore either increase the DLI or provide the same DLI by lighting over a longer photoperiod at reduced intensity, a strategy that generally results in better light use efficiency. Therefore, it would be useful to find out the photoperiod requirements of currently used genotypes. This knowledge would also be informative for greenhouse growers with respect to their blackout curtain protocols during seasons with daylength > 12-hour days.

For some long-day plants, some electric lighting technologies that contain red and far-red spectra can induce flowering with PPFDs as low as $0.02\ \mu mol \cdot m^{-2} \cdot s^{-1}$. Further, PPFDs as low as $1\ \mu mol \cdot m^{-2} \cdot s^{-1}$ may be sufficient for night interruption. However, there are currently no published studies that indicate the upper PPFD limit that will not interrupt the dark period required for cannabis to flower, which is important information for blackout curtain design and deployment.

In indoor production, it may be possible to extend the length of a typical light + dark cycle beyond the typical 24-hour day. In this way, it may be possible to induce short-day cannabis plants to flower under extended light periods, as long as the critical dark period is still regularly provided. For example, if the "day" was extended to 28 hours long, plants could be exposed to 16-hour light periods followed by 12-hour dark periods. This would compress seven normal days of lighting (i.e., 12 days + 12 nights) into six days, thus increasing total light exposure by approximately 15%. More research is needed to explore whether longer light periods (either longer photoperiods or > 24-hour days) can be provided during cannabis' flowering stage to enhance photosynthesis and increase yield. Further, some chrysanthemum research demonstrated that pure blue light can be used to extend the light period without affecting flower induction in this short-day plant. Future research should explore whether specific light spectra, such as pure blue, can be used to extend photoperiod for increasing cannabis biomass production without affecting flowering.

6.1.3 Light Spectrum and Cannabis Production

There have been some spectrum arguments made for using specific lighting technologies for the different phases of cannabis production, such as the higher blue proportion in MH and CMH promoting compact vegetative growth and higher red proportion in HPS promoting flowering. However, the relationships of photon distribution, heat distribution and efficacy characteristics of the respective technologies with respect to the cultural aspects (e.g., plant size, crop density, optimum LI) of the different production phases have also strong influencers of the lighting technology used in each phase of production. With the advent of high-output, narrow-band LED, modern horticultural LED technologies have facilitated more granular optimizations of spectrum in controlled environment crop production. Horticultural LED-spectrum research is still in its infancy, but recent publications have shown the potential for specific spectrum treatments to influence cannabis morphology, inflorescence yield and secondary metabolite composition. In North America, market research has indicated that the majority of the commercial cannabis cultivations are currently using LED and fluorescent lights for propagation and vegetative stages, and LED and HPS for flowering stage.

For the production of rooted transplants from clonal cuttings, the main objectives are to enhance root initiation and establishment, thus producing strong and healthy transplants. White LEDs with high blue spectrum and fluorescent lights are good choices. Spectral distributions with elevated proportions of blue wavelengths have been shown to enhance root initiation in cannabis and many other horticultural commodities. Some research has shown that low levels of UV at wavelengths > 310 nm may be used at the latter stage of propagation to prime transplants' resistance to different biotic and abiotic stresses.

For the vegetative stage, light spectrum can be used to steer plants to be ready for the flowering stage. For example, pure blue and far-red spectrum treatments can be used to induce stretching. Conversely, high blue content in multi-spectrum combinations (e.g., red and blue LEDs, white LEDs) can reduce internode lengths, making plants more compact. Small amounts of UV may make plants more robust and also mitigate certain foliar pathogens such as powdery mildew. Depending on the genetics of the plants and the cultivation strategies (e.g., small plants and short crop cycle vs. large plants and longer crop cycle), dynamic lighting spectrum strategies may be applied during the vegetative stage. For example, if the plants are too dense and compact, pure blue light can be applied for a few hours at the end of the day to stretch the plants, opening the canopy up for better air flow.

During the flowering stage, light can be used to enhance flower initiation, carbohydrate accumulation, production of secondary metabolites (e.g., cannabinoids and terpenes), inflorescence biomass accumulation, and so on. In the past, HPS spectra were mainly used during this stage; however, LEDs have gradually overtaken HPS to become the leading lighting technology used in this stage of cannabis cultivation. Even though insufficient research has been conducted to explore how to use light spectrum to optimize this important stage of cannabis cultivation, the general consensus is that LEDs are able to replace HPS without reducing flower yield or quality. One emerging trend with the adoption of LED technologies for the flowering stage of cannabis production appears to be increasing occurrences of terminal inflorescences (i.e., at the branch apexes) of some cultivars becoming bleached of color (Figure 6.8). This phenomenon needs more research to determine the causes and

FIGURE 6.8 An example of bleached apical buds formed in a LED-lit indoor production environment with a high proportion of red in the spectrum and high canopy intensity (> 1,000 μmol·m^{-2}·s^{-1}).

Source: Photo by David Llewellyn

mechanisms, but it is possible that the high proportions of red wavelengths found in some LED spectrum distributions, in combination with the potential for higher canopy PPFDs that are possible with LEDs (e.g., vs. HPS, due to canopy radiant heating), may have exacerbated this phenomenon. Therefore, it is advisable to test your cultivars under the selected spectrum recipes at the intended canopy intensities for negative interactions such as apical bud bleaching. Based on the current range of LED fixtures listed on the DLC website, growers now have many spectrum options to select from to match their own production environment and cultivation priorities.

Maximizing the yield of cannabinoids and terpenes in mature female inflorescences is a major objective in drug-type cannabis cultivation. Based on some earlier research and observations, it has been long believed that UV exposure can enhance both cannabinoid and terpene concentrations. However, modern cannabis genotypes generally already have far higher potency than the genotypes used when these theories were first developed. Two of our recent studies demonstrated that even low doses of UV-B had predominantly negative effects on both cannabis yield and secondary metabolite composition (Llewellyn et al. 2021a; Rodriguez-Morrison et al. 2021b). More scientific research is certainly needed to establish application protocols for using UV exposure to enhance the cultivation of modern cannabis genotypes in controlled environments.

As discussed in Section 6.1.1, there is a range of spectra used in horticultural LED fixtures, normally comprised of combinations of blue, red and phosphor-converted white LEDs (Figure 6.7). The range of spectrum distributions reflect different strategies that emphasize efficacy, color rendering or attempts to balance these factors. Generally, LED spectra primarily comprised of red and blue LEDs have higher efficacy (minimizing energy usage), but have an unpleasant magenta appearance, while fixtures with higher proportions of white LEDs can be more tolerable to work under and facilitate the observation of plant performance and pest scouting.

6.1.4 LIGHT INTENSITY AND CANNABIS PRODUCTION

This subsection and the following section address how light intensity and carbon dioxide concentration can affect cannabis biomass production and secondary metabolite composition. While these two input factors are addressed separately, it is important to be mindful of their interdependence. While it is beyond the scope of this discussion to consider the economics, both factors must be considered together to determine the optimum balance of these production inputs for individual cultivation environments and production goals.

6.1.4.1 Relationships between photosynthesis, plant growth and yield with light intensity

Photosynthetic tissues convert CO_2 under light to carbohydrates and oxygen; this process is called photosynthesis and follows the basic chemical formula outlined below.

$$6CO_2 + 6H_2O \xrightarrow{\text{light}} C_6H_{12}O_6 + 6O_2$$

Plants use these carbohydrates and oxygen to produce energy for cellular maintenance and growth in the process called respiration, which releases CO_2. Both photosynthesis and respiration (i.e., photorespiration) occur simultaneously when there is light. The photosynthetic rate ($\mu mol[CO_2] \cdot m^{-2} s^{-1}$) minus the respiration rate ($\mu mol[CO_2] \cdot m^{-2} s^{-1}$) is called net CO_2 assimilation rate (NCAR). Obviously, for cannabis growth and production, we need to create an optimal environment to promote high NCAR.

Light and CO_2 are the two most important environmental factors for photosynthesis, plant growth and the ultimate yield. How much light should you provide to your cannabis plants in controlled environment cultivation? To answer this question, let's first look at how plants respond to light intensity (LI). LI not only can affect plant photosynthesis, respiration, ultimately the NCAR; it can also influence plant morphology, which in turn affects light interception and photosynthetic efficiency.

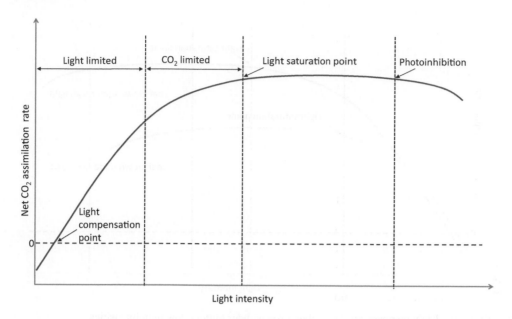

FIGURE 6.9 A typical light response curve illustrating the relationship between plant net CO$_2$ assimilation rate and light intensity.

Plant leaf and canopy-level responses to LI normally follow the pattern illustrated in Figure 6.9, and this curve is called light response curve (LRC). As LI increases, NCAR increases linearly, then the line starts to bend and gradually becomes horizontal (i.e., the light saturation point), and then begins to decrease at very high LI (i.e., photoinhibition). There is always a point near the beginning of the LRC where the photosynthetic rate equals respiration rate (i.e., the NCAR is zero); this LI is called light compensation point. Only when LI is higher than this level are plants able to grow. The linear part of the LRC is mainly light-limited. As the LRC starts to bend, the response transitions to being mainly CO$_2$-limited. Coincidentally, this means that the NCAR will be higher for a given LI when the growing environment has elevated CO$_2$ concentrations. Beyond the light saturation point, increasing LI does not result in increased NCAR. Light use efficiency (mol[CO$_2$]·mol^{-1}[PAR]) is the highest in the linear part of the LRC, called the maximum quantum efficiency. It should be noted that NCAR curves normally only provide a snapshot of how a given tissue (i.e., leaf, branch, whole plant, etc.) responds to LI at one point in time. They do not account for said tissue's light history or temporal changes in biochemistry that naturally occur as photosynthetic tissues age and newly-grown tissues adapt to their environment. For example, when plants are moved from lower to higher LIs, newly developing leaves can become thicker and smaller than the leaves that developed under the lower LI. Therefore, while plant growth and yield responses to LI generally follow similar patterns as NCAR, it is not possible to estimate growth and yield responses to LI from NCAR curves.

When constructing LRCs for a given crop or interpreting existing LRCs, it is important to keep the following points in mind:

1. The light response curves for plant tissues with different light histories (i.e., grown under different light intensities) will be different. For example, both light saturation and photoinhibition points are higher for plants grown under higher LI than those grown under lower LI (Figure 6.10). Foliar and whole canopy morphology, physiology and biochemistry can all acclimate to their growing LI. Leaf LRCs start to show obvious acclimations only a few days after plants are moved to different lighting conditions (Zheng et al. 2005).

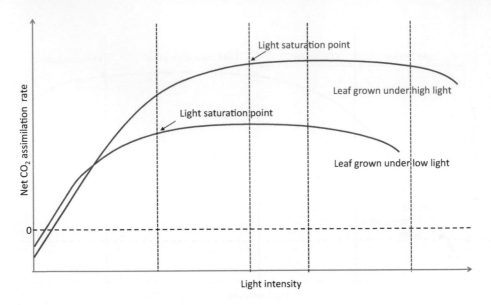

FIGURE 6.10   Light response curves of plant grown under high vs. low light intensities.

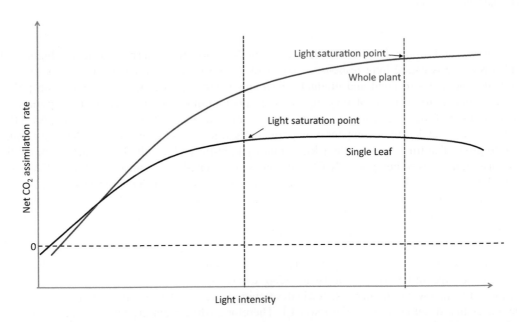

FIGURE 6.11   Light response curves of a leaf and a whole plant grown under the same lighting environment.

2. Light response curves are different depending on the level of plant tissue (i.e., single leaf vs. branch vs. entire plant). For example, the light saturation point is much lower for a single leaf than that for a whole plant (Figure 6.11). Therefore, parameters such as light saturation point obtained from a light response curve of a leaf or branch are different (usually lower) from those obtained from a whole plant.

3. Light response curves can be different for plants grown under different CO_2 levels. Plants grown under higher CO_2 levels can not only have higher NCAR, but may also have higher light saturation points (Figure 6.12).

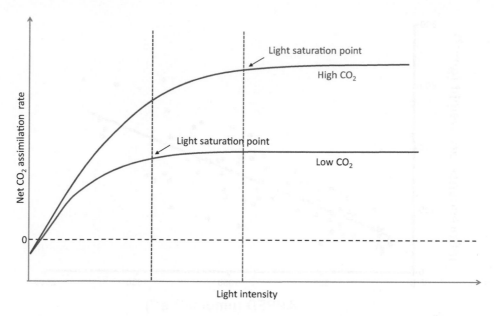

FIGURE 6.12   Light response curves of plants grown under high vs. low CO$_2$ levels.

4. Plants grown under different environmental conditions (e.g., rootzone water and nutrient levels) can have different light response curves. For example, if cannabis plants are grown in a suboptimal rootzone (e.g., lack of nutrients), then the light saturation point can be lower than normal. Regardless of the LI provided to plants, the growth and yield are limited by a nutrient imbalance.

Therefore, to construct a light response curve for a specific cannabis genotype, it is recommended to grow the plants under the cultivator's normal growing condition but under array of LIs for the entire growing period of interest (e.g., the flowering phase). Carefully measure the LI at the top of each plant—preferably twice weekly while the plants are still growing vegetatively, and then weekly thereafter. Measure yield and quality of cannabis at harvest, on a per-plant basis, and plot the data against the average measured LIs (see Rodriguez-Morrison et al. [2021a, 2021b] for more details on how to calculate average LIs, i.e., APPFD). Light response curves can also be used as a diagnostic tool. Our recent research demonstrated that cannabis is a light-hungry plant which can have a light saturation point (for inflorescence yield) higher than 1,800 μmol·m^{-2}·s^{-1} (Figure 6.13) when grown under 100% electrical light in a controlled environment. If a light response curve shows the light saturation point is much lower than 1,800 μmol·m^{-2}·s^{-1}, then it is worth investigating whether some other growing environmental factors (e.g., CO$_2$, nutrients) are limiting.

Generally, plants grown under lower light levels have larger, thinner leaves and fewer branches, and are taller with longer internodes than those grown under higher light levels. For example, our recent study (Moher et al. 2021) that grew cannabis vegetatively (i.e., under 16-hour photoperiod) under PPFDs ranging from 135–1,430 μmol·m^{-2}·s^{-1} showed that the leaf size and internode length decreased linearly with increasing LI, leaf thickness increased linearly with increasing LI, stem thickness and aboveground biomass responded quadratically to increasing LI but did not reach the light saturation point, and plant height and growth index (i.e., overall size) responded quadratically to increasing LI, reaching maxima at around 700 μmol·m^{-2}·s^{-1} and 600 μmol·m^{-2}·s^{-1}, respectively. Therefore, LI needs to be taken into consideration during the vegetative stage to prepare plants for the upcoming flowering stage. For example, for short and compact genotypes, it may be advisable to use lower LI to grow plants taller to encourage air flow and reduce foliar diseases such as *Botrytis* and

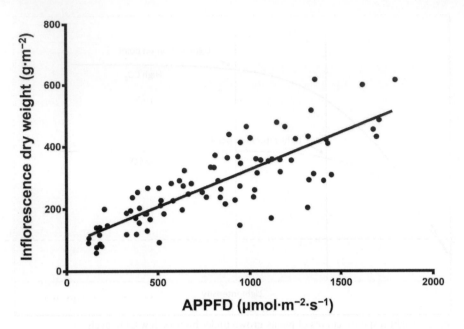

FIGURE 6.13   The relationship between average apical photosynthetic photon flux density (APPFD) applied during the flowering stage (81 days) and inflorescence dry weight of *Cannabis sativa* 'Stillwater.' Each datum is a single plant. The plants were grown under ambient CO_2 in an aquaponic system.

Source: Adapted from Rodriguez-Morrison et al. (2021a)

powdery mildew in the later plant development stages. For taller cultivars, high light intensity during vegetative growth can encourage more compact, robust growth which will improve plant structure and enable the plants to support heavier inflorescence weight during the later flowering stage.

6.1.4.2   Light Intensity and Cannabis Quality

Cannabis quality includes quite a few attributes. The major ones are inflorescence visual appearance, size, density and the cannabinoid and terpene composition. Our recent study (Rodriguez-Morrison et al. 2021a), which grew flowering-stage *Cannabis sativa* 'Stillwater' under PPFDs ranging from $120–1,800 \ \mu mol \cdot m^{-2} \cdot s^{-1}$, showed that increasing LI increases apical inflorescence size and density, as well as the proportion of the total inflorescence yield that came from (higher-quality) apical tissues. Conversely, the same study showed that LI had little impact on cannabinoid concentration, including total THC and CBD, except when the LI was extremely low (e.g., $< 200 \ \mu mol \cdot m^{-2} \cdot s^{-1}$). The concentration of total terpenes, myrcene and limonene increased linearly with increasing LI, while other terpene contents were not affected by LI. Overall, increasing LI appears to enhance the overall cannabis inflorescence quality attributes, without any known negative effects.

In summary, from the rooting of clonal stem cuttings to flower harvesting, the following light intensity schedule may be used:

For propagation of clonal cuttings

Day 1: PPFD $< 100 \ \mu mol \cdot m^{-2} \cdot s^{-1}$ to let the cutting recover from and prevent further water loss.
Day 2: PPFD $< 200 \ \mu mol \cdot m^{-2} \cdot s^{-1}$.
Days 3–7: As roots begin to initiate, gradually increase PPFD from $200 \ \mu mol \cdot m^{-2} \cdot s^{-1}$ to about $500 \ \mu mol \cdot m^{-2} \cdot s^{-1}$.
Day 8 to transplanting: Check rooting success. If roots are visible on the outer surface of the rooting medium, gradually increase the LI to match the LI at the vegetative growing room. The daily incremental increases in PPFD can be around $200 \ \mu mol \cdot m^{-2} \cdot s^{-1}$.

The PPFD for the vegetative stage may range from 400 $\mu mol \cdot m^{-2} \cdot s^{-1}$ all the way up to 1,500 $\mu mol \cdot m^{-2} \cdot s^{-1}$, depending on the genotype, cultivation strategies and the LI target for the flowering stage. For the flowering stage, the PPFD may range from 400 $\mu mol \cdot m^{-2} \cdot s^{-1}$ to almost 2,000 $\mu mol \cdot m^{-2} \cdot s^{-1}$, with higher PPFD normally resulting in higher yield and better quality. Keep in mind that the longer photoperiods in the vegetative stages naturally push the DLIs higher than the short days in the flowering stage. For example, a PPFD of 500 $\mu mol \cdot m^{-2} \cdot s^{-1}$ over an 18-hour photoperiod results in the same daily light exposure as 750 $\mu mol \cdot m^{-2} \cdot s^{-1}$ over a 12-hour photoperiod.

6.2 CO₂ IN CANNABIS PRODUCTION

The responses of photosynthetic tissues to atmospheric CO_2 concentration follow the pattern illustrated in Figure 6.14, called a CO_2 response curve. Similar to the LRC, as CO_2 increases, NCAR increases linearly then starts to bend, gradually saturates and can even decrease at very high levels. At very low CO_2 concentrations, there is a point when the photosynthetic rate equals respiration rate (i.e., the NCAR is zero); this is called the CO_2 compensation point. Plants are only able to grow when the CO_2 concentration is higher than this point. When CO_2 reaches the saturation point, further increase in CO_2 does not result in increased NCAR and excessive CO_2 can also harm plants. Excessive CO_2 can cause injuries such as chlorosis and necrosis of lower leaves and reduced growth and yield. For example, the upper thresholds are about 2,200 ppm and 1,200 ppm for tomato and gerbera, respectively. The CO_2 threshold is also LI-dependent, with lower LIs having lower CO_2 thresholds. Currently, no upper CO_2 threshold has been established in the scientific literature for cannabis.

Like LRCs, plant growth and yield responses follow similar trends as NCAR, but for the same plant the CO_2 response curves for the growth, yield and NCAR can all be different from each other.

Since plants can acclimate to their growing environments, a plant's CO_2 response curves can be different when grown under different environments. For example, when moving a plant grown under lower to higher CO_2 concentrations, stomates will start to close and leaves that developed under the higher CO_2 concentration may have reduced stomatal density. Plants also respond to CO_2 differently under different environmental conditions, such as LI and nutrient level. These types of

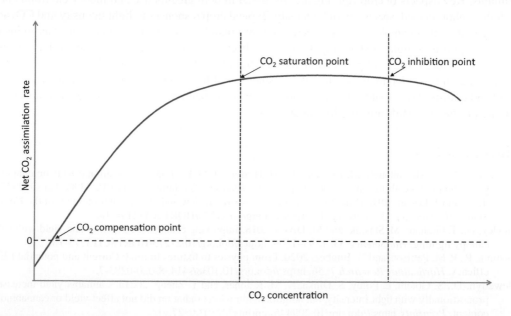

FIGURE 6.14 A typical CO_2 response curve illustrating the relationship between plant net CO_2 assimilation rate and atmospheric or leaf intercellular CO_2 concentration.

variabilities make it impossible to translate CO_2 response curves (other than basic shape) from tissues grown in one environment to another. Therefore, to construct a CO_2 response curve to reflect specific growing conditions and cannabis genotype, it is recommended to grow the plants under the cultivator's normal growing condition, but under an array of CO_2 concentrations for the whole growing period. Measure yield and quality of cannabis at harvest and plot the data against the CO_2 concentrations.

Scientists use high-tech instrumentation to construct CO_2 response curves on per-leaf and per-plant bases. For crop-level experiments, sophisticated controls are needed to maintain all environmental conditions at the same levels and also maintain CO_2 concentrations target levels. These types of research facilities are rare; therefore, so far we are not aware of any published and reliable studies that show cannabis growth, yield or quality responses to CO_2 concentration. It is believed elevated CO_2 level can improve cannabis yield compared to growing cannabis under ambient CO_2 (around 410 ppm). Therefore, CO_2 concentrations of 800–1000 ppm are commonly used in controlled environment cannabis cultivation. However, the degree of cannabis' responses to elevated CO_2 levels depends on many other factors, such as LI. Additionally, the economy of CO_2 supplementation depends on many other factors, such as the air exchange rate of the growth room or greenhouse. Note that since plants and living organisms (e.g., microorganisms) in the rootzone release CO_2 and also no photosynthesis going on in the dark, therefore, no CO_2 addition is needed during the dark period.

For stock plants and clonal propagation, elevated CO_2 concentrations may enhance the production of the number of cuttings and the cutting tissue carbohydrate content, increase rooting success rate and produce stronger transplants. During rooting, elevated CO_2 can increase auxin production to induce root initiation and formation, and enhance root growth. Therefore, elevated CO_2 environments may increase the rooting success of cannabis cuttings and enhance the root growth of transplants. Concentrations of CO_2 as high as 1,500 ppm have been recommended at the rooting stage to increase transplant quality by reducing stomatal opening, thus conserving water while enhancing photosynthesis.

6.3　SUMMARY

This chapter provides an overview of two of the most important production inputs for growing cannabis. Key aspects of crop lighting are discussed in both greenhouse and indoor environments, including photoperiod, spectrum and intensity. Typical crop responses to light intensity and CO_2 are highlighted, with an emphasis on the need to know a plant's growing environment in order to interpret its responses to light and CO_2 inputs. We also discuss recent research results, suggest target levels for the different stages of production and highlight knowledge gaps that still exist. Importantly, more research is needed to model the interdependence of light and CO_2 levels on cannabis growth, yield and quality. This would allow cultivators to balance their relative costs and benefits, thus optimizing production while minimizing input costs.

BIBLIOGRAPHY

Bauerle, W. L., C. McCullough, M. Iversen, and M. Hazlett. 2020. Leaf age and position effects on quantum yield and photosynthetic capacity in hemp crowns. *Plants* 9: 271. https://doi.org/10.3390/plants9020271.

Faust, J. E., and J. Logan. 2018. Daily light integral: A research review and high-resolution maps of the United States. *HortScience* 53: 1250–1257. https://doi.org/10.21273/HORTSCI13144-18.

Hawley, D., T. Graham, M. Stasiak, and M. Dixon. 2018. Improving cannabis bud quality and yield with sub-canopy lighting. *HortScience* 53: 1593–1599. https://doi.org/10.21273/HORTSCI13173-18.

Kusuma, P., P. M. Pattison, and B. Bugbee. 2020. From physics to fixtures to food: Current and potential LED efficacy. *Horticulture Research* 7: 56. https://doi.org/10.1038/s41438-020-0283-7.

Llewellyn, D., S. Golem, E. Foley, S. Dinka, A. M. P. Jones, and Y. Zheng. 2021a. Cannabis yield increased proportionally with light intensity, but additional ultraviolet radiation did not affect yield or cannabinoid content. *Preprints*. https://doi.org/10.20944/preprints202103.0327.v1.

Llewellyn, D., T. S. Shelford, Y. Zheng, and A. J. Both. 2021b. Measuring and reporting light characteristics important for controlled environment plant production. *Acta Horticulturae*. In press.

Lydon, J., A. H. Teramura, and C. B. Coffman. 1987. UV-B radiation effects on photosynthesis, growth and cannabinoid production of two Cannabis sativa chemotypes. *Photochemistry and Photobiology* 46: 201–206. https://doi.org/10.1111/j.1751-1097.1987.tb04757.x.

Magagnini, G., G. Grassi, and S. Kotiranta. 2018. The effect of light spectrum on the morphology and cannabinoid content of *Cannabis sativa* L. *Medical Cannabis and Cannabinoids* 1: 19–27. https://doi.org/10.1159/000489030.

Moher, M., M. Jones, and Y. Zheng. 2020. Photoperiodic response of in vitro *Cannabis sativa* plants. *HortScience* 56: 108–113. https://doi.org/10.21273/HORTSCI15452-20.

Moher, M., D. Llewellyn, M. Jones, and Y. Zheng. 2021. High light intensities can be used to grow healthy and robust cannabis plants during the vegetative stage of indoor production. *Preprints*. https://doi.org/10.20944/preprints202104.0417.v1.

Nelson, J. A., and B. Bugbee. 2014. Economic analysis of greenhouse lighting: Light emitting diodes vs. high intensity discharge fixtures. *PLoS One* 9: e99010. https://doi.org/10.1371/journal.pone.0099010.

Niu, Y., C. Jin, G. Jin, Q. Zhou, X. Lin, C. Tang, and Y. Zhang. 2011. Auxin modulates the enhanced development of root hairs in *Arabidopsis thaliana* (L.) Heynh. under elevated CO$_2$. *Plant, Cell & Environment* 34: 1304–1317. https://doi.org/10.1111/j.1365-3040.2011.02330.x.

Park, Y. G., and B. R. Jeong. 2020. How supplementary or night-interrupting low-intensity blue light affects the flower induction in chrysanthemum, a qualitative short-day plant. *Plants* 9: 1694. https://doi.org/10.3390/plants9121694.

Poorter, H., O. Knopf, I. J. Wright, A. Temme, S. W. Hogewoning, A. Graf, L. A. Cernusak, and T. L. Pons. 2021. A meta-analysis of responses of C3 plants to atmospheric CO$_2$: Dose-response curves for 85 traits ranging from the molecular to the whole plant level. *New Phytologist*. https://doi.org/10.1111/nph.17802.

Radetsky, L. C. 2018. *LED and HID horticultural luminaire testing report prepared for lighting energy alliance members and natural resources Canada*. Troy, NY: Rensselaer Polytechnic Institute.

Rodriguez-Morrison, V., D. Llewellyn, and Y. Zheng. 2021a. Cannabis yield, potency, and leaf photosynthesis respond differently to increasing light levels in an indoor environment. *Frontiers in Plant Science*. https://doi.org/10.3389/fpls.2021.646020.

Rodriguez-Morrison, V., D. Llewellyn, and Y. Zheng. 2021b. Cannabis inflorescence yield and cannabinoid concentration are not increased with exposure to short-wavelength ultraviolet-B radiation. *Frontiers in Plant Science*. https://doi.org/10.3389/fpls.2021.725078.

Shelford, T. J., and A. J. Both. 2021. On the technical performance characteristics of horticultural lamps. *AgriEngineering* 3: 716–727. https://doi.org/10.3390/agriengineering3040046.

Shelford, T. J., C. Wallace, and A. J. Both. 2020. Calculating and reporting key light ratios for plant research. *Acta Horticulturae* 1296: 559–566. https://doi.org/10.17660/ActaHortic.2020.1296.72.

Wallace, C., and A. J. Both. 2016. Evaluating operating characteristics of light sources for horticultural applications. *Acta Horticulturae* 1134: 435–444. https://doi.org/10.17660/ActaHortic.2016.1134.55.

Whitman, C. M., R. D. Heins, A. C. Cameron, and W. H. Carlson. 1998. Lamp type and irradiance level for daylength extensions influence flowering of *Campanula carpatica* 'blue clips', *Coreopsis grandiflora* 'early sunrise', and *Coreopsis verticillate* 'moonbeam'. *The Journal of the American Society for Horticultural Science* 123: 802–807.

Ying, Q., Y. Kong, C. Jones-Baumgardt, and Y. Zheng. 2020. Responses of yield and appearance quality of four Brassicaceae microgreens to varied blue light proportion in red and blue light-emitting diodes lighting. *Scientia Horticulturae* 259: 108857. https://doi.org/10.1016/j.scienta.2019.108857.

Zheng, Y., T. Blom, and M. Dixon. 2005. Moving lamps increase leaf photosynthetic capacity but not the growth of potted gerbera. *Scientia Horticulturae* 107: 380–385.

7 Canopy Management

Philipp Matzneller, Juan David Gutierrez and Deron Caplan

CONTENTS

7.1 INTRODUCTION

There are few topics in cannabis cultivation as controversial as training, pruning and de-leafing. Together termed canopy management, these interrelated practices are considered essential by most cannabis growers; however, the techniques themselves and the rationale for performing them vary drastically from one grower to the next. Managing the canopy is both an art and science, with many

DOI: 10.1201/9781003150442-7

of the current practices developed through trial and error by experienced growers. For cannabis, the scientific evidence is just beginning to emerge on canopy management and how it affects key outcomes like growth response, transpiration, yield and product quality.

This chapter's first section discusses the fundamental parameters to consider when making canopy management decisions, drawing comparisons to successfully applied strategies in well-studied crops. In the second and third sections, we describe practical applications for training, pruning and de-leafing mother (stock), vegetative and flowering plants based on our hands-on experience, consultation with experienced cannabis cultivators and recent scientific studies.

7.2 PHYSIOLOGICAL FACTORS

7.2.1 Light Interception

Light is essential for plant growth. For many fast-growing annuals, light intensity is linearly proportional to yield, with a 1% increase of light resulting in 0.5–1% higher yield (Marcelis et al. 2006). When crops are healthy and factors like nutrients and water are not limiting, dry matter accumulation is proportional to the canopy's light interception (Monteith 1977). As discussed in Chapter 6, cannabis yield can increase linearly with increasing light intensity up to 1800 $\mu mol \cdot m^{-2} \cdot s^{-1}$ or more. Maximizing light interception is an important consideration to maximize yield and quality. In modern greenhouse cultivation systems, the crop canopy captures between 75% and 95% of the light (Hand et al. 1993; Wilson et al. 1992). Without changing the light source, growers can optimize light interception through cultivar selection, planting density and canopy management. This section discusses the role of light interception in controlled environment cannabis production and how to quantify light and interpret the measurements.

7.2.2 Leaf Area Index

Plant architecture and canopy structure affect light interception and thus inflorescence yield and quality. Unfortunately, quantifying a canopy's morphological features is challenging, as they are in constant flux over the crop's lifecycle. The leaf area index (LAI) is a widely used parameter to measure canopy structure and development. It is defined as the one-sided leaf area per unit ground surface area (Figure 7.1), and can be determined using direct or indirect techniques.

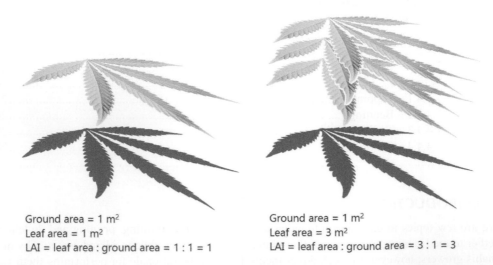

Ground area = 1 m²
Leaf area = 1 m²
LAI = leaf area : ground area = 1 : 1 = 1

Ground area = 1 m²
Leaf area = 3 m²
LAI = leaf area : ground area = 3 : 1 = 3

FIGURE 7.1 Conceptual diagram of a plant canopy with a leaf area index (LAI) of 1 (left) and 3 (right).

Direct LAI measurements involve destructively measuring plants. Each leaf in a specified area of the canopy is removed, and the leaf area is measured either manually or with a leaf area meter. High-quality leaf area meters are available from LI-COR Biosciences, CID Bio-Science and several other manufacturers. Alternatively, image analysis software (e.g., ImageJ) can be used to determine leaf area using photographs of the leaves. Although direct methods are quite accurate, they are typically labor-intensive compared to the indirect methods described in what follows.

Several commercial instruments are available for indirectly measuring LAI, such as the LAI-2200C from LI-COR Biosciences and the ACCUPAR LP-80 from METER Group. Further, a single-point or an averaging line-type photosynthetically active radiation (PAR) meter can provide the data required to estimate LAI. The principle is fairly simple: a dense canopy absorbs more light than a sparse canopy; therefore, the leaf area and the difference in PAR values above and below the canopy are related. This relationship is expressed using an inversion of the Beer-Lambert law, and methods relying on this principle are called the PAR inversion technique.

Equation 7.1 Calculating Leaf Area Index (LAI)

$$LAI = \frac{-ln\left(\dfrac{PARt}{PARi}\right)}{k} \tag{7.1}$$

PARt is the transmitted photosynthetically active radiation (PAR) measured near the growing substrate surface, *PARi* is the incident PAR above the canopy, and *k* is the canopy extinction coefficient. This parameter indicates how much radiation is absorbed by the canopy, considering solar angle and leaf angle distribution. To our knowledge, there have been no studies determining the extinction coefficient or leaf angle distribution for cannabis, but typical values for crops range from 0.7–2.5, with 1.0 being a spherical canopy angle distribution.

To measure LAI using the PAR inversion technique in controlled environments using sole-source artificial lighting (no solar radiation), follow the following instructions. Assuming an irradiance angle of 90°, a leaf angle of 1.0 and an extinction coefficient of 0.8 (until cannabis-specific research is available), we can estimate the LAI using Equation 7.1 and easy-to-collect PAR measurements.

1. Measure PARi 5–10 cm above the canopy. Take 5–10 readings per square meter.
2. Measure PARt at the upper surface of the growing media. The LAI is variable across plant canopies, so for a representative sample, use a line-type PAR meter or take 5–10 under-canopy measurements per square meter with a single-point PAR meter. Avoid taking measurements at the edge of the canopy to control for border effects.
3. Calculate the average of 5–10 readings for both PARi and PARt measurements.
4. Use the average values to solve Equation 7.1.

When the crop is exposed to solar radiation, the calculations become more complicated, as the solar angle and beam fraction must be considered. In this case, it is often best to use a commercial LAI meter, which accounts for those two parameters.

The PAR inversion technique allows for extensive and frequent sampling of the LAI, which is important for a fast-growing crop like cannabis. The frequency and intensity of pruning or de-leafing events can be informed based on the LAI. Keep in mind that only radiation that is intercepted by the crop contributes to photosynthesis. A well-maintained canopy should have dense enough foliage to prevent light from reaching the ground, but not so dense that the lowest layer of leaves is completely shaded. If other factors such as atmospheric CO_2 and temperature are not limiting, an optimal LAI will maximize net photosynthesis and increase productivity and inflorescence quality. We recommend targeting an LAI of 1.5–2.0 in as short a period as possible during the

TABLE 7.1
Optimal Leaf Area Index for Fruit, Grain, Greenhouse Flowers and Vegetable Crops

Crop	Leaf Area Index
Greenhouse tomato (Heuvelink et al. 2005)	3–4
Greenhouse cucumber (Shaikh Abdullah Al Mamun et al. 2017)	2–3
Greenhouse pepper (Cruz-huerta and De Postgraduados 2005; Peña B. and Zenner de Polanía 2015)	3–4
Peach (Faust 2000)	7–10
Apple (Patil et al. 2018)	1.5–5
Maize (Sun et al. 2019)	5–6
Soybean (Tagliapietra et al. 2018)	3.5–4
Rice (Hu et al. 2009)	7–10
Rose (Shimomura et al. 2003)	3–4

flowering stage, essentially filling the canopy quickly to maximize light interception. In our experience, an optimum LAI for cannabis during early and mid flowering is 2.5–4. More work is needed to identify specific LAI targets for different cannabis cultivars and cultivation systems.

The relationship between LAI and yield varies by crop type. For example, in leafy greens, the leaves are the harvested product, so the LAI directly affects yield; however, in fruit, vegetable and cereal crops, the plant's LAI influences yield and quality indirectly. In cannabis, both cases hold true: most of the leaves are discarded upon harvest, but the inflorescence leaves are often included in the final product. In well-studied non-cannabis crops, LAI targets vary between 1.5 and 10 (or higher), but these values depend on species, cultivar and cultivation system. For reference, Table 7.1 outlines optimum LAI values for some economically important crops.

7.2.2.1 Shade Avoidance

Plants take cues from environmental stimuli to initiate internal response mechanisms. One such mechanism involves the ability of plants to sense and react to neighboring plants. Whenever plants grow close to each other, there is mutual shading and potential competition for light. Especially when planting density is high, there are often differences in both light intensity and spectrum between the upper and lower canopy. Leaves are efficient in their use of PAR (400–700 nm) and some ultraviolet radiation (UV, 280–400 nm), but they usually reflect or transmit far-red wavelengths (700–800 nm). Plants perceive shade as a combination of a lower red–to–far-red ratio (R:FR). The plant proteins phytochrome and cryptochrome react to decreases in the R:FR ratio and blue light, and initiate the shade-avoidance response. In cannabis, a common shade-avoidance response is stem elongation (stretching), the extension of petioles, reduced branching, changes in leaf shape and increased susceptibility to pests and diseases.

Building and operating controlled environment production facilities is expensive and requires efficient use of the growing area. Selecting an optimum planting density to maximize light interception is, therefore, critical. The canopy needs to be managed from propagation through to harvest to maintain the desired plant architecture. Even during the rooting stage of cuttings or seedlings, competition in propagation environment can cause stretching (Figure 7.2). In our experience, when leaves from cuttings overlap more than 30% in the propagation environment, some cuttings outperform others, reducing the uniformity of plant height. Depending on the planting density of the cuttings, rooting time and cultivar, cuttings with 2–3 leaves with consistent LAIs result in the highest uniformity (Caplan et al. 2018). To avoid competition for light between rooted cuttings or seedlings, they should be transplanted as soon as they are established.

During the vegetative stage, canopy management is used to create a plant structure that will facilitate success in the flowering stage. Depending on the planting density, the crop must reach a

FIGURE 7.2 The shade-avoidance response is causing some rooted cuttings to outperform others. The result is a decrease in height uniformity among cuttings.

certain height, number of nodes and adequate numbers and length of branches. The plant hormone auxin plays a critical role in the regulation of these aspects of development. Auxin is synthesized in young leaves and is transported downward to the root tip through the vascular system. The shading effect causes a gradient in auxin transport, favoring elongated growth and reduced branching.

The vegetative stage ends when the photoperiod changes from long to short days. In the first 3–5 weeks of the flowering stage, plants produce new stems and leaves at a relatively high rate. Plant height can increase by 100% to more than 600% after photoperiod change; this is highly cultivar dependent, but canopy management can also influence the final size of the crop. Pruning and de-leafing at this stage influence light penetration to the lower canopy. In shaded leaves, where the transpiration rate is reduced, low delivery of the plant hormone cytokinin could inhibit leaf growth and cause stem elongation (Yang and Li 2017). Levels of abscisic acid, commonly referred to as the stress hormone, increase in shaded parts of the canopy, which is linked to a repression of branching (Reddy et al. 2013).

Plants often have a lower resistance to microbial pathogens and insects in scenarios of high competition for resources. Two important compounds regulating biotic and abiotic stress responses in plants are jasmonic acid and salicylic acid. Under shaded conditions, the synthesis and accumulation of jasmonic and salicylic acids are reduced, making plants more susceptible to diseases (Ballaré and Pierik 2017). Also, a dense lower canopy can restrict air movement, increase air humidity and cause leaf wetness. Consequently, the risk of disease is higher in dense canopies. The correct timing and intensity of pruning and de-leafing can reduce shade conditions and strengthen the resistance to diseases.

In recent years, cannabis growers have started using light spectra to manipulate plant architecture. Adding far-red into the horticultural lighting spectrum can simulate the shade response and causes stem elongation in cannabis (Campbell et al. 2019). As discussed in Chapter 6, light intensity can also be used to manipulate plant architecture. In cannabis stock plant maintenance, stretching could be a desirable outcome since cultivars with short internodes usually require above-average regrowth periods between cutting events. The addition of far-red to the light can elongate internodal length and increase cutting production.

7.2.2.2 Yield, Potency and Uniformity of Secondary Metabolites

A common objective of cannabis cultivation is to produce inflorescences of similar size and cannabinoid concentration. This can be aided by selecting a cultivation system with a short canopy, combined with proper canopy management.

Thoughtful canopy management requires a basic understanding of cannabis morphology. Cannabis produces two types of leaves; those situated within inflorescences (i.e., inflorescence or sugar leaves) and shade (or fan) leaves. The shade leaves are the first to grow during the vegetative and early flowering stage, and are typically larger than inflorescence leaves which form along with the inflorescences, acting as bracts. Four to five weeks into the flowering stage, vegetative growth slows and the crop shifts toward inflorescence development, including the multiplication and swelling of calyxes. The development of cannabis inflorescences depends on direct exposure to light. Both light intensity and spectrum play a vital role in cannabinoid synthesis and yield (Magagnini et al. 2018; Mahlberg and Hemphill 1983; Rodriguez-Morrison et al. 2021).

In some fruiting crops, de-leafing toward the end of the production cycle enhances desirable characteristics. In apples and grapevines, for example, color changes can be managed by increasing direct light exposure to the fruit (Verdenal et al. 2019). It has been demonstrated that the strategic removal of shade leaves in flowering cannabis plants can increase light penetration to the lower canopy and increase the uniformity between the top and bottom inflorescences in terms of both size and cannabinoid concentration (Danziger and Bernstein 2021). Commonly, shade leaves are removed 2–3 weeks before harvest to achieve this effect.

De-leafing during late flowering may increase photosynthetic rate in the lower canopy, which stimulates inflorescence development in the previously shaded area; however, more research is necessary to appreciate the influence of canopy management on inflorescence development and secondary metabolite production.

7.2.3   Source–Sink Dynamics

Through photosynthesis, plants use light to convert water and carbon dioxide into photosynthetic assimilates (sugars), which are shared among plant organs. During most of their lifecycle, leaves are considered source organs, as they produce more assimilates than they can consume. Most assimilates are partitioned to fruits and flowers, which are referred to as sinks. Leaves become sinks when their assimilate consumption exceeds production, often because of age or shading. Young leaves that use more energy than they produce are also be categorized as sinks. Roots are typically weak sinks because they are the first to senesce during a shortage of assimilates and grow rapidly when a surplus is available. Understanding source–sink dynamics and how to manipulate them is important to maximize the yield and quality of any crop.

7.2.3.1   Source–Sink Dynamics during Propagation

When roots develop from vegetative cuttings, it is called adventitious rooting. The process depends on complex physiological changes involving several plant hormones and carbohydrates. To induce new root growth, the base of the cutting requires elevated concentrations of hormones called auxins. Auxins are supplied naturally by the shoot apex of the cutting, but can be artificially increased by dipping the cutting in a synthetic rooting hormone (e.g., Indole-3-butyric acid). Successful rooting is achieved when leaves (the source) translocate assimilates to the base of the cutting to allow root growth (the sink).

7.2.3.2   Source–Sink Dynamics during Vegetative Growth

In the early vegetative stage, cannabis growth rates are generally relatively slow, which can likely be attributed to the source–sink dynamics of young leaves and roots. The transition from sink to source tissue is gradual, with the leaf's tip creating assimilates before its base. The gradual change allows for a balanced growth rate between roots and shoots. This transition has been studied in detail for cucumber, and can be visualized using a phloem tracer (Figure 7.3; Savage et al. 2013). Gray areas show leaf parts that have reduced dependence on external carbon (sources).

In other crops, strategies have been developed to manipulate the source–sink relationship. These include pruning, topping, de-leafing, flower and fruitlet thinning, changes to the environment and the use of hormones or nutrients. In cannabis cultivation, it is common to remove branches or

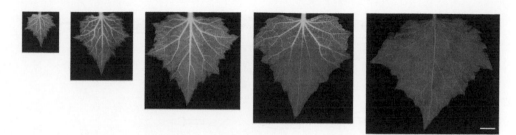

FIGURE 7.3 The transition of immature cucumber leaves from sink to source tissue. The perfusion of the phloem-mobile tracer (green) declines with transport into the leaf, indicating reduced dependence on external carbon. Cucumber leaves were imaged with a fluorescent scope during a 5-day period to visualize their source–sink transition. Scale bar = 5 mm.

Source: Savage et al. (2013)

mature leaves in the lower part of the canopy. In most cases, these leaves are shaded and have a low photosynthetic rate, having likely transitioned from sources to sinks. Correct timing and intensity of the pruning and de-leafing could increase the photosynthetic rate of the remaining canopy and increase partitioning to the inflorescences. This has been confirmed in greenhouse tomato production; however, a yield increase could only be achieved when the LAI was maintained sufficiently high (Heuvelink et al. 2005).

Thinning flowers, fruitlets and fruits are common practices to influence source–sink dynamics. In high-density orchards, small fruits are removed in the early stages of development to increase the remaining fruits' sink strength. This is a widespread practice in the cultivation of apples, pears, apricots, plums, peaches and kiwis, usually with the intent of increasing fruit size (Pawar and Rana 2019). In greenhouse tomato production, truss pruning is used to regulate fruit size and number in late summer and autumn. Light levels are high in late summer as the plant is developing a high fruit load, but when those tomatoes undergo maturation in autumn, light levels are lower. By removing some fruits, the overall yield decreases, but the individual fruit size increases. Similarly, some cannabis growers completely remove all lower inflorescences to increase the top inflorescences' size, a practice that requires more research to confirm its efficacy.

7.2.4 Air Movement

Photosynthesis requires the exchange of carbon dioxide and oxygen between the atmosphere and plants through small pores in leaves called stomata. Gas exchange occurs in a thin layer of air surrounding each leaf known as the boundary layer, and its thickness influences the rate of the exchange. A moderately thin layer encourages gas exchange and is preferable for vegetative and flowering growth, while a thicker layer is preferred for rooting cuttings to avoid moisture loss. Several factors influence boundary layer thickness, including the physical characteristics of the leaves themselves. For example, larger leaves and those with leaf hairs typically have a thicker boundary layer than small hairless leaves. Temperature, humidity and air movement also influence boundary layer thickness.

For several crops, a target air movement of 0.25–0.5 m/s is recommended to reduce the boundary layer thickness and encourage gas exchange (Nelson 2011). Lower air movement may result in high canopy humidity, reduced transpiration rate and reduced gas exchange, as demonstrated in Figure 7.4.

Air movement varies depending on where it is measured in a crop. Generally, air resistance is greater within the canopy, but that depends on plant spacing, morphology and canopy density. Research on field crops shows that when air movement is 2.5 m/s, measured at two meters above a bean crop, it is reduced to 0.9 m/s at the top of the canopy and 0.25 m/s around the middle of

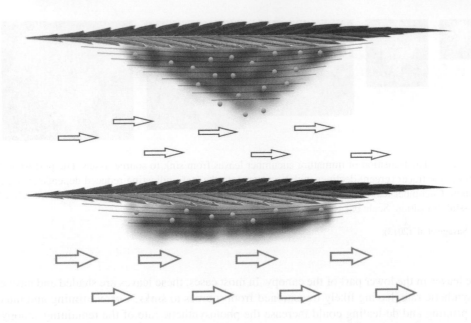

FIGURE 7.4 Conceptual diagram of cannabis leaflets under two air movement conditions. The diagram on the top has low air movement and a thick boundary layer, while the diagram on the bottom has high air movement and a thin boundary layer. A thinner boundary layer typically means higher gas exchange and transpiration.

the canopy (Thom 1971). The terms microclimate and macroclimate are often used to identify the effects of variable air movement. The macroclimate is the environmental condition in the entire greenhouse or grow room, whereas the microclimate is the condition in proximity to the plants. Regulation of the macroclimate does not guarantee uniformity of the microclimate, but in combination with cultural practices such as pruning and de-leafing, an altered macroclimate can be used as a tool to regulate air movement within the canopy.

The microclimate can have a significant effect on plant health. In a dense canopy with minimal air movement, humidity tends to rise, encouraging condensation on plant surfaces. Fungal and bacterial pathogens need water to germinate on the surface of leaves and penetrate plant tissue.

While yet unstudied in cannabis, the effect of air movement on the wetting of tomato leaves has been investigated in growth chambers (Kuroyanagi et al. 2013). Leaf wetness was completely suppressed with air movement above 0.3 m/s in the canopy. Another experiment on tomatoes investigated the effect of air circulation on the incidence of *Botrytis cinerea* (gray mold). One greenhouse had a target air movement of 0.5–1.3 m/s above the canopy and 0.1–0.2 m/s within the canopy, while a second greenhouse had no fans (control). Plants were infected with *Botrytis cinerea* and grown for one month, and the temperature and relative humidity above the canopy were similar in both greenhouses. The result was a leaf infection rate of 11.3% in the control and 0.6% with higher air movement (Sekine et al. 2007).

In our experience, an air movement of 0.5–1.0 m/s helps suppress the development of fungal and bacterial pathogens in most cannabis cultivation systems and cultivars, especially when combined with the removal of chlorotic and necrotic plant tissue and de-leafing if the canopy becomes excessively dense.

7.3 STOCK PLANTS

At present, vegetative propagation through stem cuttings is widely regarded as the easiest and most reliable method for creating new cannabis plants. Many other commercially cultivated species such as chrysanthemums, roses, poinsettias, and geraniums are propagated this way. The main advantage of propagating with cuttings is that the new plant is genetically identical to its parent or stock plant, and characteristics such as secondary metabolite production and yield remain true to type. To realize these benefits and produce uniform crops, it is critical to maintain healthy stock or "mother" plants from which cuttings are taken.

A stock plant begins life as a rooted cutting or seedling, which is then transplanted to a container or growing bed. The plant is frequently pruned during its initial stages of growth to promote lateral branching; In commercial stock plant production, this is called the scaffold development phase. When developing the scaffold, the goal is to create a structure that maximizes the number of nodes from which uniform shoots can develop (Figure 7.5). This type of canopy management maximizes the number of shoots from which cuttings can be harvested and their rooting rates. A bushy and short stock plant structure is typically favored over a tall and lean structure.

It is widely accepted that healthy and homogeneous cuttings produce uniform and vigorous flowering plants. Quality is defined by the overall size of the cutting, the number of leaves present and the thickness of the stem.

Properly timed and executed canopy management can influence cutting yield, quality and rooting success. When establishing a stock plant, topping, pruning and de-leafing can be used to shape the plants. Topping is the technique in which the active growing point of a branch is removed to disrupt apical dominance and stimulate the growth of axillary buds as lateral shoots (Figure 7.6).

FIGURE 7.5 Left: Stock plant shape and size after scaffold development. Right: Top-down view of the plant canopy.

FIGURE 7.6 Topping a young cannabis plant.

On established stock plants, topping, de-leafing and pruning may help maintain juvenility, clean overgrown stock plants or ensure new shoots consistently produce viable cuttings on a regular basis.

7.3.1 Crop Work for Stock Plant Establishment

Frequent and consistent canopy management is required to establish and manage stock plants. The activities required include topping, pruning and de-leafing. A stock plant should be encouraged to grow as a bush so that each potential cutting receives similar exposure to direct light. To achieve this structure, the plants are topped frequently, starting early in their development. It is critical to begin topping as early as possible to keep the plant compact and prevent it from allocating resources to growing shoots that will eventually be removed.

Topping can be performed using scissors, a blade, or by pinching with your fingers. Tools that can be easily sanitized between plants are preferable to prevent the spread of fungal, bacterial or viral pathogens. Pathogen outbreaks can lead to substantial cutting or stock plant losses, which can be devastating to an operation. Several sanitizers or disinfectants can be used to clean utensils; however, proven options are Virkon S and 10% Clorox bleach solutions (5.25% sodium hypochlorite). The utensils can be sanitized periodically by submerging them in a cleaning solution, following the manufacturer's use instructions. In order to withstand regular cleaning, scissors and blades with a titanium coating are typically preferred. Titanium-coated blades also stay sharp longer, resist corrosion and are more durable than uncoated stainless steel and carbon steel blades. It is important to keep all cutting tools sharp, as clean cutting ensures faster wound healing and reduces the risk of mother plant infection.

Typically, a cannabis plant has an opposite bud arrangement when grown from seed. In some instances, plants grown from cuttings have also show opposite bud arrangement. As the plant grows, it usually changes to an alternate bud arrangement; this is called a change of phyllotaxy, and is thought to occur when a plant reaches sexual maturity (Schaffner 1926). An alternate bud arrangement poses some challenges when topping stock plants to form their structure. Shoots that originate from different locations may have non-uniform light exposure, resulting in unwanted stretching and possibly an asymmetrical structure (Figure 7.7). It is therefore important to top as soon as the buds are visible and while the buds are as close to each other as possible.

FIGURE 7.7 The effect of topping on cannabis plants with opposite (top) and alternate (bottom) bud arrangement.

7.3.2 CROP WORK FOR STOCK PLANT MAINTENANCE

Once the stock plants are established and regularly producing viable cuttings, the plant canopy needs to be managed to prevent excessive undergrowth, maximize cutting quality and maintain adequate air movement. To ensure light penetration to the lower canopy, large leaves can be removed from the top of the plant. This promotes a similar lighting condition among cuttings from various locations on the plant, and thus, improved cutting uniformity.

To maintain adequate air movement within the canopy during stock plant maintenance (> 0.5 m/s), leaves and branches that are causing congestion within the canopy should be removed every 1–2 months. Further, leaves in the lower part of the canopy that are mostly shaded or show signs of necrosis and senescence should be removed regularly. During periods of low cutting demand, a heavy pruning event on the stock plant facilitates the production of new growth and restores juvenility. When such pruning events are possible, it is also a good opportunity to re-shape the plant if it has become overgrown.

Typically, juvenile cuttings root faster, possibly because they have fewer lignified cells and higher carbohydrate reserves compared to mature, woody branches. An intensive prune of a stock plant can initiate a "flush" of uniformly sized juvenile cuttings; this is a useful practice when cutting demand is high. Heavy pruning temporarily reduces cutting availability after the event, but may increase the quality and uniformity of later cuttings and can re-establish the scaffold of the plant.

7.3.3 PLANT SIZE AND SHAPE

Stock plant size affects the number of harvestable cuttings per plant, but also the way in which the canopy is managed. A small stock plant is relatively easy to maintain and produces fewer cuttings, whereas a large plant requires more crop work, water and fertilizer but may produce more cuttings. An understanding of the demand for cuttings on a weekly basis or per cutting event dictates the required number of plants.

From our experience, medium-sized stock plants (45–75 cm tall and 30–60 cm in diameter) are ideal in a commercial setting with high cutting demand. Maintaining a uniform canopy usually requires less crop work with smaller plants. With larger plants, a higher proportion of branches and leaves must be removed within the canopy to maintain light penetration and air movement compared to shorter plants; therefore, short stock plants typically require less labor to produce uniform cuttings. Shorter plants also make it possible to utilize multi-level vertical growing systems. Finally, small stock plants can be replaced more frequently as each one contributes less to the overall cutting availability and takes less time to establish than larger stock plants. If properly maintained, medium-sized stock plants can produce 20–30 quality cuttings every 14–21 days.

7.3.4 CONTAINER SIZE

A stock plant is created by transplanting a rooted cutting, explant (originating from tissue culture) or seedling into a small container; a pot 7–10 cm (3–4") in diameter and 300–500 mL in volume is typically adequate to support up to a 4-week-old, ~30 cm tall plant. Once well-rooted in the first container, the plant can be transplanted to a larger container. The size of the final container depends on the desired final size of the plant. From our experience, a container of 15–25 cm (6–10") in diameter, 1.5–6.5 L in volume, is typically sufficient for a compact plant; however, larger stock plants can be grown in a container of 19 L (5 gal) or more. The container (and therefore root mass) must be commensurate with the size of the aboveground parts. If the container is too small, it will require frequent irrigation, and drought stress can occur quickly if an irrigation event is missed. On the other hand, if the container is too large for the plant, the substrate may stay moist for extended periods and cause waterlogging, low rootzone oxygen availability and an increased risk of root disease (Zheng et al. 2007).

7.3.5 ESTABLISHING A STOCK PLANT (GENERAL METHOD)

1. Transplant a rooted cutting, explant or seedling into a 300–500 mL pot with growing media.
2. Top the plant when the main stem has reached 10–15 cm and has at least two pairs of fully expanded leaves by removing the apical meristem. This results in two distinct apical nodes.
3. Remove any shoots below the cut (Figure 7.8).
4. When the apical nodes have grown slightly taller than the two adjacent distal nodes, top the shoots that develop to create four growing points.
5. Repeat the previous step, so there are a total of eight nodes (Figure 7.9).
6. When the plant is well rooted, it can be transplanted to a larger container (e.g., 1.5–6.5 L).
7. Allow the shoots to grow until cuttings can be harvested.
8. Depending on growth rates and desired plant structure, establishing a stock plant to produce cuttings regularly takes approximately 6–8 weeks.

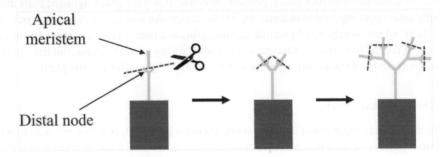

FIGURE 7.8 Topping a young cannabis stock plant.

FIGURE 7.9 Scaffold development of a three-month-old cannabis stock plant. The locations where the topping has occurred are indicated by red arrows.

7.3.6 Harvesting Cuttings

There are two common methods to harvest cuttings from a stock plant: selective harvesting and hedging (or grazing). In selective harvesting, only those cuttings that meet the specified quality requirements are harvested. In contrast, hedging involves the harvest of all viable cuttings at specified intervals. Hedging involves large reductions of biomass in a short period, which can cause stress to the stock plant and reduced vitality over time; therefore, the practice should be avoided unless cuttings demand is especially high.

When hedging, cuttings are taken every 2–3 weeks. Each cutting event involves a temporary reduction in shoot biomass, which can result in root senescence and general plant stress. After a cutting event, evapotranspiration rates decrease, along with water and nutrient and water uptake. To avoid overwatering, irrigation frequency should be reduced based on plant update. If feasible, selective harvesting is preferable, as biomass loss and its associated effects are less severe.

7.3.7 Cutting Quality

High-quality and uniform cuttings are key to the success of a flowering crop. Cuttings with a small stem diameter (< 3 mm) typically take longer to root and can be more susceptible to stem damage when placed in the rooting media. Conversely, a cutting with an overly thick stem (> 5 mm) often takes longer to root. Leaf number on a cutting also plays a role in rooting success. A cutting with too few or damaged leaves may not have enough stored assimilates and endogenous auxins for successful rooting. Cuttings with too many leaves, on the other hand, may cause overcrowding in the propagation environment and increase the potential for disease (Caplan et al. 2018).

In our experience, a quality cutting should be 10–15 cm in length, have an actively growing meristem, 2–4 fully developed and healthy leaves, and a stem that is approximately 3–4 mm in diameter.

Cutting quality may also depend on the part of the plant from which it is taken. The most vigorous growth is usually at the top of the plant, where light interception is highest. To increase cutting quality and quantity, growers can remove shade leaves at the top of the canopy to increase light penetration to lower shoots. This practice is typically carried out 2–7 days before a cutting harvest to give the previously shaded shoots a better chance to develop into quality cuttings.

Both shoot tips and axillary shoots can be used to produce cuttings; however, growth rates generally differ between the two types, which can lead to non-uniform cuttings if they are propagated together.

7.3.8 Stock Plant Turnover

A stock plant can be kept in a vegetative state indefinitely; however, there are compelling reasons to replace stock plants at regular intervals. Plant vigor will decrease as plants are constantly pruned, topped or de-leafed. It is generally accepted that plant vigor will decrease over time, leading to a reduction of cutting quality and consistency. The rate at which plant vigor decreases is not well understood for cannabis; however, accumulated stress may increase susceptibility to fungi, bacteria and viruses which can decrease plant vigor. For this reason, stock plant populations should be replaced with disease-free stock every 6–12 months.

7.3.9 Stock Plant Canopy Density

Effective use of growing space is essential while establishing and maintaining stock plants. The plants should be close enough to each other to maximize the use of the growing area, but not so close as to hinder growth or increase the risk of disease. When deciding the planting density, it is important to consider space, light, air circulation and labor availability. Table 7.2 compares some of the advantages and disadvantages for low and high planting densities.

TABLE 7.2

Comparison between High– and Low–Planting Density Stock Plant Systems

	Low Density	High Density
Advantages	• Lower cost of starting materials • Lower labor requirements	• Higher cutting uniformity and quality • Predictable canopy height • More ergonomic for workers
Disadvantages	• Plant loss has a greater impact on cutting production • Large plants are difficult to access for crop work • Lower cutting uniformity due to depth of the canopy	• Higher material cost, including pots and growing media • Greater risk of disease if plant canopy becomes overgrown • More diligent upkeep required

Light intensity must be maintained at a sufficient level to prevent plants from stretching to become tall and spindly. When cannabis stock plants receive less than 150 $\mu mol \cdot m^{-2} \cdot s^{-1}$ PPFD at canopy level, stretching is common. There should also be enough air movement around and within the canopy to prevent water vapor from condensing on the foliage. A dense and overgrown canopy with low air movement (< 0.5 m/s) can cause condensation which can encourage fungal and bacterial pathogen infection and lead to low evapotranspiration rates within the canopy.

Topping, pruning and de-leafing are all labor-intensive activities, and inadequate crop work can lead to overgrown stock plants which produce cuttings of low quality and uniformity. If labor availability is inconsistent, a low-density stock plant system may be preferred to a high-density system, as it requires less labor. The setup of the mother room should also be conducive to crop work; ergonomics and access to the plants are key considerations. A canopy situated at the waist or chest level avoids excessive reaching or bending when working with the plants.

In a commercial setting, where cutting quality and uniformity are of utmost importance and labor is typically available, a high planting density is often preferred. Plants are kept short and compact, and at a density of 6–10 plants/m^2. Each plant should be able to produce 20–30 good quality cuttings every 2–3 weeks when hedging or half as many when harvesting cuttings weekly (see Section 7.3.6 for a more detailed description of harvesting cuttings).

7.3.10 CANOPY MANAGEMENT FOR GERMPLASM PRESERVATION

Given the vast number of cannabis cultivars currently available, growers often preserve valuable genotypes using stock plants. These plants are best maintained as compact as possible to reduce inputs, labor and footprint. Ideally, these plants should be kept in 300–500 mL pots, at a height of 20–30 cm and a width of 10–15 cm; this will produce 1–4 cuttings every 2–3 weeks (Figure 7.10). It is less important to establish a scaffold but rather to prevent stem elongation by pruning and de-leafing. Frequent crop work of these small stock plants stunts their growth, keeping them in a state that is easier to maintain. Growth rates can be further slowed using low ambient temperatures and light intensity, for example, at 18–20°C and 80–120 $\mu mol \cdot m^{-2} \cdot s^{-1}$ at canopy level during the day.

7.4 VEGETATIVE AND FLOWERING PLANTS

7.4.1 THE VEGETATIVE STAGE

The cannabis lifecycle includes two growth stages: vegetative and flowering. A short-day photoperiod (~12 hours) induces flowering, lasting around 7–12 weeks depending on the cultivar and growing conditions. Generally, cannabis is an exceptionally vigorous, fast-growing plant, though the rate of dry weight increase (growth rate) varies by stage of growth, cultivar and environmental conditions. The growth rate is slow in the first few days following transplant, as the plant spreads

FIGURE 7.10   Structure of a stock plant maintained for germplasm.

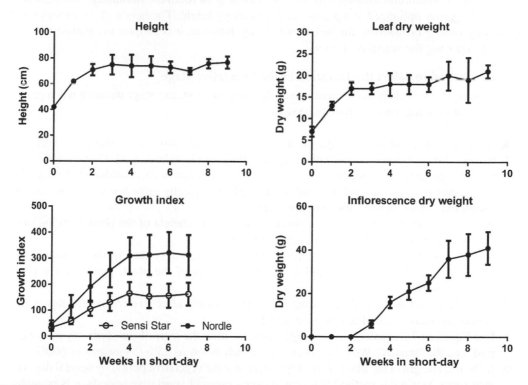

FIGURE 7.11   Mean height, leaf dry weight and inflorescence dry weight (+/− SD, n = 4) for cannabis (cv. 'Sativex') after starting a 12-hour photoperiod.

Sources: Adapted from Potter (2014). Mean Growth Index (+/− SD, n = 20) for cannabis (cv. 'Sensi Star' and 'Nordle') adapted from Yep et al. (2020) and calculated as (height [cm] × width$_1$ [cm] × width$_2$ [cm]) × 300^{-1}

roots into the new medium. Growth rates increase to their maximum during the vegetative stage and early flowering stage. As the plant matures during flowering, growth slows until the end of its lifecycle (Figure 7.11). The duration of the flowering stage is generally considered constant for cuttings from genetically identical stock plants. After initiating a short-day photoperiod, plants will be ready for harvest in 45–85 days. Conversely, the vegetative period is flexible, being easily adjusted by extending long days with artificial lighting.

7.4.1.1 Selecting the Vegetative Stage Duration

In deciding on the vegetative stage's duration, a grower should consider its effects on key outcomes such as yield (on per-crop and annualized bases), plant height and width, and production of biochemical compounds, like cannabinoids. Unfortunately, these relationships are still poorly understood. It is, however, generally accepted that a longer vegetative period will produce larger plants and higher per-plant yield. A decision needs to be made on the desired maximum size of each plant. Key considerations in this decision are planting density, cultivar growth habit and type/frequency of crop work, such as topping and pruning that are possible or desired.

In smaller-scale operations, or for those with a restricted plant quantity, longer vegetative periods may be favorable, as each harvest should be higher-yielding using fewer plants. Conversely, if the goal is to maximize both yield per unit area and consistency of the final product (inflorescence size, maturity and secondary metabolite content), then a cropping system utilizing a higher planting density and smaller (younger) plants combined with adequate crop work is advantageous.

It is important to also consider lighting uniformity when deciding on a vegetative period. Generally, the horticultural lighting fixtures in an indoor grow room are stationary, with light intensity and uniformity optimized at a pre-determined canopy height. The height of the canopy can be adjusted by raising or lowering the bench or lighting; however, it is simpler to regulate the crop's height by adjusting the vegetative period.

7.4.1.2 Method to Select the Duration of the Vegetative Stage

The following steps can help make an informed decision on vegetative stage duration when growing a new cultivar or using a new cultivation system.

1. Select a vegetative period that, on average, suits your cultivation system. Consulting breeder data is helpful. At high planting density (> 15 plants/m^2), a 10–15-day vegetative period is generally acceptable. At a lower density (< 15 plants/m^2), consider 15–30 days. If the vegetative period extends beyond 30 days, plants typically grow too large to maintain a uniform canopy in controlled environments.
2. When the short-day photoperiod is initiated, measure the height of the plant from the surface of the growing media to the uppermost node.
3. Avoid topping unless it is standard practice in the cultivation system, or the planting density is low (< 15 plants/m^2). Topping affects the height of the plant, generally favoring lateral growth.
4. Once the plant has stopped increasing in height, generally after 30 days into the flowering period, measure the final height as in step 2.
5. Use Equation 7.2 to calculate the height increase after the start of the short-day photoperiod, i.e., the "stretch." We have observed a stretch as low as 25% to upwards of 600%.
6. If the final height of the plants is too high, decrease the vegetative period by several days or vice versa. When this method is repeated over a range of vegetative periods, it is possible to estimate the relationship between the vegetative stage duration and final plant height for a specific cultivar and cultivation system.

Equation 7.2 Identifying Percent Height Increase or "Stretch" After Short-Day Photoperiod Induction in Cannabis

$$Stretch\,(\%) = \frac{(final\,height - height\,at\,flowering\,inducation)}{height\,at\,flowering\,induction} \times 100 \qquad (7.2)$$

7.4.2 PLANTING DENSITY

Planting density is one of several cultural practices that has a demonstrable effect on crop yield and quality. Greenhouse tomato growers, for example, understand that an overly dense canopy can reduce fruit size (and percentage of marketable fruit). In contrast, an overly sparse canopy can decrease yield within the growing area. For instance, in greenhouse-grown cherry tomatoes, decreasing plant spacing from 50 cm to 30 cm increased yield per unit area but led to less marketable fruit (Charlo et al. 2007). Similar research on several drug-type cannabis cultivars illustrates that, as with tomatoes, cannabis yield is linked to planting density. Lower density increases per-plant yield, and high density increases per–unit-area yield (Vanhove et al. 2011, 2012).

While the theory and basic interaction between density and yield have been documented for cannabis, there is no well-established method to optimize planting density in a particular cultivation system. Many indoor cannabis growers keep their plants small and grow at a high density (> 15 plants/m^2). Still, it is common to see cultivation systems with fewer large plants, especially when plant count is regulated.

Optimizing planting density requires a thoughtful balance between canopy light interception and the total yield of the crop. Usually, the objective is to maximize the yield in the grow area without reducing crop quality or increasing the risk of disease. The two most important considerations are light penetration and air movement within the canopy. Each inflorescence site should have enough light exposure to ensure uniform maturation among inflorescences throughout the canopy. Figure 7.12 illustrates the visually evident effects of shading on cannabis inflorescences; these include lighter colored leaves and delayed senescence of stigmas. Further, uniform air movement should be maintained within the canopy to decrease the risk of disease from moisture-loving fungal and bacterial pathogens.

Planting density affects the amount and type of crop work required, as well as the yield and uniformity of the crop. The labor associated with increased crop work and the requirement to use more plants makes overly high planting densities more costly. Conversely, a low density can reduce annualized yield and increase the need for pruning or training to maintain inflorescence uniformity within the canopy. Further, when trying to fill the canopy at a lower plant density, larger plants are required. Larger plants generally require bigger pots, more support, more fertilizer and more water

FIGURE 7.12 Effect of shading on cannabis inflorescences. The inflorescences pictured here are from similar locations on the plant. The one on the right was exposed to direct light, while the one on the left was shaded by the upper canopy.

Source: Photo by Sarah Hirschfeld

TABLE 7.3

Key Considerations in Selecting a Planting Density for Controlled Environment Cannabis Production

Factor	Consideration
Plant quantity limitations	Regulations in some areas may dictate the maximum allowable number of plants.
Cost of starting material	When starting material is expensive, a lower planting density can lead to significant cost savings.
Uniform inflorescence maturity	Shaded inflorescences generally mature more slowly than those exposed to direct light. Higher planting densities increase the proportion of the plant that is shaded, and may decrease inflorescence uniformity.
Labor	More manual pruning and de-leafing will be required at higher planting densities to maintain adequate air movement and light penetration through the canopy.
Crop duration	Higher planting density favors a lower crop duration, as a shorter vegetative period is needed to fill the canopy. Theoretically, this will lead to a higher annual yield.
Cultivar growth habits	Cultivars vary in their tendency to grow laterally (i.e., their bushiness). Bushier cultivars either require a lower planting density or more intensive pruning and de-leafing to maintain light penetration and adequate air movement within the canopy.
Facility design	At high planting densities, pockets of stagnant, humid air can develop within the canopy, facilitating fungal and bacterial pathogen infection. Circulation fans and dehumidifiers can help improve air movement and reduce inter-canopy humidity, but require capital investment.

TABLE 7.4

Comparison between Low- and High-Density Planting Systems for Controlled Environment Cannabis Production

High Planting Density	Low Planting Density
• Shorter vegetative period to fill canopy (higher annual yield)	• Lower cost of starting material
• A higher proportion of apical inflorescences	• Less crop work required to maximize light penetration
• Higher risk of immature lower flowers if pruning/de-leafing is inadequate	• Longer cultivation cycles (longer vegetative period to fill the canopy) and lower annual yield per unit area

on a per-plant basis. Table 7.3 and Table 7.4 outline some key factors to consider when deciding on a planting density for a particular operation.

7.4.3 PRUNING

Pruning is the practice of manually removing stems, leaves, buds or fruits so that only a select quantity remain. Pruning alters source–sink dynamics so that nutrients are shared between fewer fruit or flower sinks, which can increase the average fruit or flower size.

In well-studied greenhouse crops such as tomato and cucumber, optimized pruning techniques have been established to maximize parameters such as fruit quality, yield per unit area or to prevent disease. For instance, in greenhouse tomatoes, controlling the number of branches, flowers or fruits through pruning is an effective way to reduce competition between fruits. Typically, more stems indicate a higher yield but smaller, less uniform fruit.

While the effects of pruning on cannabis are largely unexplored in the literature, it is rare to see un-pruned plants in a modern controlled environment cannabis crop. Common practices include stem pruning and de-leafing (leaf pruning). Early research and experience have proven that these techniques—if used effectively—can increase the yield of uniformly sized inflorescences and decrease the variability of secondary metabolites, like cannabinoids, between upper and lower inflorescences (Danziger and Bernstein 2021).

It is important to consider pruning strategy when making irrigation and environmental control decisions. De-leafing and stem pruning both decrease the total leaf area, which is linked to canopy transpiration. More leaf area usually increases transpiration if air circulation is sufficient. The decrease in canopy transpiration after leaf or stem pruning can lower humidity in the growing environment and slow rootzone water uptake.

Pruning may also affect the spread and severity of fungal or bacterial pathogens. A congested canopy can lead to high-humidity microclimates that are prone to infection from moisture-loving pathogens. Pruning allows more air movement within the canopy, reducing the risk of pathogen infection. Canopies become most congested in high-density planting systems; therefore, pruning is essential in these systems.

7.4.3.1 Stem Pruning

Growers often prune several of the lower branches on each flowering cannabis plant. By removing lower branches, growers can increase the proportion of apical (higher) branches that tend to have higher cannabinoid concentrations (Danziger and Bernstein 2021; Namdar et al. 2018). Stem and leaf pruning make higher planting densities more manageable, as both practices increase air movement and light penetration into the canopy. Removing lower branches and leaves also limits the transmission of fungal and bacterial pathogens from the bench or growing media to the shoots.

If the priority is to optimize the quantity of larger apical inflorescences, then lower branches may be removed. Conversely, to optimize total yield per unit area with a lesser focus on uniform flower maturity and chemical composition, less stem pruning may be desirable. The latter may be the case when growing input for extraction, while the former may be better suited when producing dry flowers as a final product.

The timing and frequency of stem pruning are dependent on the intended use of the end-product (i.e., extraction of dry flower), cultivar-specific growth characteristics and the cropping system. At lower planting densities, lower branches typically have greater exposure to direct light and air movement. These branches can still produce fully developed inflorescences, which may not occur in a congested, higher planting density (depending on the extent of leaf and stem pruning). This highlights the importance of stem pruning in cropping systems using higher planting densities.

Similarly, cultivars that tend to grow with apical dominance (high height-to-width ratio) may require less pruning than bushier cultivars. Bushier plants have more congested lower canopies, which again limits light penetration and air movement to the lower branches and inflorescences.

Pruning several times may be preferential to a single intense pruning event. When large portions of the plant are removed all at once, it may cause undue stress, slowing growth rates and reducing yields. It is good practice to prune mostly before flower initiation. The vegetative stage, extending into the first two weeks of flowering, is typically a good window. When pruning later into flowering, the lower branches are typically larger, having required substantial resource allocation by the plant for their development. If they were removed earlier, those resources could be used for the development of apical branches and inflorescences.

7.4.3.2 De-Leafing (Leaf Pruning)

De-leafing (also known as leaf pruning) is the practice of manually removing leaves to increase canopy air movement and light penetration, alter source–sink dynamics, expose the inflorescence for easier harvesting and processing, or to remove drying plant material. In practice, de-leafing accelerates the development of lower inflorescences of most cannabis cultivars.

Larger shade leaves are the target for removal, as they have the greatest shading effect on lower inflorescences. To determine how many leaves to remove, select your target leaf area index (LAI; see Section 7.2.2) and remove leaves to meet this target. Leaves can either be removed using pruners

FIGURE 7.13   The epidermal tissue of this stem was stripped during de-leafing.

or by hand by bending the petiole up and down until it snaps cleanly from the stem. When de-leafing by hand, pay careful attention not to unintentionally remove the epidermal tissue of stems below the petiole (Figure 7.13). The additional wounding site is a potential entryway for pathogens.

As with stem pruning, the timing and frequency of de-leafing are dependent on cultivar-specific growth characteristics and the cropping system. High-density systems or those with fast-growing cultivars will have a higher requirement for de-leafing. Typically, de-leafing is not required until around three weeks into the flowering stage, when inflorescence sites are forming and developing. From this point until harvest, de-leafing should be carried out regularly to maintain a desirable LAI.

An intensive de-leafing event occurring 5–15 days before harvest is widespread practice for many growers. Sometimes called "pre-harvest de-leafing," the process generally takes place when leaves have stopped growing and start to show signs of senescence; therefore, subsequent de-leafing is not required. The resultant canopy has a high inflorescence-to-leaf ratio, favoring the development of lower inflorescences.

7.4.4   TRAINING

Plant training is a canopy management practice used to control the shape, size and direction of plant growth. Common training methods include topping, bending, trellising and staking. These practices can be used to manipulate canopy light penetration and air movement, to prevent plants from damage or alter the size and number of inflorescences.

Modern elite cultivars often have a high inflorescence-to-branch ratio. If left unsupported, stems may break, or plants can fall due to the inflorescence weight during late flowering. Plant supports such as trellising or stakes are common in other high-intensity cropping systems, where plants are unable to support their own weight.

7.4.4.1   Topping

Topping (or pinching) is a form of pruning in which the apical (uppermost) shoot is removed. It can be used to disrupt apical dominance and stimulate the production of lateral shoots. During the vegetative and flowering stage, topping can support the formation of an uninterrupted, uniform canopy, especially in low-density production systems or with cultivars that tend to grow with apical dominance (high height-to-width ratio). Topping is most effective during the vegetative stage or early flowering, before the plant's final form is determined. For a detailed description of topping techniques, see Section 7.3.1.

In other crops, such as tomatoes and bell peppers, topping has been shown to periodically slow growth, and it is generally accepted that the same holds true for cannabis. Therefore, it may be preferable to create a uniform canopy without topping by either adjusting the duration of the vegetative period or the planting density, or using another cultivar. When restricted by plant numbers or working with apical-dominant cultivars, topping may be an effective technique to maximize canopy uniformity.

7.4.4.2   Bending

Bending is the process of physically manipulating and securing branches to alter the structure of a plant or plant canopy. Much like topping, the purpose is to break apical dominance and facilitate better lower inflorescence development. Bending does not involve removing any plant tissue and therefore may cause less stress to the plant; however, bending is often practiced after topping, which periodically slows growth. Conversely, bending is labor-intensive and requires plant ties or supports, which are not required for topping or other types of pruning. Two commonly used methods of bending cannabis plants are the "screen of green" technique and "low-stress training."

"Screen of Green" (SCROG) is a method in which branches are spread out and bent horizontally through trellis netting, typically after topping. The SCROG method is useful in low-density production systems to shape the canopy of larger plants and increase light penetration and air movement to the lower flowering sites. This technique is labor-intensive and is not often seen in commercial production environments, since a higher planting density can achieve a similar effect with reduced labor.

"Low-Stress Training" (LST) is a bending method in which stems are tied or clipped to each other or external supports to shape individual plants. While LST and SCROG methods serve a similar purpose, the primary difference is that LST is practiced on individual plants rather than on a crop canopy.

7.4.4.3   Trellising

Trellising involves the use of one or more layers of horizontal netting to support the plant canopy. The netting keeps plants from falling and branches from snapping under their own weight. The netting is supported using upright posts, usually situated either at the edges of each bench. Typical trellis netting for controlled environment cannabis crops are square-lattices with 7–15 cm-diameter holes and are made of pliable plastics such as polypropylene (see example in Figure 7.14). Other variations are made of soft polyester cord, which is less rigid than the polypropylene variety and more prone to tangling.

Depending on the cultivar's growth rate and the height of the mature canopy, several layers of trellis netting may be required. From an operational perspective, it is simplest to install the required layers of netting when the plants are small, allowing them to grow through it. This avoids the over-handling of the plants later in the crop cycle.

7.4.4.4   Staking

Staking involves the use of vertical posts (stakes) which are secured to the stem(s) of an individual plant. Stakes can be wooden, metal or plastic. They are inserted into the growing media, then secured to the plant using a clip or tie.

Supporting individual plants through staking allows easy access and removal of individual plants from a canopy. This is practical when requiring measurements or photos of individual plants. Compared to trellising, staking requires more materials and labor since each plant requires

FIGURE 7.14   Example of a cannabis trellising system in a greenhouse.

Source: Photo courtesy of Dr. Youbin Zheng

individual attention. Therefore, staking is advisable only for low-density canopies or when removing individual plants from the canopy is required.

REFERENCES

Ballaré, C. L., and R. Pierik. 2017. The shade-avoidance syndrome: Multiple signals and ecological consequences. *Plant, Cell & Environment* 40(11): 2530–2543, November. http://doi.wiley.com/10.1111/pce.12914.

Campbell, L. G., S. G. U. Naraine, and J. Dusfresne. 2019. Phenotypic plasticity influences the success of clonal propagation in industrial pharmaceutical *Cannabis sativa. PloS One* 14(3): 1–15.

Caplan, D., J. Stemeroff, M. Dixon, and Y. Zheng. 2018. Vegetative propagation of cannabis by stem cuttings: Effects of leaf number, cutting position, rooting hormone, and leaf tip removal. *Canadian Journal of Plant Science* 98(5): 1126–1132. www.nrcresearchpress.com/cjps.

Charlo, H. C. O., R. Castoldi, L. A. Ito, C. Fernandes, and L. T. Braz. 2007. Productivity of cherry tomatoes under protected cultivation carried out with different types of pruning and spacing. *Acta Horticulturae* 761: 323–326, September. www.ishs.org/ishs-article/761_43.

Cruz-huerta, N., and C. De Postgraduados. 2005. Biomasa e índices fisiológicos en chile morrón cultivado en altas densidades. *Revista Fitotecnia Mexicana* 28(3): 287–293.

Danziger, N., and N. Bernstein. 2021. Plant architecture manipulation increases cannabinoid standardization in "drug-type" medical cannabis. *Industrial Crops and Products* 167: 113528, September. https://doi.org/10.1016/j.indcrop.2021.113528.

Faust, M. 2000. Physiological considerations for growing temperate-zone fruit crops in warm climates. In *Temperate fruit crops in warm climates*. Dordrecht: Springer Netherlands, 137–156. http://link.springer.com/10.1007/978-94-017-3215-4_7.

Hand, D. W., J. W. Wilson, and M. A. Hannah. 1993. Light interception by a row crop of glasshouse peppers. *Journal of Horticultural Science* 68(5): 695–703, January.

Heuvelink, E., M. J. Bakker, A. Elings, R. C. Kaarsemaker, and L. F. M. Marcelis. 2005. Effect of leaf area on tomato yield. *Acta Horticulturae* 691(691): 43–50, October. www.actahort.org/books/691/691_2.htm.

Hu, N., C. Lu, K. Yao, and J. Zou. 2009. Simulation on distribution of photosynthetically active radiation in canopy and optimum leaf rolling index in rice with rolling leaves. *Rice Science* 16(3): 217–225, September. https://linkinghub.elsevier.com/retrieve/pii/S1672630808600827.

Kuroyanagi, T., H. Yoshikoshi, T. Kinoshita, and H. Kawashima. 2013. Use of air circulation to reduce wet leaves under high humidity conditions. *Environmental Control in Biology* 51(4): 215–220. www.jstage.jst.go.jp/article/ecb/51/4/51_215/_article.

Magagnini, G., G. Grassi, and S. Kotiranta. 2018. The effect of light spectrum on the morphology and cannabinoid content of *Cannabis sativa* L. *Medical Cannabis and Cannabinoids* 1(1): 19–27, June 12. www.karger.com/Article/FullText/489030.

Mahlberg, P. G., and J. K. Hemphill. 1983. Effect of light quality on cannabinoid content of *Cannabis sativa* L. (Cannabaceae). *Botanical Gazette* 144(1): 43–48.

Marcelis, L. F. M., A. G. M. Broekhuijsen, E. Meinen, E. M. F. M. Nijs, and M. G. M. Raaphorst. 2006. Quantification of the growth response to light quantity of greenhouse grown crops. *Acta Horticulturae* 97–103. www.actahort.org/books/711/711_9.htm.

Monteith, J. L. 1977. Climate and the efficiency of crop production in Britain. *Philosophical Transactions of the Royal Society of London. B, Biological Sciences* 281(980): 277–294, November 25. https://royalsocietypublishing.org/doi/10.1098/rstb.1977.0140.

Namdar, D., M. Mazuz, A. Ion, and H. Koltai. 2018. Variation in the compositions of cannabinoid and terpenoids in *Cannabis sativa* derived from inflorescence position along the stem and extraction methods. *Industrial Crops and Products* 113: 376–382, March. https://doi.org/10.1016/j.indcrop.2018.01.060.

Nelson, P. V. 2011. *Greenhouse operation and management*, 7th ed. London: Pearson.

Patil, P., P. Biradar, A. U. Bhagawathi, and I. S. Hejjegar. 2018. A review on leaf area index of horticulture crops and its importance. *International Journal of Current Microbiology and Applied Sciences* 7(4): 505–513.

Pawar, R., and V. S. Rana. 2019. Manipulation of source-sink relationship in pertinence to better fruit quality and yield in fruit crops: A review. *Agricultural Reviews* 40(3), August 7. http://arccjournals.com/journal/agricultural-reviews/R-1934.

Peña B., F., and I. Zenner de Polanía. 2015. Growth of three color hybrids of sweet paprika under greenhouse conditions. *Agronomía Colombiana* 33(2): 139–146, May 1. https://revistas.unal.edu.co/index.php/agrocol/article/view/49667.

Potter, D. J. 2014. A review of the cultivation and processing of cannabis (*Cannabis sativa* L.) for production of prescription medicines in the UK. *Drug Testing and Analysis* 6(1–2): 31–38.

Reddy, S. K., S. V. Holalu, J. J. Casal, and S. A. Finlayson. 2013. Abscisic acid regulates axillary bud outgrowth responses to the ratio of red to far-red light. *Plant Physiology* 163(2): 1047–1058, October 1. https://academic.oup.com/plphys/article/163/2/1047-1058/6110994.

Rodriguez-Morrison, V., D. Llewellyn, and Y. Zheng. 2021. *Cannabis inflorescence yield and cannabinoid concentration are not improved with long-term exposure to short-wavelength ultraviolet-B radiationno*, June. www.preprints.org.

Savage, J. A., M. A. Zwieniecki, and N. Michele Holbrook. 2013. Phloem transport velocity varies over time and among vascular bundles during early cucumber seedling development. *Plant Physiology* 163(3): 1409–1418, November.

Schaffner, J. H. 1926. The change of opposite to alternate phyllotaxy and repeated rejuvenations in hemp by means of changed photoperiodicity. *Ecology* 7(3): 315–325, July. https://doi.org/10.2307/1929314.

Sekine, T., M. Aizawa, T. Nagano, and T. Takahashi. 2007. Suppression of gray mold and leaf mold of tomato by ventilation using fans and analysis of mechanism. *Annual Report of the Society of Plant Protection of North Japan* 58: 46–53.

Shaikh Abdullah Al Mamun, H., W. Lixue, C. Taotao, and L. Zhenhua. 2017. Leaf area index assessment for tomato and cucumber growing period under different water treatments. *Plant, Soil and Environment* 63(10): 461–467, November 2. www.agriculturejournals.cz/web/pse.htm?volume=63&firstPage=461&type=publishedArticle.

Shimomura, N., K. Inamoto, M. Doi, E. Sakai, and H. Imanishi. 2003. Cut flower productivity and leaf area index of photosynthesizing shoots evaluated by image analysis in "arching" roses. *Journal of the Japanese Society for Horticultural Science* 72(2): 131–133.

Sun, J., J. Gao, Z. Wang, et al. 2019. Maize canopy photosynthetic efficiency, plant growth, and yield responses to tillage depth. *Agronomy* 9(1): 1–18.

Tagliapietra, E. L., N. A. Streck, T. S. M. Da Rocha, et al. 2018. Optimum leaf area index to reach soybean yield potential in subtropical environment. *Agronomy Journal* 110(3): 932–938.

Thom, A. S. 1971. Momentum absorption by vegetation. *Quarterly Journal of the Royal Meteorological Society* 97(414): 414–428, October.

Vanhove, W., T. Surmont, P. Van Damme, and B. De Ruyver. 2012. Yield and turnover of illicit indoor cannabis (*Cannabis* spp.) plantations in Belgium. *Forensic Science International* 220(1–3): 265–270. http://doi. org/10.1016/j.forsciint.2012.03.013.

Vanhove, W., P. Van Damme, and N. Meert. 2011. Factors determining yield and quality of illicit indoor cannabis (*Cannabis* spp.) production. *Forensic Science International* 212(1–3): 158–163. http://doi. org/10.1016/j.forsciint.2011.06.006.

Verdenal, T., V. Zufferey, A. Dienes-Nagy, et al. 2019. Timing and intensity of grapevine defoliation: An extensive overview on five cultivars in Switzerland. *American Journal of Enology and Viticulture* 70(4): 427–434, October.

Wilson, J. W., D. W. Hand, and M. A. Hannah. 1992. Light interception and photosynthetic efficiency in some glasshouse crops. *Journal of Experimental Botany* 43(3): 363–373, March. https://academic.oup.com/ jxb/article-lookup/doi/10.1093/jxb/43.3.363.

Yang, C., and L. Li. 2017. Hormonal regulation in shade avoidance. *Frontiers in Plant Science* 8: 1527, September 4. http://journal.frontiersin.org/article/10.3389/fpls.2017.01527/full.

Yep, B., N. V. Gale, and Y. Zheng. 2020. Comparing hydroponic and aquaponic rootzones on the growth of two drug-type *Cannabis sativa* L. cultivars during the flowering stage. *Industrial Crops and Products* 157: 112881, December. https://linkinghub.elsevier.com/retrieve/pii/S0926669020307986.

Zheng, Y., L. Wang, and M. Dixon. 2007. An upper limit for elevated root zone dissolved oxygen concentration for tomato. *Scientia Horticulturae* 113(2): 162–165, June. http://linkinghub.elsevier.com/retrieve/ pii/S0304423807001203.

8 Management of Diseases on Cannabis in Controlled Environment Production

Cameron Scott and Zamir Punja

CONTENTS

DOI: 10.1201/9781003150442-8

8.1   INTRODUCTION

Plant pathogens have been impactful to humans throughout history, damaging and destroying crops and causing hunger, malnutrition and starvation. Some diseases have caused people to uproot their families in search of other work or food. Initially, blighted and diseased crops were seen as punishment from wrathful or unhappy gods, and ancient Romans and other peoples sought to manage plant diseases by appeasing deities with sacrifices and prayer. Eventually, through advances in science and inventions, such as the microscope, scientists were able to determine that microscopic organisms were responsible for these diseases. As a result, several hundred thousand plant diseases have been described to date. Management approaches for plant diseases have also shifted over time. Approaches have progressed from the use of compounds like Bordeaux mixture (lime and sulfur) to control downy mildew of grapes, and sulfur or copper to manage blights, to our modern synthetic chemicals, as well as biological and cultural control methods. Even with modern diagnostic tools and management strategies, plant diseases still cause hardship and upwards of $200 billion USD in crop losses worldwide every year (FAO 2021).

   In this chapter, general diagnostic approaches and management strategies for plant diseases will be discussed. How these management strategies apply to the most prominent diseases of cannabis produced in controlled environments (*Fusarium* crown and root rot and damping off, *Pythium* crown and root rot, powdery mildew and bud rot), and the biology of the pathogens that cause these diseases, will be the main focus. As the production of hemp and cannabis increases throughout the world, new diseases are emerging or less common diseases are spreading. Being able to identify and manage these new potential threats will also be discussed to assist growers.

8.2   PLANT DISEASES AND THEIR DIAGNOSIS

8.2.1   Definition

The definition to be used for disease in this chapter is: a change from a plant's normal development and appearance caused by a living (biotic) entity that reproduces and spreads to adjacent plants i.e., it is infectious. This definition applies to most pathogens, including fungi, bacteria, viruses, viroids, phytoplasmas and nematodes. Other definitions of disease may encompass nonliving (abiotic) factors, such as nutritional deficiencies, pH of the growing media, physical or mechanical damage, drought, overwatering, light damage or any persistent damage to the plant; i.e., they are noninfectious and do not spread. Determining whether a plant is affected by a disease or an abiotic cause is the first step in the diagnosis of a problem. Improper diagnosis—and subsequently, inadequate treatment methods—can be costly in terms of time, money (labor, cost of products, etc.) and the potential loss of yield and quality of the crop.

8.2.2 Diagnosis

To properly manage any plant disease affecting a crop, one must first identify what the problem is. Accurate diagnosis of a disease requires the investigation and consideration of many interacting factors, such as what signs and symptoms are displayed on plants in the affected areas, potential patterns of symptoms that may be present and the various biotic factors present at the time, as well as the interplay of abiotic and environmental influences.

The problem should be described in terms of specific symptoms on the affected plant. Symptoms are defined as the visible manifestation—both externally and internally—of the effects of a disease. Examples of symptoms include necrosis, yellowing of the foliage (chlorosis), wilting of the plant, stunted growth, rotting of roots, darkening of internal tissues, blighting and cankers. It should also be determined where symptoms occur on a plant, and whether they occur on all tissues or only on a specific part of the plant. For example, chlorosis may occur only on new growth, only on older leaves or over the entire plant.

In addition to symptoms, signs of plant disease can be used in diagnosis. Signs are the observable physical presence of a pathogen and are most commonly seen in fungi. Examples include conidia (spores), mycelium and reproductive structures such as pycnidia or sclerotia. Signs are generally characteristic of a particular pathogen and therefore are useful for the diagnosis of a disease.

The pattern of symptom appearance on affected plants may also be indicative of whether it is caused by a biotic or abiotic factor. Are the affected plants found throughout the growing area, or only in one particular location? If all the plants along a wall or a particular tray of cuttings are affected, it could be indicative of an abiotic cause. Uniform damage on a plant, or over a larger number of plants, is typically associated with abiotic factors. Damage that is random or unevenly distributed on the plant is more likely to be caused by a pathogen or a biotic factor. If the problem appears to be spreading to other plants, it is likely to be caused by a biotic factor. These differences are summarized in Table 8.1.

8.2.3 Abiotic Factors

Abiotic and environmental stresses can complicate attempts to diagnose plant diseases as they can cause similar symptoms to those caused by some pathogens. For example, nutrient deficiencies that

TABLE 8.1

A Comparison of Abiotic and Biotic Causes of Symptoms on Plants

Abiotic Factors	Biotic Factors
Uniform symptoms	Symptoms appear uneven
Symptoms observed in a uniform or discrete area	Affected plants are spread throughout the area
Symptoms do not spread or progress	Symptoms progress and may spread to neighboring plants or growing areas
No signs associated with pathogens are observed	Signs associated with pathogens are observed

cause yellowing of leaves, such as nitrogen or magnesium deficiency, may appear similar to chloro-sis caused by a root rot pathogen such as *Fusarium*. The effects of overwatering may appear similar to symptoms of root rot caused by *Pythium*. Environmental and abiotic factors that can affect plant health include extremes of pH of the growing medium, nutrient excesses and deficiencies, improper drainage, lack of water or overwatering, extremes of temperature and humidity, improper light levels, physical damage from severe pruning and girdling from trellises. All of these need to be considered when a diagnosis is being made. In addition, extraneous materials (sprays, fertilizer, etc.) applied to plants should be considered with regard to potential for damage to plants if applied incor-rectly or at a higher dose than recommended. If other crops are grown nearby, drift from sprays applied to those crops should also be considered as a potential source of damage or contamination to your plants.

The characteristics of the cannabis strain (genotype) being grown, and its normal appearance at that stage in its growth, should be known. Knowing the features of a healthy plant allows for more accurate comparisons to be made to the suspected symptoms on diseased plants. Chlorosis and leaf drop are normal at the end of the plant's growth cycle, cuttings may wilt if not hardened off and some strains may appear phenotypically differently from others—but that does not suggest they may be affected by a pathogen.

Confirmation of disease diagnosis is important and can be provided by a commercial testing lab, a university or government diagnostic lab, or by qualified consultants. Growers are advised to seek the help of someone who is proficient in disease diagnosis. This person should conduct a site visit, if possible, as it is often very difficult to diagnose an issue by phone or other correspondence, especially without access to samples to examine. Growers may also consult previous descriptions of these types of problems from reliable sources that can be found online or in institutional libraries to try to associate the symptoms seen with previously reported symptoms of the suspected disease.

8.3   HOW DISEASE DEVELOPS

8.3.1   DISEASE DEVELOPMENT

Many factors are required for the development of a disease. The first is a susceptible host plant. The vigor of the plant, the age of the plant, the conditions under which the plant is growing and its genetic composition all play a role in determining whether or not it may be susceptible to disease. The second requirement is the presence of a pathogenic organism (fungi, bacteria, virus, etc.), which also determines the incidence and severity of disease. The third requirement is an environment that is conducive to the development of a pathogen. High humidity, leaf wetness and an anaerobic or overwatered rootzone are examples of environments that are conducive to the development of some pathogens. Conversely, factors such as extreme temperatures and dryness may inhibit plant disease by reducing pathogen growth, even if the plant is susceptible to infection. The environment may also play a role in the growth, reproduction and dispersal of the pathogen, as well as affect the susceptibility of the host plant.

8.3.2   DISEASE TRIANGLE

The interactions between a plant, a pathogen and the environment can be described in a disease triangle (Figure 8.1). The disease triangle is a fundamental concept in plant disease development, which illustrates all three elements that must be present in order for a disease to develop. However, even when all requirements of the disease triangle are present, disease may fail to develop due to inappropriate timing. Therefore, "time" may be considered as the fourth element of this model, creating a disease tetrahedron, or disease pyramid. The pathogen, the host and a conducive environ-ment must exist together for a certain period of time in order for disease to develop. The length of time during which these three elements occur together also plays a role in the severity and incidence

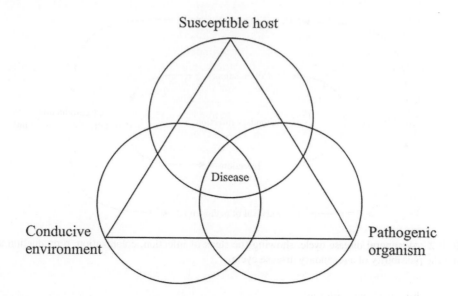

FIGURE 8.1 The disease triangle, highlighting the three main factors necessary for the development of disease.

of disease. For example, increasing wetness duration on flowers and fruits on strawberries can increase the incidence of *Botrytis* infection. A similar phenomenon may also apply to cannabis and bud rot development.

By altering any individual component of the disease triangle, the incidence or severity of the disease could be altered. Therefore, disease management practices by growers should be targeted to disrupt one or more components of the disease triangle. Planting a less susceptible host genotype, altering the humidity in the growing environment or spraying a product that reduces spore production by a pathogen, are all examples of disruption of one aspect of the disease triangle that can reduce disease development.

8.3.3   The Disease Cycle

Diseases progress over time according to a series of specific events known as the "disease cycle" (Figure 8.2). The events that occur in succession, resulting in the development of disease include the following.

1. Infection of the host by a pathogen (from primary inoculum such as fungal spores, bacterial cell or virus particles) that establishes itself in the plant tissues. Symptoms may be apparent here.
2. Colonization of the tissues as the pathogen grows and develops on the host. Symptoms will be present.
3. Reproduction by the pathogen, creating more inoculum (secondary inoculum) that can re-initiate infection.
4. Spread of the pathogen, sometimes in several life cycles during one growing season.
5. Survival of the pathogen.

Knowledge of the disease triangle, disease cycles and the various biotic and abiotic factors that influence them can be used to identify opportunities to disrupt the cycle, leading to disease management. These types of general management approaches include exclusion or avoidance, eradication,

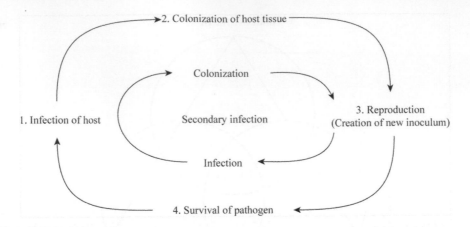

FIGURE 8.2   A general disease cycle, showing the steps of infection, colonization, reproduction and survival, and the possibility of a secondary disease cycle.

protection and resistance (genetic and induced). These approaches are discussed in more detail in the following section.

8.4   DISEASE MANAGEMENT PRINCIPLES

8.4.1   EXCLUSION

In this disease management strategy, the aim is to prevent the introduction of the pathogen into the growing environment. For example, exclusion can be achieved by placing a quarantine on any plant or plant material before it is allowed to enter the facility in an area separated from the main growing environment. The plants are examined regularly for signs or symptoms of disease, or insect pests, before they are released. The use of pathogen- or pest-free plant material is another important aspect of exclusion. Producers should verify from their source of cuttings or seeds that they were produced in an environment free of pathogens. This is particularly important as at the present time, as there are no government-enforced certification requirements for propagated cannabis plants. Testing services could be used to ensure the status of plants with regard to presence of pathogens before they are introduced into a facility. Other examples of exclusion of pathogens from the growing environment include biosecurity measures such as proper sanitation of equipment entering the area where plants are being grown, implementation of footbaths and ensuring workers move through rooms considered to be most "clean" to least "clean." These should be placed at all points of potential entry into a growing facility. The use of high-efficiency particulate air (HEPA) filters and ultraviolet (UV) lights can also preclude introduction of pathogen spores into a clean facility.

8.4.2   AVOIDANCE

This strategy aims to prevent establishment of a pathogen by creating an environment that is not conducive to disease development. For example, cultural control methods are practices that attempt to alter the growing environment to prevent infection by the pathogen. Improved drainage of soil, growing susceptible varieties only when disease pressure is low, minimizing excess nutrients, pruning plants to improve air circulation and reduce areas of high humidity, and providing proper post-harvest storage conditions are examples of cultural control methods that can reduce disease development and spread.

8.4.3   ERADICATION

The principle of eradication is based on eliminating or reducing the amount of the pathogen even after it has entered the growing environment, but before it spreads and becomes established. This can be achieved through cultural control methods, such as removal of infected plant material or entire plants.

Physical control methods to eradicate a pathogen include treatment of seeds with disinfectants or hot water, pasteurization of irrigation water, sterilization of growing media and cleaning of tools, equipment, benches and other surfaces during a growing cycle and between cycles. Tissue culture methods utilizing meristem tissue can be used to eradicate a virus from stock plants, but it has not been widely used yet in cannabis.

8.4.4   PROTECTION

Protective measures are aimed to protect the plants before they become infected, through the application of a treatment that can kill or severely reduce growth of the pathogen, e.g., fungicide sprays, drenches and dusts. An example of a protective measure that can be used on cannabis in Canada are sulfur-based products to manage powdery mildew. When fungal spores come in contact with sulfur-treated leaves or flower tissues, they do not germinate, thus disrupting the disease cycle. Some products may act to eradicate a pathogen, as well as protect against future infection. One example of this is potassium bicarbonate products (MilStop, Armicarb), as they may eradicate a pathogen through their direct fungitoxic characteristics, as well as offer some degree of protection as they reduce the development of fungal mycelium and spores on plant tissues by altering their pH and osmotic pressure for a time. Traditional chemical fungicides are not registered for use on cannabis currently, but are a form of eradication or protection used in other crops.

The application of biological control agents can also provide protection against infection by root-infecting pathogens and foliage or flower-infecting pathogens. Biological control agents (living organisms formulated in commercial products for control of pathogens or insect pests) can be used to reduce the level of pathogen inoculum in soils or on plants when applied as drenches or sprays, respectively. These microbes compete with the pathogen for resources by colonizing the soil or tissues before pathogens can do so, or in some cases, they will directly parasitize and destroy pathogens.

8.4.5   GENETIC RESISTANCE

The use of plant varieties that are resistant or tolerant to one or more diseases is a key strategy in disease management. A resistant variety is able to prevent the pathogen from infecting tissues or reduce its growth and infection through the use of a genetically determined component, such as an enzyme or toxin, resulting in little or no infection. A tolerant variety is still susceptible to disease, but when infected, it will show less damage and still yield well. A variety is said to be immune if it never becomes diseased. At the other extreme, a susceptible variety allows infection to proceed to where the symptoms become severe and the plant could be killed. In most agricultural crops, the use of resistant varieties can be inexpensive, effective and safe. They can reduce both crop losses from disease, as well as reduce disease management costs.

In cannabis, detailed information on resistant varieties is scarce and is based mostly on grower observations. The experience of growers suggests that there are differences among strains in susceptibility to a number of important diseases, including powdery mildew and *Botrytis* bud rot. However, insufficient knowledge of the source(s) of strains and their genetic background can make it difficult to confirm if resistance genes are present. Genetics derived from known breeders or seed-banks, or cuttings originating from a reliable source, should be used to manage diseases. Further characterization of the basis of the observed resistance is ongoing, and efforts to develop new resistant strains of cannabis are in progress. The evaluation of landraces and strains from diverse geographical origins should prove to be useful in the search for resistance genes.

8.4.6 INDUCED RESISTANCE

The basal resistance that plants have against a range of pathogens can be induced through a variety of methods. For example, plant hormones such as salicylic acid can induce a series of defense responses following application that can lead to systemic acquired resistance. This form of resistance is finite in its activity, but can provide protection from infection over several weeks after treatment is made. Some biocontrol agents or plant extracts are reported to be able to induce resistance by inducing expression of proteins that can enhance defense against diseases, provided that they are applied in advance of the onset of infection. These defenses are typically not long lasting, and their importance in disease management in cannabis needs to be researched.

8.5   DISEASE MANAGEMENT PRACTICES FOR CANNABIS PRODUCTION

8.5.1   CULTURAL CONTROL

- Clean plant materials. For the prevention of any potential disease, incoming plant material that may be infected should be quarantined and inspected. *Pythium* and *Fusarium* may be introduced into a growing facility on infected roots or media that plants are grown in. Powdery mildew and viral diseases can also be introduced into a growing facility on plants if not monitored. Plants can carry pathogens but may not express symptoms for some time, depending on the pathogen. Footbaths, sanitary clothing and sanitation of tools and equipment should all be used to reduce the introduction of pathogens.
- Clean environment. *Pythium* can spread through free or standing water, as well as irrigation systems, and extra care must be given to maintaining cleanliness in these areas. Drip lines, plumbing, hoses, water storage tanks, nutrient tanks and other irrigation equipment should be regularly cleaned. Untreated water sources such as rivers or ponds may also be a source of *Pythium* in the growing environment, and should be avoided. Reusing growing media between cycles should be avoided, as this allows for the transfer of pathogens between production cycles and into new environments. Sterilization of these media between cycles can reduce the inoculum carryover. Air filters and purifiers—e.g., photocatalytic, UV or ozone—may be used to reduce the spread and intake of pathogen spores. The dehumidifiers and air filters should be regularly cleaned to reduce inoculum buildup.
- Care should be taken to thoroughly clean and sanitize all tools and equipment, as well as all surfaces in the growing environment during the turnaround period after harvest. Special attention should be given to equipment used for harvesting plants and trimming buds. Trimming machines may trap plant debris internally, and as resin builds up on surfaces such as belts, spores of molds such as *Penicillium* may accumulate. Wounding from harvest and trimming may increase the incidence of infection, as these wounds provide an opportunity for contaminants to colonize flower tissues post-harvest.
- Hydrogen peroxide or peracetic acid (peroxyacetic acid) contained in ZeroTol or SaniDate 5.0, and products containing bleach or alcohol, can also be used for sanitation. Heat from a flame or bead sterilizer may also quickly and effectively sanitize tools. Products containing dodecyl dimethyl ammonium chloride, such as Chemprocide or Kleengrow, may also be effectively used to sanitize equipment and surfaces, or in footbaths (Figure 8.3). In our studies, when tested at concentrations of 0.4% and 1%, Chemprocide completely inhibited growth of *Penicillium olsonii*, *Fusarium oxysporum* and *Botrytis cinerea* in liquid culture. ZeroTol was also effective at reducing the growth of these fungi, but only when used at a 1% concentration. These findings indicate both compounds have fungitoxic properties

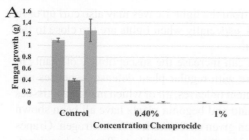 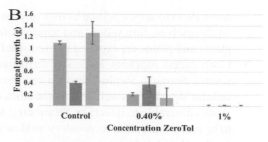

FIGURE 8.3 The effects of three concentrations of A) Chemprocide or B) ZeroTol on the growth of *Penicillium olsonii, Fusarium oxysporum* and *Botrytis cinerea* in potato dextrose broth. Cultures were grown on a shaker table at 125 rpm for seven days before being strained, dried and weighed (n = 4). Error bars are 95% confidence intervals.

against various fungi commonly found in cannabis growing facilities. For cleaning equipment used in harvesting and trimming, other options may also be worth considering, such as food-safe degreasers and steam.

- Removal of diseased plants. Removal of infected plants or symptomatic plant parts is important to reduce the spread of pathogens in the growing environment. Plants should be removed as soon as they show symptoms of infection by disposing of them outside the facility to reduce potential spread of inoculum. Plants affected by *Fusarium* and *Pythium*, and foliage and buds infected by powdery mildew and *Botrytis* bud rot, should be discarded.

- Manage irrigation. Maintaining appropriate moisture levels in the growing media or soil is key to managing root pathogens. Overwatering can encourage the spread of *Pythium* by providing free water, which helps to spread inoculum, as well as by creating anaerobic conditions in the rootzone which weakens roots and promotes infection. Low spots that lead to pooling of water on the ground or on tables should be avoided. The nutrient solution should also be well aerated to reduce *Pythium* growth and infection, especially in systems like deep water culture. Well-draining media will also help to reduce *Pythium* and *Fusarium* infection by reducing anaerobic conditions and increasing aeration in the rootzone. Extreme cycles of wetting and drying should be avoided, as these can cause roots to die back and predispose them to infection.

- Compartmentalization of plants and their rootzones can reduce spread of pathogens such as *Fusarium* and *Pythium* between plants.

- Stress avoidance. Plants that are growing vigorously and not subjected to any form of stress are generally better able to tolerate or resist infection by pathogens. Forms of nutrient-related stress in the soil or nutrient solution include excessive salinity from high salt levels, and ionic stress from high concentrations of specific ions such as Na^+, Mg^{2+} and Cl^-. These can directly reduce root growth, as well as potentially cause damage or dieback of lateral roots, making them more susceptible to *Pythium* infection. Some *Pythium* species can tolerate conditions of high salinity.

- Care should be taken to avoid excessive damage to roots during transplanting, and as previously mentioned, overwatering and cycles of extreme wetting and drying should also be avoided, as these factors may also cause stress on plants and promote disease.

- Lighting intensity should also be appropriate to the needs of the plant to avoid stress. Cuttings and younger plants growing vegetatively do not require light as intense as plants in flower, and providing them with too much light—or too much light too quickly—may cause them to appear wilted as they turn away from the light. Leaves may also curl up or in on themselves, turn yellow and appear burned, especially when lights are too close to the plants and plants are exposed to excessive heat.

- Limit excess nutrients. Excessive fertilization can increase the susceptibility of plants to several diseases. Surplus nitrogen can interfere with regular plant defense responses and cellular signaling, as well as limit the silica content of leaves, all of which may make tissues more vulnerable to infection. Strawberry, tomato and begonia plants have all been shown to be more susceptible to powdery mildew when given excessive levels of nitrogen. Grapes fertilized with a higher rate of nitrogen were also more susceptible to *Botrytis* bunch rot. The concentration of nitrogen in a plant also plays a role in the growth and development of the pathogen, with higher levels of nitrogen resulting in increased *B. cinerea* sporulation on basil, increased powdery mildew sporulation on barley and tomato plants, and the production of more virulent *B. cinerea* spores on tomato. This may result in a more rapid spread of the pathogen, as it is able to produce more inoculum for subsequent infections.

- Nutrient levels also have an effect on bacterial pathogens, such as *Xanthomonas campestris* pathovars. Fertilization with nitrogen, or excess nitrogen, was found to delay the onset and severity of black rot on cabbage and leaf spot on tomato, respectively. Although it is unclear if this is a practical solution for the management of *Xanthomonas campestris* pv. *cannabis*, which has been reported to cause leaf spot on hemp, due to the costs of inputs and potential trade-offs in disease susceptibility to other pathogens and plant health.

- The form of nitrogen fertilizer used, as well as the amount, have been shown to have an effect on the disease severity of pathogens such as *Fusarium* spp., as well. These factors directly affect the virulence of the *Fusarium*, as well as the susceptibility of the host to this pathogen.

- The effects of different forms and levels of nutrients—such as nitrogen, potassium, magnesium and others—on the susceptibility of cannabis to disease has yet to be determined.

- Climate management and air movement. Relative humidity in the growing environment should be kept low (< 50–60%) to reduce powdery mildew and *Botrytis* infections. Lower relative humidity is especially important later when inflorescences are mature and most susceptible to disease. The severity and incidence of powdery mildew and *Botrytis* infections have been shown to increase with increased humidity on crops such as grape and tomato.

- Relative humidity is the amount of water vapor the air is holding compared to what it can hold, at a specific temperature. As the temperature of the air increases, so does its capacity to hold water. When the relative humidity reaches 100%, also known as the saturation point, water can condense on plant surfaces. As relative humidity is dependent on temperature, large fluctuations in temperature should be avoided to manage humidity and condensation. For example, a growing environment which is quickly cooling or heating—such as at sunset, sunrise or when lights are turned on or off—can result in a rapid change in relative humidity and increase the possibility of condensation. This can lead to increased disease and stress on plants. Proper heating, cooling and ventilation can minimize fluctuations in temperature at critical times of the day.

- When considering relative humidity and temperature for disease management, it is also important to consider the vapor pressure deficit being created and the effect this can have on plants. The vapor pressure deficit (VPD) is a measure of pressure created by the difference between the amount of water vapor in the air and the amount it can hold. The VPD of the growing environment can also inform growers about the rate at which plants are transpiring. Too low of a VPD can cause plants to reduce transpiration, resulting in reduced movement of water and nutrients, and increased disease. Conversely, a high VPD may also increase the rate of transpiration too much, leading to stress, drought and nutrient deficiencies. Throughout the production cycle, cannabis plants will require different VPDs to maximize their growth, and it is important that growers balance these needs with their potential effects on disease development. For more information about VPD, relative humidity and environmental control, see Chapter 2.

- Having adequate air circulation by utilizing numerous well-placed fans, pruning and trellising plants, de-leafing plants and providing appropriate plant spacing can also help to create a consistent climate throughout the growing area. This will in turn help to reduce powdery mildew and *Botrytis* bud rot, and potentially other foliar diseases. Providing adequate air movement, cool temperatures and appropriate levels of humidity in drying rooms can also help to reduce post-harvest decay.

- Control insect pests. Management of insect pests (shore flies, cannabis aphids, fungus gnats, rice root aphids, etc.) is an important aspect of managing diseases in the growing environment, as these insects may be vectors for plant pathogens, enabling their spread. Insects may also damage root tissues and make them more susceptible to infection, as well as increase overall stress on the plants. Root aphids, for example, can potentially increase damage to roots and increase the likelihood of root pathogens such as *Fusarium* and *Pythium* infecting and causing disease.

8.5.2 Physical Control

- Pasteurization. Recirculated irrigation water is a significant way that many plant pathogens are spread. In order to eliminate these potential pathogens, water may be treated with heat in a process known as pasteurization. This is an effective and safe method to reduce the amount of inoculum present, although the potentially large amount of energy—and subsequently high cost of treatment—may make it prohibitive on the scale required for some facilities. Generally, protocols recommend bringing water to 95°C (203°F) for 10–30 seconds. Research has shown that water treatment may effectively eliminate plant pathogens at lower temperatures (42–48°C) as long as the exposure to these temperatures is longer (6–12 hours). One study showed that *F. oxysporum* conidia may be inactivated by heating water to 54°C for as little as 15 seconds. This may help to make pasteurization a more environmentally and economically sound practice, although it is still most likely only a realistic option for smaller producers.

- Irradiation. Treatment with UV light may be used to eradicate pathogens in irrigation water. Propagules or inoculum of *Pythium*—such as zoospores, oospores and mycelium, and conidia, chlamydospores and mycelium of *F. oxysporum*—can be killed using this treatment method. The degree to which the ultraviolet light is transmitted through the water plays an important role in the dosage of radiation required to eliminate these pathogens. Factors such as the turbidity of water can affect the transmittance of ultraviolet light. The duration of treatment required will vary with the intensity of the irradiation and the volume of water. Physical control methods applicable to irrigation water such as ultraviolet light and filtration are discussed in more detail in Chapter 5.

- Ultraviolet radiation in the form of UV-B or UV-C can also be used to manage powdery mildew on cannabis. There are several factors that can affect the efficacy of this treatment, including the intensity of the UV light plants receive, the duration of the treatment, how often treatments are applied, when treatments are applied during the day and the amount of foliage to be treated. Typically, plants will receive one "pass"—or treatment—per day, with treatments during either the day or at night showing efficacy at managing powdery mildew on cucumber plants. Comparable results have also been reported by applying a higher dose of UV-C every fourth day. This approach works well at preventing new fungal infections, but may have less activity on established infections. An excessive dose of UV can cause plant injury and stress, which could increase the incidence of other diseases, such as *Botrytis* gray mold. UV treatments can be made on a smaller scale using hand-held devices. Boom mounted lamps for treatments on a larger scale are also available.

- The use of gamma radiation or electron beam radiation may be allowed to reduce post-harvest contamination on cannabis where it has received regulatory approval.
- Filtration. Pathogens may be removed from irrigation water through physical filtration. The most common methods use either membranes of differing pore sizes or materials like sand and gravel (slow sand filtration). These methods each have their own associated costs, benefits and scalability issues. Physical control methods applicable to irrigation water such as ultraviolet light and filtration are discussed in more detail in Chapter 5.

8.5.3 Biological Control

8.5.3.1 Root-Infecting Pathogens

There are several commercial biological control products that can be used on cannabis for disease management. The biocontrol products contain fungi such as *Trichoderma* spp. (Rootshield, Asperello, Trianum, etc.) or *Gliocladium catenulatum* (Prestop, Lalstop). In addition, bacterial products contain *Bacillus* spp. (Rhapsody, Stargus or Double Nickel) or actinomycetes such as *Streptomyces griseoviridis* (Mycostop) (Figure 8.4). These products contain spores of fungi or

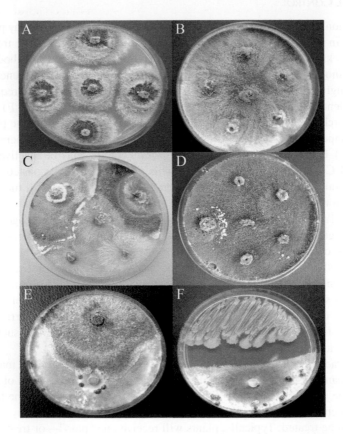

FIGURE 8.4 Fungi and bacteria utilized for biocontrol of plant pathogens and their action against select pathogens on potato dextrose agar. A) *Gliocladium catenulatum* (Prestop, Lalstop) growing from colonized cannabis stems; B) *Trichoderma harzianum* from Rootshield growing from colonized cannabis stems; C) A *Trichoderma* spp. and a *Fusarium* spp., seen at the bottom, growing from cannabis stems; D) *Trichoderma asperellum*, from the product Asperello, growing from colonized cannabis stems; E) *Trichoderma asperellum* overgrowing *Sclerotinia sclerotiorum* in a dual culture assay; F) *Bacillus amyloliquefaciens* (Stargus) inhibiting the growth of *Sclerotinia sclerotiorum*. All stem tissues shown in this figure were treated with the biocontrol agent, incubated for seven days, surface sterilized and plated.

bacterial cells at a high concentration, usually in excess of 10^8 cells/ml, in a formulation that provides stability and longevity to the microbe. These biocontrol products prevent infection by *Fusarium* or *Pythium* on cuttings or rooted plants, and should be applied as a drench before infection takes place.

Based on research conducted on many crops, these biocontrol agents can produce antibiotics or enzymes that inhibit the pathogen, or they may compete for nutrients or space on the root system. Fungal products based on *Trichoderma* spp. or *Gliocladium catenulatum* are also known to actively destroy (through mycoparasitism) *Pythium* and *Fusarium* hyphae. Although these biocontrol agents—especially the fungal biocontrols—are known to persist in media for extended periods of time, reapplications may still be needed. On other crops, these microorganisms have also been shown to have the added benefit of stimulating plant root and shoot growth. These bacteria and fungi have this effect, as they may colonize plant tissues and interact with the plants through the production of hormones and growth enhancers.

The use of soils that naturally contain a complex of these beneficial microorganisms is another potential biological control method. These soils are often referred to as suppressive soils or "living soils," and they can reduce the growth of a pathogen and the disease it causes. These types of soils may also have the added benefit of stimulating plant growth compared to soils with less numerous or diverse microbial communities, such as sterilized soil or media like rockwool. Mature composts are a source of many of these microorganisms. Research has shown that composts may be inoculated and used as a substrate for specific biocontrol organisms such as *Trichoderma harzianum*.

The general suppressiveness of the soil seems to be associated with the overall biomass and microbial activity of the soil, whereas the ability to suppress more specific pathogens or organisms is associated with the presence and relative abundance of specific organisms. Substrate respiration, amendments added to the soil, soil physiochemistry and other abiotic conditions also alter the microbial contents and suppressiveness of soils. These approaches should be considered by organic producers.

8.5.3.2   Foliar- and Flower-Infecting Pathogens

Several biocontrol products that contain bacteria have been shown to reduce powdery mildew on crops such as cucumber, zucchini, strawberry and grape. As with root pathogens, production of antibiotics and enzymes from the microbes in these products provides protection when they are applied prior to establishment of infection. On cannabis plants, applications of Rhapsody or Stargus, containing *Bacillus* spp., made at weekly intervals was shown to reduce powdery mildew development (Figure 8.5). Actinovate SP (*Streptomyces lydicus*) was shown to have a limited effect on powdery mildew on cannabis, and only at low disease pressure. The use of these biocontrol products for disease management will likely require they be applied frequently and/or used in a rotation with other products during periods when plants are most susceptible to disease.

8.5.4   Biorational Products

8.5.4.1   Root-Infecting Pathogens

Products containing plant extracts such as neem oil or powder, clove oil, thyme oil, tea tree oil extract (TIMOREX GOLD), orange oil extract (PREV-AM) and others reportedly have some ability to reduce fungal growth and manage disease on other crops. Biorational products such as these still need to be evaluated for disease management on cannabis or hemp plants.

8.5.4.2   Foliar Pathogens

The use of plant extracts to effectively manage foliar diseases such as leaf spots and powdery mildew has been demonstrated on crops such as tomato, cucumber, pea and okra. Two effective products include Regalia Maxx (containing extracts from the giant knotweed *Reynoutria sachalinensis*) and products containing extracts from the neem tree (as oils). These products are most effective when applied preventatively or under low disease pressure. Regalia Maxx is registered for use on

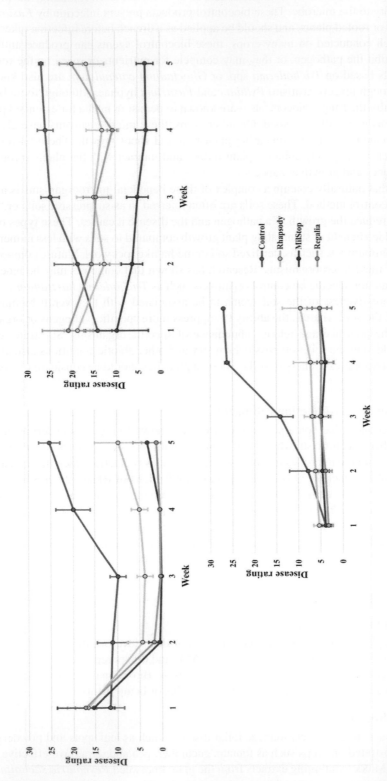

FIGURE 8.5 The effects of Rhapsody ASO, Regalia Maxx and MilStop on powdery mildew disease progression under various disease pressures. Treatments were made once weekly and disease ratings were calculated by rating the 30 most diseased leaflets per plant. There were four replicates per treatment per trial. Error bars represent 95% confidence intervals. Further details can be found in Scott and Punja (2020). For control of *Botrytis* infection, Stargus is registered to manage bud rot for indoor and outdoor hemp production by the EPA. On cannabis, there is no data available yet to demonstrate the efficacy of these products at managing *Botrytis*.

cannabis and hemp in Canada, while neem-based products and Regalia Maxx are approved for use on hemp in many U.S. states. In other crops, Regalia Maxx has been shown to promote resistance to pathogens by increasing antimicrobial compounds or enzymes; however, the biochemical responses of cannabis plants following treatment have not been studied. Regalia Maxx has been shown to significantly reduce powdery mildew development when applied weekly (Figure 8.5).

8.5.5 REDUCED-RISK CHEMICALS AND CONVENTIONAL FUNGICIDES

In addition to the biocontrol and biorational products previously mentioned, there are several other options that growers can consider as a part of their disease management strategy. These products are based on chemistries that are considered safe for plants and the environment and are described in what follows. However, applications of these products, as well as Regalia Maxx, made during flower development can cause damage to stigmas on cannabis plants (Figure 8.6).

- Chlorine: The addition of compounds containing chlorine to irrigation water can effectively eradicate the majority of propagules of pathogens such as *Fusarium* and *Pythium*. A concentration of 5 ppm active chlorine is recommended to reduce the spread and survival of propagules of these pathogens. Factors such as the biology of the pathogen or propagule, water quality and exposure time all affect the efficacy of chlorine. This treatment is discussed in more detail in Chapter 5.
- Potassium bicarbonate: Products such as MilStop, when applied as a foliar spray, can reduce the development and spread of powdery mildew on cannabis and other crops (Figure 8.5). These products act by altering the pH and osmotic pressure of the surface of tissues on which they are applied, which then disrupts the growth of mycelium and spores. These products appear to have both curative and preventative effects when used to manage powdery mildew on cannabis.

FIGURE 8.6 Damage from foliar sprays on stigmas.

- Hydrogen peroxide: ZeroTol shows some efficacy in reducing powdery mildew on cannabis and cucumber when applied as a foliar spray. The effect may be due to direct toxicity to the pathogen. On other crops, these products have been shown to play a role in inducing resistance to disease on treated plants. Data on the efficacy of hydrogen peroxide products and their mode of action is limited.
- Silicon: Foliar applications of the silicon-based product Silamol reduced powdery mildew on cannabis at low disease pressures. Silicon sprays can also be effective for disease management on crops such as grape, wheat and cucumber. The use of silicon in soil or nutrient solution can also reduce powdery mildew development on crops such as rose, cucumber, zucchini and wheat. Silicon is taken up by plants and utilized as part of the plant's defense response. Silica is deposited in plant tissues and limits powdery mildew infection. Application of silicon to the roots seems to be more effective than foliar sprays, since it results in higher levels of silicon accumulation within the plants. The use of root-applied silicon to cannabis plants for disease management has not been tested.

At the present time, applications of conventional fungicides are not approved for cannabis or hemp growers. While fungicides with active ingredients such as fludioxonil, thiabendazole, fluopyram, prothiconazole, and others such as propamocarb, metalaxyl-M and cyazofamid, have been shown to reduce *Fusarium* or powdery mildew and *Pythium* infections, respectively, on other crops, future research may determine the safety parameters around which they could be used on cannabis and hemp crops.

8.5.6　Disease Resistance

Most agricultural crops utilize disease-resistant varieties that have been developed through selective breeding. Genes that confer resistance to pathogens such as *Fusarium* and powdery mildew–causing pathogens such as *Podosphaera xanthii* have been identified in other crops and can provide stable disease reduction. In cannabis, breeding for the purpose of selecting disease-resistant strains has not yet been undertaken on a scale comparable to other crops, although there exists a broad diversity of germplasm from which selections could be made. Concerted efforts are being made by cannabis and hemp producers that will begin to formally identify such sources of resistance. Screening methods using pathogens known to be of importance in specific areas of production will begin to identify the most suitable strains that combine traits of commercial interest with disease resistance traits.

8.6　INTEGRATED DISEASE MANAGEMENT

The utilization of several different approaches for managing diseases in a cohesive plan will allow producers to implement an integrated disease management plan (IDM). This may be combined with management of insects and other pests into an integrated pest management (IPM) program. Continuous monitoring of environmental and crop variables, as well as scouting for pest and disease signs and symptoms, are key to a well-designed IPM strategy. This allows growers to act proactively when disease levels are low and manageable, i.e., below the economic threshold (ET). The implementation and efficacy of an IPM program should be adjusted as needed based on repeated observations and consideration of good horticultural practices, the scale of the operation and costs (labor, pesticides, etc.).

Economic thresholds are based on the extent to which the disease or insect pest can potentially reduce the yield or quality of the crop if left untreated. This includes a cost-benefit analysis to establish at which point the damage caused by a pathogen or insect pest requires management intervention and the associated costs. The economic injury level, or EIL, is the lowest level of damage that will cause economic impact. The ET is also known as the Action Threshold, as when pest pressures reach this point, action should be taken to avoid reaching the EIL and losing crop value.

Below the ET, the costs of controls exceed their possible benefits, whereas the EIL is a break-even point whereby the cost of control equals the benefits. For other crops, ET and EIL values have been established, whereas for cannabis, these values need to be determined by individual growers.

Factors that affect EIL and ET values include the cost of inputs (labor, pesticides, equipment, etc.), the efficacy of the inputs and the value of the harvest. The ET value should account for the time it may take to treat the affected areas and how extensively the disease may progress during that time. The effects of diseases or pests on subsequent production cycles and the facility as a whole should be considered.

8.7 PROMINENT DISEASES OF CANNABIS

Described in the following subsections are the most important disease affecting cannabis crops, based on the frequency of occurrence and the damage caused to the crop.

8.7.1 Crown Rot, Root Rot and Damping Off Caused by *Fusarium* Spp

Fusarium species are prevalent and potentially devastating soilborne fungal pathogens that cause vascular wilts and crown rot on field and greenhouse crops around the world. These crops include cucumber, tomato, ornamental flowers, legumes, pulses, *Brassica* spp. and banana. On cannabis plants, *Fusarium* has been shown to be able to infect at all stages of growth, from propagation through to flowering. Affected plants develop crown and root rot symptoms, and damping off on cuttings may occur. Infection of flowers by *Fusarium* has also been observed, and will be discussed later.

8.7.1.1 Causal Agent

On cannabis, the most prevalent *Fusarium* species is *Fusarium oxysporum*, although other species such as *Fusarium proliferatum* and *Fusarium solani* can also cause similar symptoms. The mycelium, which is the vegetative form of the fungus, is visible on plants and can be white, light orange or light pink in color. When grown on agar medium (such as potato dextrose agar [PDA]) in the laboratory, cultures of *Fusarium* tend to display shades of purple, red or light pink (Figure 8.7). The spores of *Fusarium* spp. are colorless and recognizable under a microscope due to their canoe-like shape. Smaller spores called microconidia may also be present.

8.7.1.2 Disease Cycle, Symptoms and Signs

Fusarium spp. grow best at temperatures between 28°C and 34°C, although growth at lower temperatures can occur. The fungus can be present in soil, rockwool, peat-based growing media, perlite and other substrates, and grows best at pH 5–7. If soil is contaminated with *Fusarium*, the spores and mycelium can infect and colonize the developing roots of the plant (Figure 8.8). The fungus secretes enzymes that degrade the cell walls of roots, causing root rot. Mycelium is then able to grow intercellularly (between the cells) within the roots until it reaches the xylem vessels of the plant. Wounds from transplanting or other forms of damage. e.g., insect feeding injury, can increase infection by *Fusarium*. Once in the xylem vessels, the pathogen colonizes these tissues, which then collapse, reducing uptake of water and nutrients. Hence, symptoms of yellowing and wilting are observed. Production of toxins by the fungus can also cause symptoms of yellowing and wilting. High temperatures during the summer can cause increased wilting of diseased plants.

The roots of affected plants appear brown and necrotic (root rot). In cannabis, dark sunken lesions on the crown of the plant (crown rot) also may develop. Mycelium may be visible on the infected tissues, especially at the soil line or on areas affected by crown rot. When the stems of infected plants are cut open, the pith and xylem tissues appear brown or black. Sometimes, diseased plants may remain asymptomatic until plants are stressed.

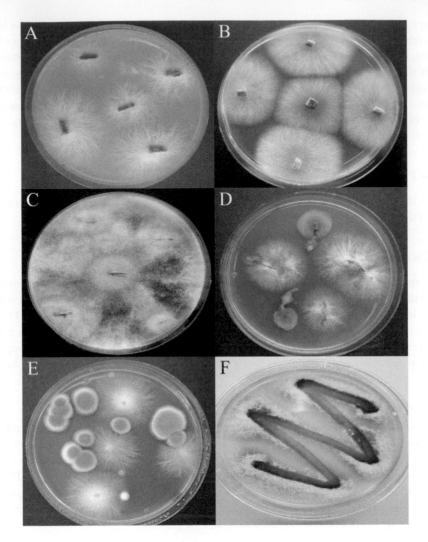

FIGURE 8.7 Various species of *Fusarium* growing on potato dextrose agar medium. *Fusarium oxysporum* growing from infected cannabis stems from plants sampled in A) vegetative growth or B) from infected cuttings; C) *Fusarium* and *Pythium* isolated from co-infected cannabis roots; D) *Fusarium* and other fungi growing from infected hemp seeds; E) A sample of the air in a propagation room showing the presence of *Fusarium oxysporum* and a *Penicillium* species; F) A swab from a damped off cutting that produced a pure culture of *Fusarium*.

Fusarium spp. can also cause damping off on cannabis cuttings. These symptoms start as soft, dark waterlogged areas at the base of stems, which progress upward and eventually cause the cutting to collapse (Figure 8.9). The humid conditions present during propagation of cannabis increases development of this disease. As the infection spreads, the cutting will rot, and often times the mycelium can be seen growing on the affected regions. Secondary organisms such as soft rot bacteria or *Penicillium* spp. may also be present. Poor air circulation, as well as overly wet media, may increase the severity of damping off. Cuttings which originate from infected stock plants, even if they are free of symptoms associated with *Fusarium*, are at particular risk of developing damping off, as these cuttings may already have the pathogen present inside them. This is a common way by which *Fusarium* has the ability to spread. Stock plants that have been grown for longer than approximately six months have a higher chance of becoming infected by *Fusarium*.

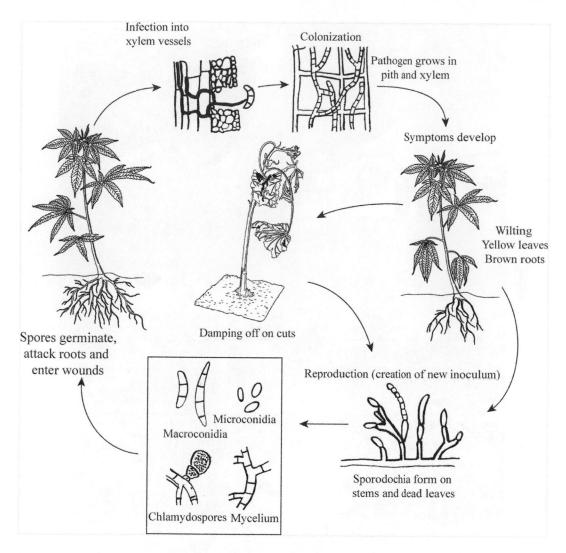

FIGURE 8.8 The disease cycle of *Fusarium* on cannabis.

Source: Adapted from Agrios (2005)

Spores of *Fusarium* can be spread by water, in air and on equipment. A diseased plant on a flood table or in a tray with healthy plants may release spores that spread to adjacent plants. Contaminated rooting or growing medium should not be reused, as spores can survive and initiate disease in subsequent growing cycles.

Plant debris (leaves, stems) infected by *Fusarium* may also spread the pathogen to other plants. Spores and mycelium can survive on plant residues in the growing environment and in soil for several weeks, and potentially initiate disease in subsequent cropping cycles. Equipment or tools—including pots, trays, domes and shears—that come into contact with infected plant material—especially when there is visible mycelial growth—may spread the pathogen to other plants or other areas of the growing environment. Seeds of crops such as wheat are known to transmit *Fusarium*, although the degree to which *Fusarium* spp. are spread through cannabis seeds is not yet known.

FIGURE 8.9   The signs and symptoms of *Fusarium* diseases on cannabis plants. A–B) Damping off on cuttings caused by *Fusarium*, with visible white mycelium on the outside of the cuttings; C) Damping off affecting numerous cuttings; D) Root and crown rot on plants in vegetative growth; E) Stunted plants (right) next to otherwise healthy plants; F) Chlorosis and stunting on plants in flower, with healthy plants around them; G) Intense chlorosis throughout an infected plant; H) Chlorosis, stunting, premature leaf drop and wilt on a mature cannabis plant; I) Close-up highlighting the characteristic chlorosis caused by *Fusarium* infection; J–K) Cross sections of cannabis stems showing blackened and necrotic vascular tissue.

8.7.1.3   Management Approaches

1. Start with planting material that is free of *Fusarium*. Incoming plant material should be examined for symptoms (yellowing, wilting, root rot) and placed under quarantine and tested for the presence of the pathogen. Stock plants should be replaced approximately

every six months to avoid a buildup of *Fusarium* internally. Suspected plants should be tested for the presence of the pathogen by a plant disease diagnostic lab.

2. Minimize introduction or spread of the pathogen through contaminated soil, plant debris or potentially by spores on shoes or clothing of personnel and visitors. This requires that shoe covers, gloves and footbaths with disinfectant be placed at entrances to growing rooms. Filters and UV lights in air intakes may also reduce the number of spores being brought into the facility.

3. Reduce plant-to-plant spread. When taking cuttings from a mother plant, preparing cuttings for propagation or pruning plants, shears should be regularly cleaned with isopropyl alcohol, bleach or heat. Avoid movement of workers and equipment from a diseased area to a clean area. Regularly scout and remove symptomatic plants and the growing media as soon as possible and destroy them.

4. Sanitize equipment and tools regularly. Previously used trays, domes, pots, flood tables, humidifiers, dehumidifiers, and fans should be cleaned thoroughly and regularly with a detergent and water, as well as a disinfectant. Irrigation equipment such as drip lines or pipes may be cleaned by flushing them with hydrogen peroxide or other cleaning agents. For growers who recirculate nutrient solutions, filtration through various membrane systems or sand (slow filtration), heat treatment (pasteurization) or UV radiation may be required. Addition of chlorine-containing compounds, such as sodium hypochlorite or chlorine dioxide, to achieve levels up to 5 ppm chlorine, may reduce survival of *Fusarium*.

5. Avoid damage to roots from transplanting. Avoid excessive dry-back of the growing media or overwatering of plants. Insect pests such as fungus gnats, shore flies and root aphids may also create wounds on roots or act as vectors of *Fusarium*.

6. Apply biological control products such as Rootshield and Prestop as drenches at recommended rates when propagating plants. Reapplication may be beneficial later in production.

7. Provide adequate—but not excessive—humidity and good air movement in the propagation environment. Avoid overwatering.

8.7.2 CROWN AND ROOT ROT CAUSED BY *PYTHIUM* SPP.

Pythium spp. are fungal-like organisms called oomycetes (also known as water molds). Oomycetes are protists that differ from fungi like *Botrytis cinerea* or *Fusarium oxysporum* because their cell walls contain beta glucans and cellulose rather than large amounts of chitin, and they produce flagellated motile spores. These root-infecting pathogens occur worldwide, and they are commonly found in soil. They are favored by wet conditions and grow over a range of temperatures, with many favored by temperatures over 30°C in hydroponic cultivation systems. *Pythium* species cause damage to many greenhouse crops, including pepper, tomato, cucumber, and ornamentals, as well as field-grown crops such as soybean, cotton, strawberry and turfgrass. On cannabis, *Pythium* spp. cause crown rot and root rot, as well as pre- and post-emergence damping off on hemp.

8.7.2.1 Causal Agent

The *Pythium* spp. reported to infect cannabis are *Pythium myriotylum*, *Pythium dissotocum*, *Pythium ultimum* and *Pythium aphanidermatum*. When grown on PDA, they produce cottony white aerial mycelium which quickly covers the plate (Figure 8.10). Unlike *Fusarium*, they do not produce any pigment. Some species grow with a particular undulating or radiating pattern to their mycelium. Spores (zoospores), if present, are produced in structures called sporangia. Other spore types that may be produced are thick-walled oospores that allow for long-term survival.

8.7.2.2 Disease Cycle, Symptoms and Signs

Pythium can be introduced into the growing environment on rooted cuttings, contaminated soil, water and plant debris. Once inside the growing environment, *Pythium* begins to spread and infect

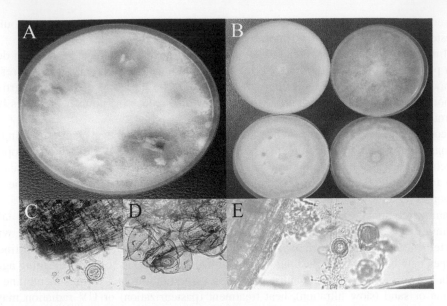

FIGURE 8.10   The causal agents of *Pythium* diseases on cannabis and their biology. A) *Pythium myriotylum* growing out of surface sterilized cannabis stem tissues; B) Four different *Pythium* spp. isolated from infected cannabis plants; C–D) Stained oospores produced by *Pythium myriotylum* on infected cannabis roots; as well as E) unstained oospores and sporangia (top right structure).

the roots of plants, particularly if roots are damaged during transplanting or by insect feeding. Infection is achieved mainly by the zoospores, although other species of *Pythium* may rely more on structures like mycelia to spread (Figure 8.11). Zoospores are spores that are able to actively travel through water using their flagella. These mobile spores come from germinating oospores (and the subsequent zoosporangium) and sporangium. When zoospores find their way to a host root, they start the process of infection by transforming into thick celled immobile structures called cysts. These cells adhere to roots and penetrate them, allowing growth within the root tissues to begin. *Pythium* spp. are able to infect nonwounded roots, as well as wounded tissues, with root tips and young root hairs being most susceptible to infection.

Colonization of root tissues by mycelium causes browning, with *Pythium* spreading both between and inside root cells. As the destructive necrotrophic phase of infection continues, roots may appear stubby, with absence of feeder roots (Figure 8.12). The outer region of the root, the epidermis and cortex, may slough off, leaving only the pith and vascular bundle. The crowns of infected plants may also appear sunken and dark, which can extend several centimeters up the stem. The *Pythium* disease cycle includes formation of sporangia on the root surface, while oospores may be formed inside the root tissues. These propagules can spread to other plants, causing additional infections. The pathogen may also survive between growing cycles on plant debris, in growing media and in water.

Foliar symptoms will become visible as the disease progresses, especially during warm weather (Figure 8.12). Plants infected by *Pythium* appear stunted and may begin to wilt as their root systems are destroyed. Infected plants may also appear moderately chlorotic, but otherwise look healthy. Initially, wilted plants may recover, but symptoms generally become more severe as the disease progresses. Wilting caused by *Pythium* occurs very rapidly, with plants drying out and dying over a few days. This is especially true in hot dry weather and on larger plants during flowering.

Damping off caused by *Pythium* is more commonly reported to be a problem on hemp rather than cannabis. Infection occurs if seeds are planted in media infested with the pathogen or seedlings are infected after germination of propagules in water. If seeds germinate, they may still be killed

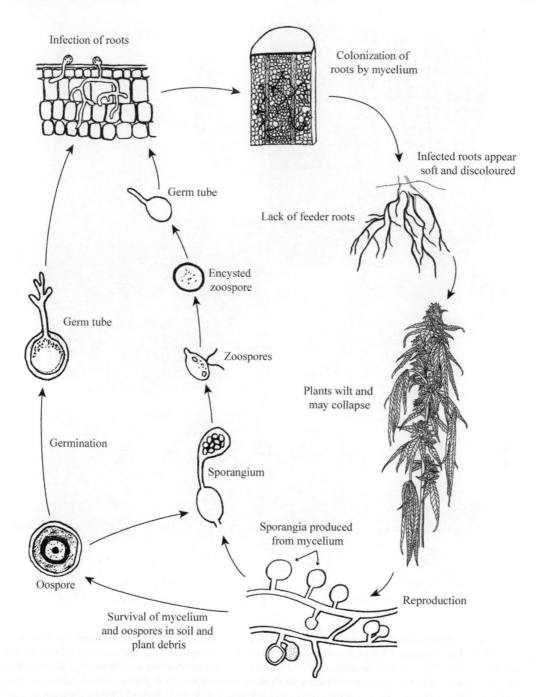

FIGURE 8.11 The disease cycle of *Pythium* on cannabis.

Source: Adapted from Agrios (2005)

by the pathogen before they emerge from the media. This is known as pre-emergence damping off. Seedlings may also be destroyed by *Pythium* in post emergence damping off, as the pathogen causes root and stem tissues to collapse. Symptoms such as stunted growth, yellowing and rot near the soil line may occur on older seedlings, causing wilting and plant collapse. Damping off can cause uneven stands and a notable reduction in plant counts.

FIGURE 8.12 Symptoms of root and crown rot on cannabis plants caused by *Pythium* species. A–B) Root rot caused by *Pythium* species; C) A mature cannabis plant showing symptoms of severe crown rot; D) A healthy plant (left) compared to a plant inoculated with *Pythium myriotylum* showing root browning; E) A healthy plant (left) compared to two plants inoculated with *Pythium myriotylum*; F) A plant in flower starting to show symptoms of wilt and chlorosis. G) More advanced symptoms of wilt; H–I) Plants killed by *Pythium*.

8.7.2.3 Management Approaches

1. Starting planting material should be free of *Pythium*. Incoming plant material should be examined for symptoms and placed under quarantine and tested for the presence of the pathogen.
2. Minimize introduction or spread of the pathogen through contaminated soil, plant debris or potentially spores on shoes or clothing of personnel and visitors. This requires that shoe

covers, gloves, hair and beard nets, and footbaths with disinfectant be placed at entrances to growing rooms.

3. Reduce plant-to-plant spread by minimizing sharing of irrigation water, such as on flood tables. Avoid movement of workers and equipment from diseased areas to clean areas. Regularly scout and remove symptomatic plants and the growing media as soon as possible and destroy them.

4. Sanitize equipment and tools regularly. Previously used trays, domes, pots, flood tables, humidifiers, dehumidifiers and fans should be cleaned thoroughly and regularly with a detergent and water, as well as a disinfectant. Irrigation equipment such as drip lines or pipes may be cleaned by flushing them with hydrogen peroxide or other cleaning agents. For growers that recirculate nutrient solutions, filtration through various membrane systems or sand (slow filtration), heat treatment (pasteurization) or UV radiation may be required. Addition of chlorine-containing compounds, such as sodium hypochlorite or chlorine dioxide, to achieve levels up to 5 ppm chlorine may reduce survival of *Pythium*.

5. Avoid damage to roots from transplanting. Insect pests such as fungus gnats, shore flies and root aphids may also wound roots or act as vectors of *Pythium*.

6. Avoid overwatering. This creates an anaerobic environment which is conducive to *Pythium* infection. Well-draining media should also be used. Nutrient solution should be aerated, and excessive salt levels should be avoided.

7. Apply biological control products such as Rootshield and Prestop as drenches at recommended rates when propagating plants. Reapplication may be beneficial later in production.

8.7.3 POWDERY MILDEW

Powdery mildew diseases affect a large number of field- and greenhouse-grown crops around the world, including tomato, cucumber, pepper, rose, hops, and grape. On cannabis, powdery mildew appears as white colonies on the upper surface of leaves, making it one of the more easily recognizable diseases. Powdery mildew occurs in all production environments and may affect plants at all stages of growth; although it will rarely kill plants, it can significantly reduce their growth.

8.7.3.1 Causal Agent

Powdery mildew on cannabis is caused by several *Golovinomyces* species: *G. cichoracearum*, *G. ambrosiae* or *G. spadiceus*. The hop powdery mildew pathogen *Podosphaera macularis* can occasionally infect cannabis and hemp and cause similar symptoms, but it is not considered a major pathogen.

8.7.3.2 Disease Cycle, Symptoms and Signs

Golovinomyces spp. are obligate pathogens, which requires living tissues to grow and reproduce. Infections on cannabis plants start when spores germinate on leaves, floral tissues or—less commonly—on the stems of plants. These airborne spores (conidia) are released from conidiophores from overwintering mycelium or neighboring infected plants (Figure 8.13). The spores produce a germ tube which penetrates the host cells to form haustoria, structures which absorb nutrients and water from the cells without killing them (Figure 8.14). Further mycelium and conidiophores develop as distinct white powdery patches on the upper surface (adaxial side) of leaves and release more spores. This can occur within seven days of initial infection. Powdery mildew mycelium does not invade further into the plant from the epidermis and is not known to spread systemically within cannabis plants by infecting vascular or pith tissues. Spores adhering to the outside surface of plants can spread the disease.

Generally, *Golovinomyces* species grow best in warm climates with minimal leaf moisture, but high levels of humidity encourage pathogen growth. If conditions are unfavorable, the pathogen may produce chasmothecia (overwintering fruiting bodies containing sexual spores) on leaves

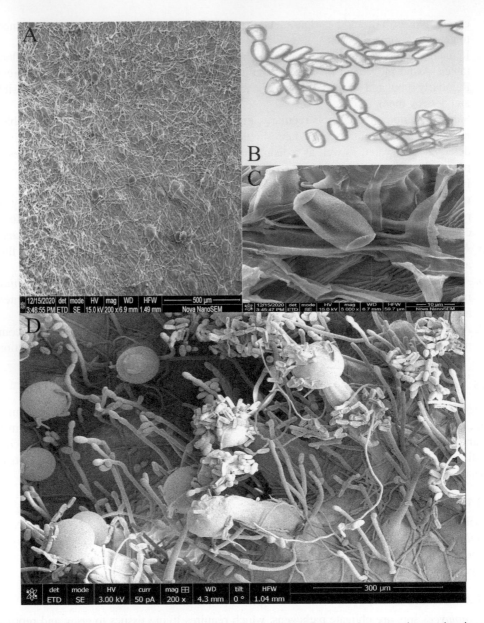

FIGURE 8.13 The biology of *Golovinomyces* spp. A) A scanning electron microscope image showing powdery mildew growing on cannabis leaves and covering trichomes; B–C) Light microscope and scanning electron microscope images, respectively, of *Golovinomyces* spp. conidia; D) A scanning electron microscope image of powdery mildew growing over trichomes. Conidiophores can also be seen extending from the tissues.

and plant debris. Chasmothecia of *Golovinomyces spadiceus* appear as very small brown or black dots on the undersides of leaves on hemp, although they have not been observed on cannabis plants.

Cannabis plants infected with powdery mildew may appear stunted, leaves may appear brown and drop prematurely, and flower quality and yields may decrease (Figure 8.15). Cannabis flower infected with powdery mildew will be of lower quality and be less marketable due to the visible growth of the fungus.

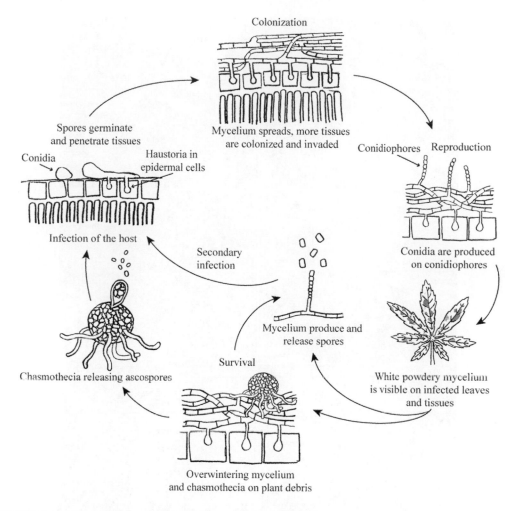

FIGURE 8.14   The disease cycle of powdery mildew on cannabis.

Source: Adapted from Agrios (2005)

8.7.3.3   Management Approaches

1. Starting planting material should be free of powdery mildew. Incoming plant material should be examined for symptoms, placed under quarantine and tested for the presence of the pathogen.
2. Minimize introduction or spread of the pathogen through plant debris or potentially spores on shoes or clothing of personnel and visitors. This requires that shoe covers, gloves, hair and beard nets, and footbaths with disinfectant be placed at entrances to growing rooms. Equipment or tools should be cleaned with a detergent and water, as well as a disinfectant. Avoid movement of workers and equipment from diseased areas to clean areas.
3. Reduce plant-to-plant spread through the use of air purification and filtration. Humidifiers, dehumidifiers and fans should be cleaned regularly. Replace filters regularly.
4. Maintain relative humidity at appropriate levels (< 50–60%) to limit infection. Minimize fluctuations in temperature and humidity. Ensure air movement in the growing environment with fans and convection tubing. Pruning, training and de-leafing plants may help to improve air movement, as well. Space plants appropriately. This will help to reduce pockets of higher humidity.

FIGURE 8.15 Signs of powdery mildew infection on cannabis plants in different stages of growth. A–C) Powdery mildew infection on leaves and flower tissues; D–E) Severe powdery mildew infection.

5. Remove infected leaf material. Limit the spread of inoculum by placing infected leaves in a sealed bag or container in the growing environment before removing them from the facility.
6. Apply products such as MilStop, ZeroTol or Regalia Maxx at rates and intervals as per the labels. Apply when disease pressure is low. Reapplication may be necessary. Sulfur-based products, biocontrols or UV light exposure may also be used to manage powdery mildew.
7. Cannabis strains that are less susceptible to powdery mildew should be used, especially when seasonal disease pressure is high.

8.7.4 BUD ROT CAUSED BY *BOTRYTIS CINEREA*

Botrytis cinerea causes diseases on a wide range of economically important crops. These include greenhouse crops such as tomato, strawberry, basil and ornamentals such as roses, as well as field or orchard crops such as cherry, grape, raspberry, apple and legumes. *Botrytis* affects almost all types of plant tissues and causes symptoms that include stem cankers, blossom blights, damping off, leaf spots, fruit rot and storage rot.

8.7.4.1 Causal Agent

On cannabis, *Botrytis* primarily infects the flowers, especially in high humidity environments. This results in soft, rotten, discolored flowers and is commonly referred to as bud rot. If this disease is not managed effectively, it can quickly spread through a growing environment, resulting in severe losses. *Fusarium* species and *Sclerotinia* may also cause bud rot, although *B. cinerea* is the predominant cause of this disease. The conidia and conidiophores of *Botrytis* have a distinct and characteristic gray color, giving it the common name "gray mold" (Figure 8.16).

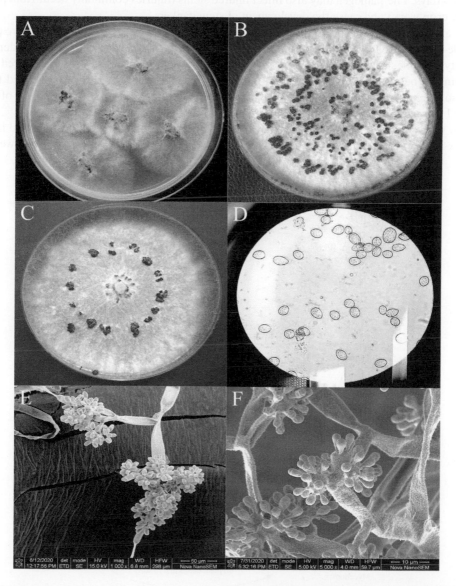

FIGURE 8.16 The biology of *Botrytis cinerea*, the main cause of bud rot on cannabis. A) *Botrytis cinerea* growing from infected flower tissues on PDA; B) *Botrytis cinerea* on PDA producing sclerotia; C) *Botrytis cinerea* on PDA producing sclerotia and characteristic gray conidiophores; D) *Botrytis* spores under a microscope; E–F) Scanning electron microscope images of *Botrytis cinerea* conidiophores.

8.7.4.2 Disease Cycle, Symptoms and Signs

Botrytis cinerea survives as mycelium in plant debris or sclerotia (small dark hardened spheres of mycelium). From these structures, conidiophores are produced which release spores in the growing environment. Spores are predominantly spread by air and germinate on flowers or other tissues if there is high humidity or free moisture. Overhead watering, pruning shears, workers and other equipment may also spread *Botrytis* spores.

Once spores have germinated, they penetrate into the tissues and release enzymes which begin to degrade the cells (Figure 8.17). Tissues may appear soft, discolored and rotten. As the pathogen continues to colonize plant tissues, it produces more spores which cause secondary disease cycles and the rapid spread of the pathogen in the growing environment. This can occur over a period of just a few days. The pathogen may also infect injured stems (injuries commonly occur from pruning or other crop work) and cause the development of stem cankers.

On cannabis plants, bud rot may start on the outside of flowers, especially if wounds are present (from de-leafing, etc.), or from the interiors of flowers, near the nodes where they are most dense and humidity is highest. Flowers infected with *Botrytis* will initially appear soft and discolored before advancing to being crisp, light brown and desiccated (Figure 8.18). Leaves near affected flowers may also appear brown and dry, and can be a characteristic symptom to scout for. Gray or off-white mycelial growth may also be observed on affected flowers.

Even with consistent scouting and removal of infected buds throughout cultivation and harvest, *Botrytis*-infected flowers may still be present in the post-harvest environment. These flowers may

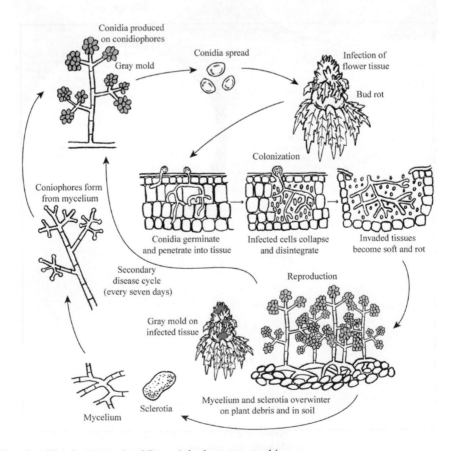

FIGURE 8.17 The disease cycle of *Botrytis* bud rot on cannabis.

Source: Adapted from Agrios (2005)

FIGURE 8.18 Symptoms of bud rot on cannabis. A) Bud rot on immature buds; B–C) Bud rot with noticeable mycelium production; D) *Botrytis cinerea* infecting multiple sites on one branch, highlighting how infections may start in areas of dense tissue and grown outwards; E–F) Severe bud rot causing complete destruction of affected tissues.

not have signs and symptoms that are the same as infections pre-harvest. Mycelial growth may spread over buds placed closely together if the relative humidity and air movement in the drying room is not managed appropriately.

8.7.4.3 Management Approaches

1. Minimize the introduction and spread of the pathogen in contaminated plant debris, or potentially spores on shoes or clothing of staff and visitors. This requires that shoe covers, gloves, hair and beard nets, and footbaths with disinfectant be placed at entrances to growing rooms. Equipment or tools should be cleaned with a detergent and water, as well as a disinfectant. Avoid movement of workers and equipment from diseased areas to clean areas.
2. Reduce spread of spores in the growing environment through the use of air purification and filtration systems. Humidifiers, dehumidifiers and fans should be cleaned regularly. Replace filters regularly.

3. Maintain relative humidity at levels below 50–60% to limit infection. Minimize fluctuations in temperature which can cause rapid changes in humidity and condensation on the surface of leaves and flowers. Provide air movement in the growing environment with fans and convection tubing. Pruning, training and de-leafing plants may help to improve air movement, as well. Space plants appropriately. This will help to reduce areas of higher humidity.

4. Remove infected flowers as soon as they appear. Limit the spread of inoculum by placing flowers in a sealed bag or container in the growing environment before removing them from the facility.

5. Select strains that are not known to produce overly dense inflorescences, as they tend to trap moisture and maintain a high humidity that allows spores to germinate. Pruning and training may also help to reduce the size of flowers. Strains that are susceptible may be planted when seasonal disease pressure is low.

8.7.5 POST-HARVEST DECAY

8.7.5.1 Causal Agents

Cannabis flowers are also susceptible to rot or decay post-harvest from a variety of fungi including *Botrytis cinerea*, *Penicillium* species, *Fusarium* species, *Aspergillus* species and *Cladosporium westerdijkieae* (Figure 8.19, Figure 8.20). Post-harvest decay affects yields and the quality of flowers, and may cause batches to exceed the acceptable limits for total yeast and mold counts.

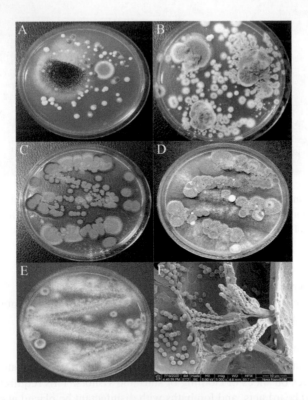

FIGURE 8.19 Fungi that are found post-harvest on cannabis flowers. A) *Aspergillus niger*, *Fusarium oxysporum*, a *Penicillium* species, bacteria and yeast on PDA from a swab of a post-harvest bud; B) *Penicillium* and *Aspergillus flavus* growing from post-harvest flower tissues; C) A swab of post-harvest flower showing the presence of *Penicillium* and *Cladosporium*; D) A swab of post-harvest flower showing the presence of *Botrytis cinerea*, *Aspergillus flavus* and *Penicillium*; E) *Fusarium* and *Penicillium* growing from a swab taken from infected post-harvest flower tissue; F) A scanning microscope image of the conidiophores of *Penicillium*.

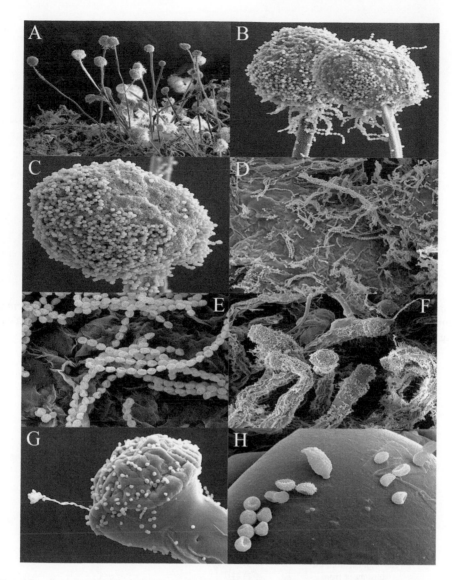

FIGURE 8.20 Scanning electron microscope images of *Aspergillus* and *Penicillium* on cannabis flowers. A–C) The spore heads of *Aspergillus* on conidiophores; D) *Penicillium* colonizing and sporulating on flower tissues; E) Chains of *Penicillium* spores; F) *Penicillium* spores on trichomes; G–H) Close-up of *Penicillium* spores on trichomes.

8.7.5.2 Symptoms and Signs

Wounds from harvesting and trimming flowers, especially excessive wounding from automated trimming machines, provide openings for these fungi to colonize flower tissues. Typically, post-harvest infections may begin to be observed within 3–6 days of starting the drying period, with species such as *Penicillium*, *Cladosporium* and *Aspergillus* causing discoloration and decay of tissues. Small patches of white mycelium may also be observed (Figure 8.21).

Fungi may spread during trimming and drying through the air as spores, through contaminated equipment or tools, or on the hands of workers (Figure 8.22). During the wet trim process, wounding of flowers can cause a buildup of spores that can progress to mold development during the drying process. Dry trimming or hand trimming may reduce the buildup of spores. Unsanitary

FIGURE 8.21 Signs of post-harvest decay. A) *Fusarium* growing on drying buds; B) *Botrytis cinerea* and *Penicillium*, indicated by an arrow, growing on drying buds; C) White spots of *Penicillium* growing on numerous drying buds; D) *Penicillium* on a drying bud.

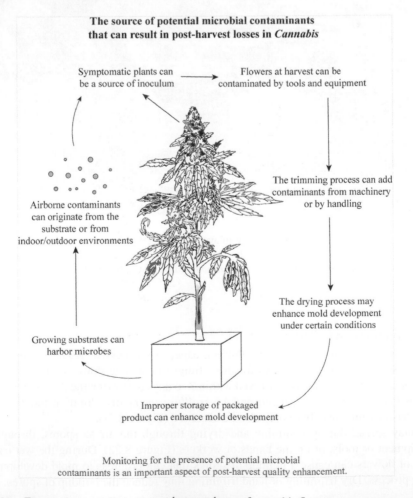

FIGURE 8.22 Factors that contribute to post-harvest decay of cannabis flowers.

equipment or excessive aerial contamination and inappropriate drying conditions all contribute to post-harvest losses of cannabis.

8.7.5.3 Management Approaches

1. Harvested flowers should examined and be free of visible bud rot.
2. Minimize introduction or spread of contaminants. This requires that shoe covers, gloves, hair and beard nets, and footbaths be utilized. Equipment or tools should be cleaned with a detergent and water, as well as a disinfectant. Special attention should be given to trimming machines, including regularly cleaning inside in areas where plant debris may build up. Food-safe degreasers may be effective to clean areas where resin and plant debris build up. Air filtration and purification should also be used in trimming rooms, as the buildup of spores and particulate matter can be high during these operations.
3. Avoid excessive physical damage to buds during harvesting and pruning.
4. Maintain optimal humidity, temperature and airflow in the drying area. Air filtration and purification should be used in drying rooms.
5. Removed tissues that show signs of infection post-harvest.

TABLE 8.2
The Most Important Pathogens Currently Affecting Cannabis Production and Their Management Options

Common Name of Disease	Pathogen(s)	Management Options
Damping off	*Fusarium oxysporum*	Have appropriate humidity levels
	Fusarium proliferatum	Improve air circulation
	Fusarium solani	Apply biocontrols during propagation
		Remove diseased cuttings
		Replace stock plants regularly
***Fusarium* root and crown rot**	*Fusarium oxysporum*	Test mother plants to ensure they are free of disease
	Fusarium proliferatum	Replace stock plants regularly
	Fusarium solani	Apply biocontrols during propagation and vegetative stages of growth
		Avoid injury to roots
***Pythium* root and crown rot**	*Pythium myriotylum*	Avoid overwatering
	Pythium dissotocum	Avoid injury to roots
	Pythium aphanidermatum	Apply biocontrols during propagation and vegetative stages of growth
		Treat irrigation water with UV or chlorine
Powdery mildew	*Golovinomyces* spp.	Improve air circulation
		Manage temperature and humidity
		Prune and de-leaf plants
		Apply products such as Regalia Maxx, MilStop, ZeroTol or others
		Treat plants with UV-C light
		Remove and destroy diseased leaves
		Plant resistant varieties
Bud rot	*Botrytis cinerea*	Improve air circulation
	Fusarium spp.	Manage temperature and humidity
		Remove diseased buds
		Avoid varieties with dense inflorescences
Post-harvest decay	*Botrytis cinerea*	Maintain appropriate humidity, temperature and air movement in drying rooms
	Penicillium spp.	Avoid damage to buds during harvesting and trimming
		Removed infected post-harvest buds

8.8   EMERGING PATHOGENS IN CONTROLLED ENVIRONMENTS

In addition to the major pathogens of cannabis discussed previously, there is the potential for other pathogens reported on hemp or cannabis to become more prevalent and widespread. It is unclear to what extent diseases reported on hemp produced in the field may cause disease in indoor or greenhouse growing environments, but growers should be aware of alternate sources of inoculum from crops grown in proximity to cannabis.

8.8.1   *SCLEROTINIA* HEMP CANKER AND BUD ROT

Sclerotinia species have been reported to cause hemp canker and crown rot on hemp in California, Kentucky, New York, Alberta and New Brunswick. Major hosts of *Sclerotinia* spp. include soybean, canola, potato, sunflower and other vegetable crops. Symptoms of infection on hemp include wilt, browning or bleaching of foliage, and dry discolored lesions on the stem or crown of the plant. Small, black sclerotia may be present externally on lesions or internally when tissues are broken open. Cottony white mycelium may also be observed, which is a characteristic feature of *Sclerotinia* infections and gives it the common name "white mold." Infections on inflorescences from *Sclerotinia* can also occur on hemp and cannabis (Figure 8.23).

FIGURE 8.23   Some of the emerging and less common pathogens of cannabis and hemp. A) Symptoms of hop latent viroid on a cannabis plant growing vegetatively; B) An example of a mosaic type symptom commonly caused by viruses; C–D) Infections caused by *Botrytis* and other fungi on cuttings; E) *Botrytis* infecting the stem of a cannabis plant; F) *Sclerotinia* growing from infected cannabis flower tissues.

8.8.2 Southern Blight

Sclerotium rolfsii is the causal agent of southern blight on hemp and numerous other crops such as apple, peanut, potato, tomato, beans, cereals and cotton. Southern blight has been reported on hemp in Kentucky, Alabama, Virginia, Tennessee and North Carolina, as well as in parts of Italy.

Symptoms include wilting and yellowing of mature plants, discoloration and rot of the crown and death of plants. White mycelium can often be seen growing out from the base of infected plants and along the soil surface. Tan or off-white sclerotia may also be produced on diseased tissues. Symptoms of southern blight are worsened by warm dry weather.

8.8.3 *Rhizoctonia* Root and Stem Rot

Root and stem rot, caused by *Rhizoctonia solani*, has been reported on hemp in North Carolina and Kentucky. Symptoms include root rot, necrotic lesions on the stems of plants at the soil line, and wilting. *Rhizoctonia* also can reportedly cause damping off, as well as web blight, which causes foliage and flowers to die back rapidly and may leave a tan or off-white mycelial webbing on affected tissues. Alternate hosts of *Rhizoctonia solani* include rice, beans, soybean, corn, cotton, wheat, turf grass and greenhouse crops such as tomato, pepper and eggplant.

8.8.4 *Neofusicoccum* Stem Canker and Dieback

Hemp plants in Italy and the United States (Arkansas) have been reported to be affected by *Neofusicoccum parvum*, which causes symptoms of leaf curl and leaf discoloration, as well as stem canker and dieback on the main stem and branches. Stem cankers appeared as sunken tan lesions. White mycelium and black pycnidia were also visible on lesions. *Neofusicoccum* often caused the death of plants. These symptoms have been observed and replicated on a range of cultivars of hemp, including Carmagnola, Bejko, Cherry Wine, Cherry Blossom and Berry Blossom. This pathogen has a broad host range and also infects crops such as grape, strawberry, stone fruits and ornamental trees.

8.8.5 Fungal and Bacterial Leaf Spots

Leaf spots on hemp have been reported to be caused by fungal pathogens belonging to the genera *Septoria*, *Bipolaris* and *Cercospora*, and by bacterial pathogens such as *Xanthomonas* spp. These pathogens predominantly occur on hemp in the warm and humid southeastern states of the United States, including Kentucky, Florida and North Carolina. Symptoms may appear as circular or angular spots, with or without a distinct border on leaves. There may also be a yellow halo around the spots. Some species produce pycnidia on infected tissues, which appear as black flecks on the diseased area. As the disease becomes more severe, spots may coalesce. Alternate hosts of these pathogens and possible sources of inoculum include cereals and grains, as well as citrus and soybean.

8.8.6 Viruses and Viroids

Beet curly top virus (BCTV) has been reported on hemp in western Colorado. Symptoms start as a fading of leaf color to a pale green, which extends from the base of the leaves to the tips. A yellow and green mosaic pattern also develops. As the disease progresses, symptoms affect the entire plant, including new growth, and leaves will take on a curled, narrow appearance. The plant may appear stunted, curled and distorted. This virus has been reported to affect numerous different cultivars of hemp at various stages of growth. Alternate hosts of BCTV include sugar beet, beans, pepper, spinach and tomato. BCTV is spread only through leafhoppers.

Another virus that affects cannabis plants is lettuce chlorosis virus (LCV), which to date has been reported in Israel and Canada. Affected plants have symptoms of interveinal chlorosis, which can cause leaves to appear chlorotic and bright yellow throughout, as well as partial necrosis and brittleness. Transmission of LCV was confirmed to be through *Bemisia tabaci* whiteflies and not through seed. Vegetative cuttings taken from infected mother plants showed symptoms that were often more severe than the symptoms observed on the mother plants. This particular isolate of LCV, currently named LCV-Can, was confirmed to infect two different varieties of lettuce, as well as rose periwinkle. Other plants susceptible to LCV include lettuce, beets, tobacco and weeds such as shepherd's purse and hemlock.

Hop latent viroid was first reported on cannabis plants in California, and has spread to growing regions in North America and likely elsewhere. A viroid, unlike a virus, does not have a protective outer coating (a capsid or envelope) and are only RNA. Affected plants may have chlorotic and distorted leaves, brittle stems, an abnormal or stunted pattern of growth, and reduced yield (Figure 8.23). Flowers may also "dud" as the viroid can cause a reduction in flower development.

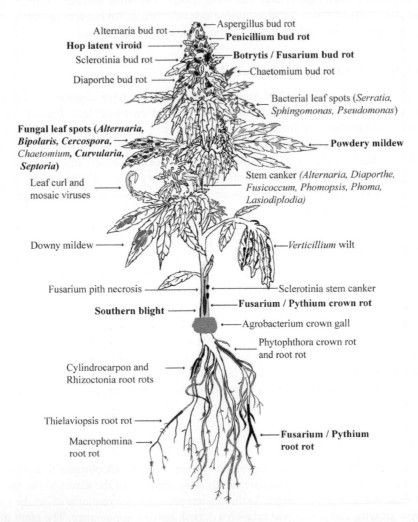

FIGURE 8.24 The currently recognized pathogens on cannabis and hemp plants. Pathogens shown in bold are the most damaging.

Source: (Punja 2021)

Infected plants may appear asymptomatic for prolonged periods of time before showing symptoms. Hop latent viroid is known to spread mechanically and through seed. It is believed to have originated from hop plants, in which it causes no symptoms. The extent of damage caused by HPLv is currently unknown, but it is of concern to the industry at large.

8.9 ACKNOWLEDGMENTS

We thank Samantha Lung for assistance with the preparation of disease cycle figures, and Alastair Roberts for his assistance with assessing the efficacy of ZeroTol and Chemprocide in Figure 8.3. This research was funded by the Natural Sciences and Engineering Research Council of Canada.

BIBLIOGRAPHY

Agrios, G. N. 2005. *Plant pathology*. Amsterdam: Elsevier Academic Press.

Bains, P. S., H. S. Bennypaul, and S. F. Blade. 2000. First report of hemp canker caused by *Sclerotinia sclerotiorum* in Alberta, Canada. *Plant Disease* 84: 372.

Bektaş, A., K. M. Hardwick, K. Waterman, and J. Kristof. 2019. Occurrence of hop latent viroid in *Cannabis sativa* with symptoms of cannabis stunting disease in California. *Plant Disease* 103: 2699.

Chatterton, S. S., and Z. K. Punja. 2009. Chitinase and β-1,3-glucanase enzyme production by the mycoparasite *Clonostachys rosea* f. *catenulata* against fungal plant pathogens. *Canadian Journal of Microbiology* 55: 356–367.

Cummings, J. A., C. A. Miles, and L. J. du Toit. 2009. Greenhouse evaluation of seed and drench treatments for organic management of soilborne pathogens of spinach. *Plant Disease* 93: 1281–1292.

Food and Agriculture Organization of the United Nations (FAO). 2021. New standards to curb the global spread of plant pests and diseases. https://www.fao.org/news/story/en/item/1187738/icode/

Gauthier, N., K. Leonberger, and C. Bowers. 2019. *Production and Pest Management*. Proceedings of the 1st Annual Scientific Conference of the Science of Hemp, Lexington, Kentucky, October 10–11.

Giladi, Y., L. Hadad, N. Luria, W. Cranshaw and O. Lachman. 2020. First report of beet curly top virus infecting *Cannabis sativa* in Western Colorado. *Plant Disease* 104: 999.

Hadad, L., N. Luria, E. Smith, N. Sela, O. Lachman, and A. Dombrovsky. 2019. Lettuce chlorosis virus disease: A new threat to can nabis production. *Viruses* 11. https://doi.org/10.3390/v11090802.

Koike, S. T., H. Stanghellini, S. J. Mauzey, and A. Burkhardt. 2019. First report of sclerotinia crown rot caused by *Sclerotinia minor* on hemp. *Plant Disease* 103: 1771.

Majsztrik, J. C., R. T. Fernandez, P. R. Fisher, D. R. Hitchcock, J. Lea-Cox, J. S. Owen Jr., L. R. Oki et al. 2017. Water use and treatment in container-grown specialty crop production: A review. *Wat er, Air & Soil Pollution* 228. https://doi.org/10.1007/s11270-017-3272-1.

Mersha, Z., M. Kering, and S. Ren. 2020. Southern blight of hemp caused by *Athelia rolfsii* detected in Virginia. *Plant Disease* 104: 1562.

Orr, R., and P. N. Nelson. 2018. Impacts of soil abiotic attributes on fusarium wilt, focusing on bananas. *Applied Soil Ecology* 132: 20–33.

Punja, Z. K. 2018. Flower and foliage-infecting pathogens of marijuana (*Cannabis sativa* L.) plants. *Canadian Journal of Plant Pathology* 40: 514–527.

Punja, Z. K. 2021. Emerging diseases of *Cannabis sativa* and sustainable managem ent. *Pest Management Science*. https://doi.org/10.1002/ps.6307.

Punja, Z. K., D. Collyer, C. Scott, S. Lung, J. Holmes, and D. Sutton. 2019. Pathogens and molds affecting production and quality of *Cannabis sativa* L . *Frontiers in Plant Science*. https://doi.org/10.3389/fpls.2019.01120.

Punja, Z. K., C. Scott, and S. Chen. 2018. Root and crown rot pathogens causing wilt symptoms on field-grown marijuana (*Cannabis sativa* L.) plants. *Canadian Journal of Plant Pathology* 40: 528–541.

Scarlett, K., L. Tesoriero, R. Daniel, and D. Guest. 2014. Sciarid and shore flies as aerial vectors of *Fusarium oxysporum* f. sp. *cucumerinum* in greenhouse cucumbers. *Journal of Applied Entomology* 138: 368–377.

Scott, C., and Z. K. Punja. 2020. Evaluation of disease management approaches for powdery mildew on *Cannabis sativa* L. (marijuana) plants. *Canadia n Journal of Plant Pathology*. https://doi.org/10.1080/0 7060661.2020.1836026.

Sutton, J. C., C. R. Sopher, T. N. Owen-Going, W. Liu, B. Grodzinski, J. C. Hall, and R. L. Benchimol. 2006. Etiology and epidemiology of pythium root rot in hydroponic crops: Current knowledge and perspect ives. *Summa Phytopathologica*. https://doi.org/10.1590/S0100-54052006000400001.

Szarka, D., B. Amsden, J. Beale, E. Dixon, C. L. Schardl, and N. Gauthier. 2020. First report of hemp leaf spot caused by a *Bipolaris* species on hemp (*Cannabis sativa*) in Kentucky. *Plant Health Progress* 21: 82–84.

Tian, X., and Y. Zheng. 2013. Evaluation of biological control agents for fusarium wilt in *Hiemalis begonia*. *Canadian Journal of Plant Pathology* 3: 363–370.

Warren, J. G., J. Mercado, and D. Grace. 2019. Occurrence of hop latent viroid causing disease in *Cannabis sativa* in California. *Plant Disease* 103: 2699.

Willsey, T., S. Chatterton, and H. Cárcamo. 2017. Interactions of root-feeding insects with fungal and oomycete plant pathogens. *Frontiers in Plant Science*. https://doi.org/10.3389/fpls.2017.01764.

9 Management of Insect Pests on Cannabis in Controlled Environment Production

Jason Lemay and Cynthia Scott-Dupree

CONTENTS

DOI: 10.1201/9781003150442-9

9.1   IPM IN CANNABIS

Like many specialized agricultural topics, insect pest management in cannabis is significantly understudied. There is very little validated and cannabis-specific information for managing pests. While there are many knowledgeable legacy and legal market growers, many sources of information—including online forums and personal websites—are littered with misidentified insects, inappropriate biological control recommendations and ineffective, unregistered or even dangerous chemical control suggestions. In this chapter, we will introduce the concept of integrated pest management (IPM) and common arthropod (insect and mite) pests associated with cannabis production in controlled environment. Unfortunately, due to a lack of cannabis-specific knowledge at this time, information presented in this chapter is largely generalized based on what is known from other crops in controlled environment. As cannabis-specific knowledge is developed, the information and recommendations presented within this chapter may soon be outdated. Always consult a biological control supplier, pest management consultant or local extension agent for the most current information.

Integrated pest management (IPM) is a framework developed in the 1950s in response to the growing insecticide resistance problem in agriculture. A commonly accepted definition of IPM as per the Food and Agriculture Organization of the United Nations (2022) is:

> The careful consideration of all available pest control techniques and subsequent integration of appropriate measures that discourage the development of pest populations and keep pesticides and other interventions to levels that are economically justified and reduce or minimize risks to human health and the environment. IPM emphasizes the growth of a healthy crop with the least possible disruption to agro-ecosystems and encourages natural pest control mechanisms.

Integrated pest management is a decision-based system that requires both a biological understanding of the pest and specific monitoring protocols. With IPM, the idea of *eradication* is replaced by the idea of *management*. Integrated pest management strives to maintain pest populations below levels that cause losses greater than the cost of managing that pest (Figure 9.1). This is known as the economic injury level (EIL). Management actions should be taken before a pest reaches the EIL, and this is called the economic or action threshold (ET). A management action is a reactive measure and could include the release of a biological control agent (i.e., beneficial insects), a pesticide application or a physical control measure. Developing specific ETs or EILs can be difficult and are not universally applicable. It is up to growers to determine when pest levels are causing economic damage in their facility. Due to a low tolerance for contamination in the cannabis flower, as well as the high value of the crop, the ET for most insect pests in cannabis is low. A management action should therefore be initiated upon the initial detection of the pest of concern.

Monitoring, or scouting, is the cornerstone of any IPM program. Before you can undertake a management action, you need to know what pests and beneficial insects are present in your crop. Scouting is divided into two components: *inspection* and *identification*. Inspection of the crop needs to occur at frequent intervals, at least weekly. Insect populations can change rapidly, especially when temperatures are high. Simply using traps to monitor for insects is not sufficient, as many pests are not trapped reliably by sticky cards or other traps. Rather, traps should be used to supplement frequent crop inspections. How and when scouting occurs should be defined in a standard operational procedure (SOP). This ensures that scouting remains consistent, regardless of who is scouting the

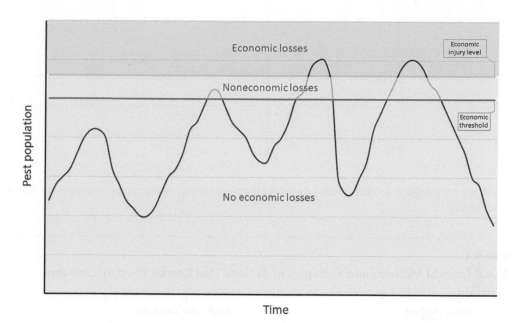

FIGURE 9.1 Pest populations over time based on scouting data that indicates when pest populations reach the economic threshold (blue line) and economic injury level (yellow line). Management action would be undertaken when pest populations reach the economic threshold to prevent pests from causing economic losses.

Source: Copyright Jason Lemay, University of Guelph

crop. When a pest is found, the location should be marked, either with a colored flag or with colored flagging tape. Effective scouting allows you to find pests and track their population growth, as well as monitor the effectiveness of any management actions initiated. At a minimum, scouts should be equipped with a 10–20× hand lens, and ideally also have access to a 40–60× microscope. These tools allow for the second critically important component of scouting: identification.

IPM relies upon using information about specific pests in the crop to determine if and when a management action is needed to prevent economic losses. This chapter will provide information to help identify pests and understand the biology of the insects you may find in a controlled environment cannabis production facility. In most cases, a species-level identification is necessary to determine the appropriate management action. When scouting, taking detailed notes is very important. For both pests and beneficial insects, important data to record includes the date, precise location, cultivar, location on the plant, number of insects and their life stage, current environmental conditions, recent management actions and any observations on plant damage or stress. A standardized form should be developed and used to record information when scouting (see Figure 9.2 for an example). Alternatively, various electronic scouting applications have been development, facilitating the digitization of scouting records and data. Using electronic scouting applications digitizes the data at the point of collection, which helps minimize input errors. They can also archive the data, preventing data losses and facilitating historical comparisons. This information can help you identify trends in insect populations and help predict when or where pest outbreaks can occur—and consequently, help adapt future management actions. If an IPM program is only as good as its scouting program, then a scouting program is only as good as the details in the recorded observations.

Management tactics used in IPM can be placed into various categories: cultural, physical, biological and chemical. Genetic control is an underused management tactic which is experiencing a renaissance in terms of interest by IPM experts. Novel technologies like RNA interference, sterile insect technique, and CRISPR are all considered genetic controls, and their application in agriculture is receiving a lot of attention. Their potential application in cannabis IPM is still being investigated.

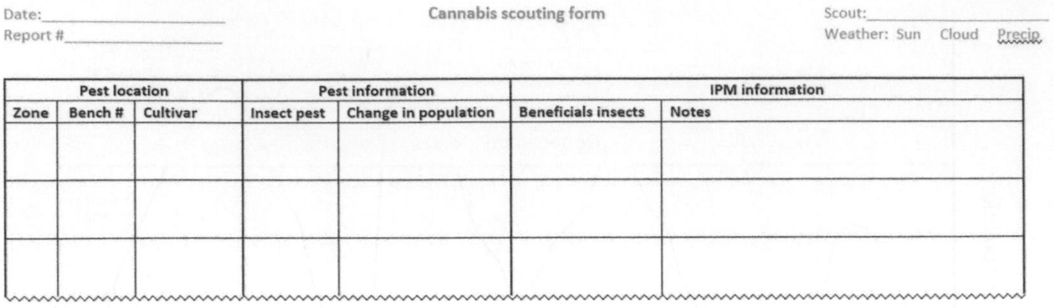

Date:__________ Cannabis scouting form Scout:__________

Report #__________ Weather: Sun Cloud Precip

Pest location			Pest information		IPM information	
Zone	Bench #	Cultivar	Insect pest	Change in population	Beneficials insects	Notes

FIGURE 9.2 Example of a form that can be used to take notes when scouting to make sure relevant information is recorded.

TABLE 9.1
Cultural Control Methods and Examples of Actions That Can be Used in Cannabis Production

Cultural Control Method	Applicable Example
Establishing biosecurity protocols	Quarantining incoming plant material until it can be inspected for the presence of pests or pathogens
	Having clean, workplace-only clothes and foot baths
	Controlling access to stock plants and the production environment to prevent introducing pests
	Organizing tasks to work from clean areas to infested areas
Proper sanitation both during and between crops	Removing heavily infested, dead or dying plant material
	Sweeping debris off the floor or growing area
	Disinfecting the growing area between crops
	Using a crop-free period to kill pests
	Ensure that no weeds or other plants that can harbor pests are present
	Ensure cannabis waste is disposed of immediately
Following proper production methods	Producing a healthy crop capable of tolerating biotic stress
	Proper rootzone management (e.g., irrigation) to prevent root diseases and fungus gnat development
	Not overfertilizing the crop, as this can lead to increases in pest populations
Maintaining environmental conditions less favorable for pest development	Maintaining higher relative humidity to suppress spider mites
	Managing soil moisture and eliminating puddles from forming to prevent fungus gnat development
	Remove weeds that can harbor pests
	Consider longer photoperiods (> 12 hours) to prevent beneficial insects from diapausing
Selecting resistant or less susceptible plant varieties	There are no cannabis cultivars with known resistance to insect pests at this time
	Some pests show preference for certain cultivars (i.e., cannabis aphid prefers cv. 'Blueberry' and other fruity or citrus smelling cultivars)

9.1.1 Cultural and Physical Control

Cultural controls are preventative measures taken to ensure a healthy crop that has increased tolerance to pests or made the crop environment less favorable for pests or more favorable to beneficial insects. Examples of cultural control are listed in Table 9.1. Cultural controls are often seen as the first steps in preventing pests from entering a facility or becoming established in the crop.

Biosecurity and *sanitation* are two of the most important cultural controls to establish in a production facility and can reduce the potential for pest outbreaks and further management actions. All incoming plant material should be inspected under a microscope for the presence of insects, including eggs, and any signs of pathogens. Plant material should then be kept isolated for 24–72 hours and re-inspected before being moved into production areas. Screening vents with the appropriately sized mesh can prevent pests from entering the facility through open vents. This can also reduce airflow, so plan appropriately. Humans can also bring pests into the facility. Clean workplace-only clothing and restricted access to production areas can prevent the introduction of pests brought in inadvertently by employees. These practices will greatly reduce—but not eliminate—the possibility of pests entering the production facility. Proper workflow is also important as a cultural control. Always work from areas free of pests and move toward infested areas to minimize the dispersal of pests and pathogens by workers and equipment.

Sanitation is essential to prevent movement of pests between crops, including when multiple crop cycles occur simultaneously. Avoid placing new batches of plants near older plants, as pests can quickly disperse to the new plants and perpetuate the pest cycle. Keeping each batch of plants isolated from each other and using a crop-free periods between each crop can prevent pest populations from continuing in the next batch. A crop-free period should be scheduled at least once a year. For most insect pests of cannabis, a week without a host plant can cause them to die off. Weeds or other non-crop plants can harbor pests between crop cycles. Maintaining an ambient temperate above 40°C and relative humidity below 50% for 3–4 days during the crop-free period can further reduce pest populations. When a prolonged crop-free period is not used, use the time between crops to disinfect as much of the growing area as possible. This includes cleaning troughs or tables, drippers and lines, as well as training wires, trellises or any plant supports. Disinfection is an important part of the overall facility sanitation. Equipment such as pruners, or shears, and sprayers should be sanitized frequently, and especially before moving to the next production area in the greenhouse.

Cultivar selection can be an important part of pest management. Unfortunately, there are currently no cultivars with known resistance to pests. Some cultivars may be more susceptible to pests. Proper note taking while scouting can help you determine what cultivars are more prone to pest outbreaks.

Physical controls can be preventative or reactive but involve physically killing, removing or preventing pests. This includes using screens on vents or openings, aspirating or vacuuming pests, removing pests by hand, de-leafing, and mass trapping (including sticky cards or tape traps, pheromone baited traps or electric insect traps such as UV lights and bug zappers). When de-leafing, place the removed foliage directly into bags to prevent pests from returning to the crop. Physical controls are often labor-intensive and are rarely efficient, with the exception of vent screening, which may be the most efficient physical control option. Physical controls also provide a rapid reduction in pest numbers, which can in turn improve the efficacy of other management actions.

9.1.2   BIOLOGICAL CONTROL

Biological control is the use of living organisms to control pests. These organisms are referred to as biological control agents (BCA). They are predators, parasitoids, pathogens or competitors of pests. Biological control recommendations, such as selection of BCA or release rates, are often crop- and pest-specific, and the crop production methods and environment can further influence these recommendations and their effectiveness. Due to a lack of cannabis-specific information, we will not provide specific release rates in this chapter. It is best to seek this information directly from the BCA supplier you will be utilizing in your facility. Biological control is not a "one-size-fits-all" solution, nor is it static. Even when validated recommendations or strategies have been developed, biological control programs must be tailored for specific facilities. Successful biological control programs are constantly adapting to changes in pest communities, environmental conditions, production methods, etc.

How BCAs perform on cannabis is also mostly unknown. For example, the effects of trichome densities on predatory mites has been evaluated in crops such as tomatoes and chrysanthemums, but not in cannabis. Studies have found that trichome densities greater than 100–200 trichomes/cm^2 can decrease the walking speed in various predatory mites (BCAs). Similarly, the volatile terpenes produced by cannabis may interfere with the prey-finding ability of BCAs. This knowledge is crucial to determine optimal release rates for BCAs in cannabis. The effectiveness of BCAs is dependent on many factors such as pest population, production system, use of supplemental lighting, plant stage, cultivar, environmental conditions and pesticide use. Novel pests such as cannabis aphid and hemp russet mite pose even greater challenges, as we have no recommendations from other crops on which to base our control strategies.

Most biological control of arthropod pests (i.e., insects and mites) involve the use of predators and parasitoids. Predators consumer more than one prey, while parasitoids consume just one individual over the course of their development. Predators with chewing mouthparts will typically consume the entire prey, while predators with piercing/sucking mouthparts only consume the soft insides of their prey, leaving behind the exoskeleton carcass. Parasitoids differ from parasites by killing their host during development, while parasites only weaken their host.

There are two approaches to biological control in controlled environments: the inoculation approach and the inundation approach. The inoculation approach can be thought of as a preventative program or proactive response to pests that could be in the crop. The goal is to establish BCAs populations in the growing facility prior to pests being present. This preventative approach gives you the best chance at maintaining pests below the EIL. Certain predatory mites such as *Amblyseius swirskii* and *Amblyseius andersoni* are available in controlled release sachets also called breeding system sachets. These are small sachets with the predatory mite, a food mite and a carrier that serves as a food source for the food mite. This way, the predatory mite has prey available and a safe place to breed. The predatory mites leave the sachets and disperse into the crop over time. This system is helpful when pest populations in the crop are too low to sustain the predatory mites. This release method produces thousands of mites from a single sachet over the course of 4–6 weeks. Similarly, some aphid parasitoids (e.g., *Aphidius colemani* and *Aphidius matricariae*) can be released preventatively and maintained on banker plants in the absence of prey. Banker plants are non crop plants infested with insects that are not a pest of the crop being grown. Banker plants for *Aphidius* typically consists of wheat, oat or barley infested with bird cherry-oat aphid (*Rhopalosiphum padi*). This aphid has a restricted host range, so it will not feed on most crop plants, but provides a host for parasitoids in the absence of pestivorous aphids in the crop. Other factitious foods such as pollen, nonviable moth eggs and shrimp cysts can be used to provide food to support generalist predators when pests are not present in the crop. Using factitious food in cannabis production can be difficult, as they are often applied directly on the plant, possibly contaminating the cannabis flowers. Not all BCAs can be used in inoculative biological control programs. For example, *Phytoseiulus persimilis* cannot be maintained on factitious food sources in the absence of spider mites.

Conversely, the inundation approach is a reactive response to pests that are establishing in the crop. Inundative biological control involves releasing BCAs in response to pest hotspots. A greater selection of BCAs are available for inundative biological control programs, as the pests necessary to sustain them are present in the crop. The goal of this approach is to rapidly overwhelm the pests with BCAs. However, this leads to unsustainable populations of BCAs as pest populations decline, and subsequent BCA releases are required to manage future pest outbreaks.

Occasionally, biological control will fail to maintain pests below the ET and further management is required. This can occur for a variety of reasons, but seasonality is a common factor. Many BCAs enter diapause, a state of suspended activity, development, or reproduction, when day length or temperatures get too low. In most cases, insects in the adult stage will enter diapause when day lengths are < 12–14 hours and/or temperatures drop too low (the specific critical temperature varies between species, but is generally below temperatures experienced in cannabis production). In many cases, only the adult life stage will diapause, but this can still result in reduced population growth or a lack of subsequent generations. Juvenile insects will continue to feed and develop. Biological control suppliers try to select strains or rearing methods that minimize diapause, but it can be very difficult to eliminate entirely. Diapause is not a binary state across the entire population; as day length and temperature near the critical point, a greater percentage of the adult population enters diapause. Furthermore, the lack of pollen or nectar produced by cannabis plants in drug-type production systems can reduce the longevity and fecundity, or prevent oviposition, in some BCAs. When possible, source BCAs from local producers, as long-term storage and extended travel with fluctuating environmental conditions can reduce the effectiveness and longevity of BCAs compared to fresh and locally reared BCAs. Another reason biological control programs can fail is due to poor quality in the BCAs. These are living organisms and can be sensitive to adverse conditions. Incoming shipments of BCAs should be inspected to confirm their quality and viability. Detailed guides on how to complete quality assurance checks have been compiled by Buitenhuis (2017) in *Grower Guide: Quality Assurance of Biocontrol Products*, and by van Lenteren (2003) in *Quality Control and Production of Biological Control Agents: Theory and Testing Procedures*.

9.1.3 CHEMICAL CONTROL

When cultural, physical and biological controls fail to maintain pests below the ET, chemical control is often the next step. However, most products registered for use on cannabis should not be relied upon to save a crop from pest outbreaks. Due to the variability between and frequent changes within different jurisdictions, we will not be providing specific chemical control recommendations for the management of insect pests in cannabis. Very few chemical pesticides are registered for use on cannabis. Insecticidal soaps, horticultural oils and microbial biopesticides make up most of the registered insecticides at this moment in Canada, and elsewhere. Chemical control in jurisdictions without legal cannabis production (i.e., the United States) can be more complicated. Furthermore, before spraying any pesticides on cannabis, it is important to: 1) check local regulations to ensure it is legal to apply the product; 2) read the product label to know the rate, application method, and safety requirements (re-entry interval, pre-harvest interval, required PPE, etc.); 3) determine the compatibility of the product with BCAs in the crop—many suppliers offer compatibility charts; and 4) test the product on a small number of plants of each cultivar to determine if it damages the plant (i.e., phytotoxicity).

Most chemical control options in cannabis require excellent coverage, including on the underside of the foliage. Sufficient pressure from the spray nozzle is necessary to penetrate the canopy. De-leafing can help the pesticides reach further into the canopy. Water sensitive paper can be used to test the coverage of sprays in different areas of the canopy.

Microbial biopesticides are chemical controls that use living organisms to infect and kill pests. In this chapter, we will place them in the chemical control section to separate them from macrobial BCAs (i.e., insects, mites, nematodes). Common microbial biopesticides include *Bacillus thuringiensis* (commonly known as Bt), *Beauvaria bassiana*, *Isaria fumosorosea*, *Lecanicillium lecanii* and *Metarhizium anisopliae*. Other than Bt, these are all fungi. These products have differing compatibility with BCAs, so be sure to check the compatibility charts offered by most suppliers. These products are also incompatible with most fungicides, and compatibility charts for these products are also available. Many microbial biopesticides require warm and humid environments to be effective.

9.2 ARTHROPOD PESTS OF CANNABIS

A common urban myth is that cannabis is not susceptible to insect pests due to the insecticidal properties of its essential oil. Unfortunately, this is far from true. In fact, there are even insects that are not known to feed on any plants other than cannabis. Due to the historical legal nature surround cannabis, very little is known about the pest complex associated with this plant. With changes in the 2014 Farm Bill in the United States and the 2018 legalization of recreational cannabis in Canada, there is an increased opportunity to study aspects of cannabis production, including insect pests. While some there is considerable overlap between pests of cannabis and hemp, due to major differences in cultivation methods, morphological and biochemical properties, and end-use products, the

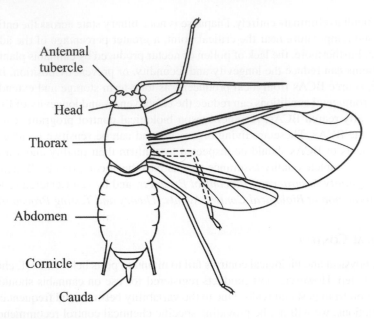

FIGURE 9.3   Aphid morphology displaying the three segments of an insect body, as well as important body parts used to identify aphid species in this chapter.

Source: Copyright Tiffany Yau

two crops may need to be managed differently. For example, cannabis grown for dried flower (i.e., smokable flower) has a much lower tolerance for insect residue or other contaminates than cannabis grown for cannabinoid extraction or industrial hemp.

Most of the arthropods mentioned in this chapter are insects, yet there are a few that are not insects (i.e., mites). However, for the purpose of this chapter, we are grouping the non-insect arthropod pests with the insect pests. Most insect pests can be placed into one of two categories based on their mouthparts: chewing insects or piercing/sucking insects. In a controlled environment production system, chewing insect pests are less common than piercing/sucking pests; however, the occasional caterpillar or beetle will find its way into the crop and can become problematic if not managed appropriately. Pests in controlled environments are therefore predominantly of the piercing/sucking type, leading to fewer obvious signs of damage. Juvenile insects are called either larva or nymphs based on what type of development they undergo. All juvenile insects will molt as they grow, and the stages between each molt are referred to as instars. Insect bodies are divided into three main parts: the head, thorax and abdomen. The head contains the mouthparts, eyes and antenna. The thorax contains the locomotive parts of the insect, including the three pairs of legs and two pairs of wings (when applicable). The abdomen contains the digestive, excretory and reproductive organs (Figure 9.3).

9.2.1   ARTHROPOD VIRUS TRANSMISSION

Viruses remain significantly understudied in cannabis, and our understanding of arthropod vectors is rapidly changing. Currently, most known viruses in cannabis are transmitted mechanically, suggesting that insects are not the primary vector of viruses in cannabis. Yet, there are a few instances of confirmed insect vectors. Giladi et al. (2020) confirmed beet curly top virus in industrial hemp and its vector beet leafhopper (*Circulifer tenellus*) is an occasional pest of cannabis grown outdoors. Hadad et al. (2019) found that sweetpotato whitefly (*Bemisia tabaci*) can transmit lettuce chlorosis virus to cannabis, and it is possible that greenhouse whitefly (*Trialeurodes vaporariorum*) can also transmit this virus. Schmidt and Karl (1970) reported that cannabis aphids can transmit cucumber mosaic virus and alfalfa mosaic virus to cannabis. Cannabis aphid may also be a vector of hemp streak virus and hemp mosaic virus, but this has not yet been confirmed. Green peach aphid and melon can also transmit cucumber mosaic virus and alfalfa mosaic virus. However, both are non-persistent viruses, so an aphid is only infectious for minutes—maybe hours—after feeding on an infected plant.

9.2.2 APHIDS

Aphids (Aphididae) are one of the most common insect pests in cannabis production. As a group, aphids are highly polyphagous, making them a significant agricultural pest on many crops worldwide. They feed on plants by inserting their needle-like mouthpart called a stylet into the plant tissue, allowing them to ingest phloem sap. Aphids are quite small (generally < 4 mm in length) but easily identifiable by their round or oval shape, relatively long legs and antenna, and the presence of a pair of cornicles at the terminal end of the abdomen (tubes resembling tailpipes) (Figure 9.3). Most aphids excrete a defensive wax-like compound from their cornicles in response to a threat, such as a predator. Aphids also produce honeydew which is excreted from their rectum. Phloem sap is rich in sugar but poor in other nutritional qualities, including nitrogen. Aphids consume large amounts of phloem sap due to the low nutritional quality, as well as the high hydrostatic pressure, in the phloem, which allows for passive feeding once the phloem has been penetrated. Aphids concentrate the necessary sugars and nitrogen and expel the excess sugars as a sticky liquid called honeydew. A black mold called sooty mold can grow on honeydew, causing aesthetic damage, and can reduce photosynthesis and leaf transpiration in plants. Aphid honeydew can be a food source and a host location cue for aphid predators and parasitoids, but it may also impede the movement of beneficial insects due to its sticky nature. Some species of ants are known to guard aphids from predators to protect the supply of honeydew that the ants use as food.

Aphid damage can be hard to see, especially at low population densities. They do not cause visual feeding marks, such as stippling or chewing damage. As the aphid population increases, plants may begin to wilt or show signs of nutrient deficiencies. Aphids rarely cause economically significant yield loses in cannabis, but severe infestations can result in stunted plants and even death. However, just the presence of aphids, their caste skins or sooty mold can contaminate the flowers and significantly affect the harvestable yield of cannabis.

Aphids have a complex lifecycle. Most species (90%) of aphids are monoecious, meaning their entire lifecycle occurs on a single plant species. The remaining aphid species are heteroecious, or host-alternating, meaning they have a primary host—often a tree species on which they overwinter—and a secondary host, usually an herbaceous species on which they spend the summer. Aphids can also be separated by holocyclic or anholocyclic lifecycles. Aphids that are holocyclic have a lifecycle whereby a period of sexual reproduction occurs, often in the fall to produce eggs to overwinter. These egg-producing aphids are called oviparae. Conversely, aphids with anholocyclic lifecycles reproduce by viviparous parthenogenesis (live birth and asexual reproduction) year-round. Other than when oviparae produce eggs for overwintering, aphids give birth to live and genetically identical offspring. Aphids giving live birth are called viviparae. Holocyclic and heteroecious life cycles often—but not always—occur together. However, the lifecycle of an aphid species can differ geographically. Warm locations such as tropical regions and controlled environments such as greenhouses tend to produce anholocyclic and monoecious aphid populations. All the aphid species mentioned in the following subsections reproduce by viviparous parthenogenesis year-round in controlled environments—with the exception of cannabis aphid, which will produce some males and sexual oviparous females. This means that almost all aphids in cannabis are females. Aphids also produce telescoping generations, whereby aphids are born pregnant with the next generation offspring already developing inside of them. This is why aphid populations can increase so rapidly.

Aphids can be either apterous (without wings) or alate (winged). Typically, the apterous form predominates. Alate aphids are produced due to overcrowding, in response to changes in the host plant quality or in heteroecious aphids to facilitate migration to the primary host for overwintering. Furthermore, morphological differences also exist between viviparae and oviparae and between males and females. More than 1,500 species of aphids have been recorded in North America, and correct species level identification of aphids can be very difficult due to their highly variable morphology. However, only a very small number of aphid species are known to be pests in cannabis. We will provide basic diagnostic criteria to help you identify the known aphid species that could be present in a cannabis production system. It is strongly advised to confirm the identity of any aphids before proceeding with management actions. The following diagnostics serve only as guidelines to start the identification process.

Due to their size and poor flight ability, aphids should be easy to exclude from controlled environments if proper cultural and physical controls are established and followed. Still, there are two species of aphids that plague most cannabis production facilities: cannabis aphid and rice root aphid. Another four species of aphids that are common pests in controlled environments have been recorded on cannabis.

When scouting for aphids, the first sign of an infestation is often cast skins or honeydew. As aphids grow, they shed their exoskeleton in a process called molting. Most aphids molt four times before becoming adults, and each time, a white flaky skin is left behind. Aphids feeding and developing on the underside of leaves and stems will leave skins or honeydew droplets on the upper surface of foliage below them.

9.2.2.1 Cannabis Aphid

Cannabis aphid (*Phorodon cannabis*), also known as hemp aphid, hemp louse or bhang aphid, is monophagous, feeding only on cannabis. Attempts to transfer cannabis aphid to other hosts such as hops (*Humulus lupulus*) have been unsuccessful. Unlike other species within the genus *Phorodon*, cannabis aphid does not use *Prunus* species as an alternate host to overwinter.

Cannabis aphid are typically pale yellow or green, but can also be pink or red in color. Adults can have three darker green stripes running lengthwise down the dorsal surface of their abdomen (Figure 9.4). They are a slightly elongated oval shape and appear slightly flattened dorsoventrally.

FIGURE 9.4 Cannabis aphid (*Phorodon cannabis*) adults and nymphs on a cannabis leaf. Both alate and apterae adults are present. Red and green color morphs of cannabis aphids can be present on the same leaf. Beige aphids have been parasitized by *Aphidius* sp. parasitoids.

Source: Copyright Jason Lemay, University of Guelph

Winged forms have a dark brown or black head and thorax, as well as dark patterning on their abdomen. Early instar nymphs are very small, roughly the size of a pin head, while apterous adults can reach 1.9–2.7 mm in length. Cannabis aphid has a similar appearance to hop aphid (*Phorodon humuli*), but the two can be differentiated by distinct morphological criteria (Cranshaw et al. 2018). However, host transfer experiments indicate that both pests are likely restricted to the host plant in their name, so host association should be sufficient to differentiate the two species (i.e., hops aphid on hops and cannabis aphid on cannabis) (Cranshaw et al. 2018; Carmichael 2020). Projections on the antennal tubercles (colloquially referred to as horns—growths point in the part of the head where the antenna start) (Figure 9.3) help to differentiate cannabis aphid from the other aphids mentioned in this chapter.

Cannabis aphid females begin to reproduce within 1–2 weeks and have a 3–4-week life span. Cannabis aphid is likely monoecious, completing its entire lifecycle on cannabis. However, cannabis aphid has a holocyclic lifecycle, as females have been observed to oviposit on cannabis foliage in the fall. We are unsure what initiates oviposition, but photoperiod is likely a factor. Cannabis aphid eggs are ovate green and turn black. Cannabis aphid maintains its holocyclic lifecycle in greenhouses, as eggs-laying has been observed in greenhouses in Ontario, Canada. Cranshaw et al. (2018) report one such instance four weeks after the photoperiod was changed from 16:8 hr to 12:12 hr. This photoperiod change would simulate the shorter days in the fall when cannabis aphid in natural outdoor settings would be preparing to overwinter. It is likely that cannabis aphid eggs require a cold overwintering period before hatching. Their viability without this cold period (e.g., in a greenhouse) is currently unknown.

9.2.2.2 Rice Root Aphid

Rice root aphid (*Rhopalosiphum rufiabdominale*) is the other aphid species commonly found in cannabis production facilities. Unlike cannabis aphid, rice root aphid is a cosmopolitan pest with many known hosts. It was first recorded on cannabis in 2016, but by 2020 has been found in cannabis production facilities throughout North America due to a particularly strong association with controlled environment cannabis production. Most cannabis production systems—including soil, coconut coir and rockwool, as well as deep water culture, nutrient film technique and aeroponics—can support rice root aphid colonies. However, it has yet to be recorded on cannabis grown outdoors. This aphid is primarily associated with the roots of grasses, including small grains and sedges. Rice root aphid has been recorded on various broadleaf horticultural crops in North America, such as cotton and celery grown outdoors, as well as peppers and squash grown in controlled environments.

Apterous rice root aphid are approximately 2.0–2.5 mm in length and have a dark green or brown color that transitions to a rusty red color around the cornicles (Figure 9.5). A heteroecious and holocyclic lifecycle has been reported in Asia, with *Prunus* spp. as a primary host. However, Cranshaw and Wainwright-Evans (2020) report that a monoecious and anholocyclic lifecycle occurs on cannabis in North America (i.e., completes entire lifecycle on cannabis roots through parthenogenesis). Alate aphids will climb to the upper foliage of cannabis plants to disperse through the crop. When rice root aphids are present, it is not uncommon to find dead alate aphids trapped by the sticky trichomes on cannabis foliage. If you see a group of only alate aphids on the foliage or stems, that is a sign to look at the roots for rice root aphids.

As rice root aphid spend very little time outside of the rootzone, damage can be difficult to quantify. They cause very little contamination to the foliage or inflorescence. However, if inadequately managed, rice root aphid can reach populations that reduce plant growth and vigor due to the amount of phloem sap consumed. A decline in root health is also possible, and an increase in susceptibly to root diseases is also possible but has yet to be confirmed.

9.2.2.3 Other Aphid Species

Other aphid species have been recorded on cannabis, but are not as common as cannabis aphid and rice root aphid in cannabis crops at this time. Most notably are the cotton/melon aphid (*Aphis gossypii*), green peach aphid (*Myzus persicae*), foxglove aphid (*Aulacorthum solani*) and black bean aphid (*Aphis fabae*). These are polyphagous aphid species often found in controlled environment

FIGURE 9.5   Rice root aphids (*Rhopalosiphum rufiabdominale*) on the roots of cannabis sativa rooted into a rockwool block.

Source: Copyright Jason Lemay, University of Guelph

and field production systems. Their biology and biological control on other crops are well known, but how cannabis affects the development and control of these species is unknown. These aphids can contaminate the flowers and create honeydew like cannabis aphid. They are all monoecious and anholocyclic in controlled environment cannabis production facilities.

Melon aphid (*Aphis gossypii*), also known as the cotton aphid, is a highly polyphagous aphid found on over 700 species of plants worldwide. *Aphis gossypii* is a small aphid with apterous adults measuring 1–1.8 mm in length. It is highly polymorphic with several color morphs ranging from yellow to a dark (almost black) green color. However, the dark green mottled with lighter green morph is the most common. This color variability can make visual identifications difficult. However, the cornicles are entirely black from base to tip in all color morphs, while the cauda, a tail-like appendage, is comparatively paler. These characteristics can be used to distinguish cotton/melon aphid from other aphids found on cannabis (Figure 9.6).

Green peach aphid (*Myzus persicae*) is highly polyphagous and is found on more than 400 species of plants worldwide. Apterous adults measure 1.7–2.0 mm in length. There are three known color morphs of green peach aphid: green, orange and red. The green morph is more common in the spring and early summer, while the red morph is more common in the late summer and fall due to its tolerance for higher temperatures (> 25°C). The red morph is also known to be more tolerant to insecticides. The orange morph is believed to be an intermediate between the red and green morph, but is not common outside of the mid-Atlantic region of the United States. This species can be differentiated from other aphid species on cannabis by its convergent antennal tubercles (Figure 9.7).

FIGURE 9.6 Melon aphid (also called cotton aphid) (*Aphis gossypii*) apterae. The dark cornicles, black from base to tip, help differentiate melon aphid from other aphids in cannabis.

Source: Copyright Jim Baker, North Carolina State University, Bugwood.org

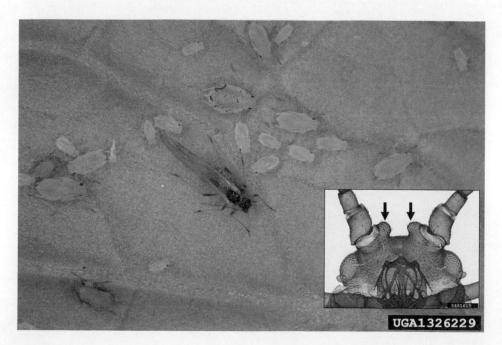

FIGURE 9.7 Multiple life stages of green peach aphid (*Myzus persicae*), including apterous adults and nymphs, as well as an alate adult. Image in insert displays the convergent antennal tubercles that are diagnostic for green peach aphid.

Sources: Main image copyright Whitney Cranshaw, Colorado State University, Bugwood.org; insert image copyright Brendan Wray, AphID, USDA APHIS PPQ, Bugwood.org

Foxglove aphid (*Aulacorthum solani*), also known as the glasshouse-potato aphid, is extremely polyphagous, being recorded on over 82 different families of plants worldwide. It is a larger aphid with apterous adults roughly 1.5–3.0 mm long. Foxglove aphids are green and have a darker green patch at the base of each cornicle. Their antennae are longer than their body and marked with dark bands at the end of each antennal segment. It is a cool weather aphid, preferring temperatures from 15–25°C. It is rarely found in greenhouses during the summer months, when temperatures are commonly above 25°C. Foxglove aphid has an active defensive behavior and is prone to drop from the plant when disturbed (Figure 9.8).

Back bean aphid (*Aphis fabae*) has a cosmopolitan distribution, but of the aphids mentioned in this chapter, it is the least commonly found in controlled environments. It is polyphagous and found on more than 200 wild and cultivated herbaceous plants. Apterous adults are 1.2–2.9 mm long, and have a dark green or black color. The antenna are roughly two-thirds the length of their body, and like their legs, are pale yellow with black bands at the joints. The cornicles on black bean aphid are shorter than other aphids found on cannabis. Unlike the cotton/melon aphid, both the cornicles and cauda will be similarly dark. Black bean aphid will sometimes have white waxy spots on their abdomen, especially as immature alates. These aphids feed predominantly on new foliage, but will colonize the entire plant over time. Black bean aphid forms large colonies that are usually protected by ants. (Figure 9.9)

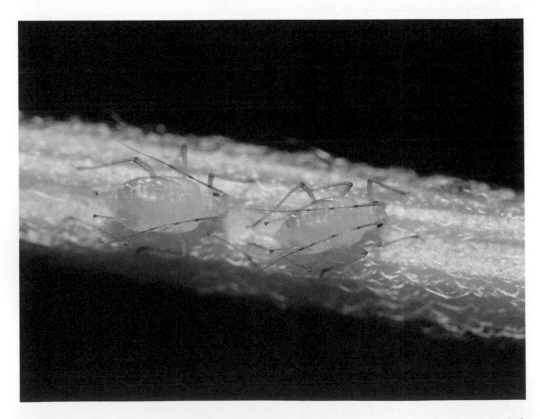

FIGURE 9.8 Adult foxglove aphid (*Aulacorthum solani*). The long antenna with dark bands and the darker spots at the base of each cornicle help differentiate this aphid from other aphids on cannabis.

Source: Copyright Alexis Tinker-Tsavalas

FIGURE 9.9   Adult and nymph black bean aphids (*Aphis fabae*).

Source: Photo by Peter Gabler

9.2.3   Aphid Control

9.2.3.1   Biological Control

The development of effective biological control options for cannabis aphid and rice root aphid are still in their infancy. Further work is needed to determine the most effective parasitoids and predators for control of these pests.

9.2.3.1.1   Parasitoids

It is very important that you accurately identify the aphid species you are managing, as this will impact the parasitoid you choose as a BCA. Parasitoids work best as a preventative measure against aphids. Once aphids are established, it is unlikely that parasitoids alone will help you regain control. Aphid parasitoids lay eggs inside their host, and when they hatch, the larva feeds on the living aphid. After some time, the aphid dies and becomes mummified, while the larval parasitoid pupates and finally emerges as an adult. Aphids parasitized by *Aphidius* become golden brown, and parasitism by *Aphelinus* turns them black. Both *Aphidius* and *Aphelinus* are effective searchers and can locate aphid colonies just as they are starting. It is important to released parasitoids randomly through the crop, and not just in aphid hot spots. Foraging parasitoids can locate small aphid colonies before they are found through scouting. *Aphidius colemani* and *A. matricariae* can be maintained in cannabis crops on banker plants. Unfortunately, there are no commercial banker plants developed for *Aphidius ervi* in North America. English grain aphids (*Sitobion avenae*) are available commercially as an *A. ervi* banker plant host in Europe, and it can be collected from wild populations in North America. Parasitoid development from egg to adult takes ten days at 25°C, with the aphid becoming "mummified" approximately seven days after oviposition. *Aphelinus* development is a little longer than *Aphidius* and can take upwards of two weeks to emerge after the mummy has formed. *Aphidius* and *Aphelinus* adults live for 2–3 weeks, but the lack of nectar resources in cannabis could result in a decrease in performance (e.g., longevity, fecundity, searching ability) by parasitoids in comparison to other crops. Adult female *Aphelinus* can feed directly on smaller aphids for nutrients (host feeding) as opposed to laying an egg in them. It is currently unknown if they will host feed on cannabis aphid. Importantly, these parasitoids do not enter a reproductive diapause under short-day lengths like some other aphid predators.

Choosing what parasitoids to utilize in a cannabis crop can be difficult (Table 9.2). All three *Aphidius* species, as well as *Aphelinus abdominalis*, will parasitize cannabis aphid, but *Aphidius ervi* and *Aphidius matricariae* likely provide the best control. *Aphidius matricariae* will parasitize black bean aphid, green peach aphid and cotton/melon aphid, although the latter is better controlled by *Aphidius colemani*, especially when temperatures are above 25°C. Both *Aphidius ervi* and *Aphelinus abdominalis* will parasitize foxglove aphid. *Aphidius matricariae*, *A. colemani* and *Aphelinus abdominalis* may parasitize alate forms of rice root aphids when they are in the plant canopy but will not provide any control of this pest since most of the population is restricted to the rootzone. Many BCA suppliers sell a mixed-species product—multiple species of *Aphidius* together. Due to morphological similarities, it can be very difficult to differentiate the species and

---

TABLE 9.2

List of Aphid Species Found on Cannabis in Controlled Environments and Their Associated Parasitoids

Aphid	Associated Parasitoid
Cannabis aphid	*A. ervi*[*]
	A. matricariae[*]
	A. colemani
	Aphe. abdominalis
Cotton/melon aphid	*A. colemani*
	A. matricariae[+]
Green peach aphid	*A. matricariae*
	A. colemani[1]
Foxglove aphid	*A. ervi*[*]
	Aphe. abdominalis
Black bean aphid	*A. matricariae*[*]
	A. colemani[+]
Rice root aphid	No parasitoids will control this species

Notes:

[*] Indicates that the aphid is more effectively parasitized by this BCAs than others when multiple species are capable of parasitizing it.

[+] Indicates poor control of the aphid by this BCA.

[1] *Aphidius colemani* is a more effective parasitoid of green peach aphid than *A. matricariae* when temperatures are above 25°C.

FIGURE 9.10 Emergence holes on parasitized mummies. The parasitized aphid on the left has a jagged emergence hole with no lid indicating a hyperparasite emergence. The aphid on the right has a smooth emergence hole with a lid, indicating an *Aphidius* sp. emergence.

Source: Copyright Jason Lemay, University of Guelph

consequently to determine which *Aphidius* species is providing the best aphid control in your production system.

The effectiveness of some parasitoids may decrease during the summer due to the presence of other wasps that are parasites of *Aphidius*. These hyperparasites can be monitored by checking the emergence holes on the parasitized aphid mummies. *Aphidius* emerges from a smooth round hole at the rear of the abdomen on the mummified aphid. There is often a lid or flap remaining on *Aphidius* emergence holes. Hyperparasites emerge from a jagged hole that is often off to the side of the abdomen and does not have an attached lid (Figure 9.10). Controlling hyperparasite populations can be difficult, especially when crop production is continuous as is often the case for cannabis production. Parasitoids should be used in combination with a predator, but intraguild predation can occur (i.e., ladybird beetles eating parasitized aphids).

9.2.3.1.2 Predators

Predators provide better control when aphid populations are high and have the advantage of not leaving behind a mummified aphid. Furthermore, the predators mentioned in what follows are largely generalist and will predate on any of the aphids found on cannabis, with the possibility that there may be some preference to a certain prey species if multiple species are present. Some of these predators will also provide control against some non-aphid pests such as thrips, mites and whiteflies.

Aphidoletes aphidimyza is a midge with a predatory larval stage. They prefer cooler temperatures (21–25°C) and high humidity (> 70%). Adults are nocturnal, searching for aphids at dusk. *Aphidoletes aphidimyza* is an effective searcher and will locate small aphid colonies in the same manner that the parasitoid *Aphidius* does. Adults appear similar to fungus gnats or mosquitos and can be found hanging from spider webs where they mate. Eggs are laid directly in aphid colonies. Larva are orange maggots that grow to 3 mm in length. Larva feed for 1–2 weeks and then drop to the ground to pupate for two weeks. When the aphid density is high, larva will kill more aphids than they consume. *Aphidoletes aphidimyza* may be difficult to establish in some cannabis facilities, as they require moist soil for pupation and the pupa are prone to desiccation. Therefore, constant re-introductions may be required. *Aphidoletes aphidimyza* will diapause under shorter daylengths (12–14 hours), so they may not be suitable for control of aphids during the flowering stage of cannabis when short daylengths are used. Very low–intensity lights, such as garden walkway lights, have been used to prevent diapause initiation in *A. aphidimyza*.

Green lacewing (*Chrysoperla carnea* or *Chrysoperla rufilabris*) larvae will provide localized control of aphids, but do not disperse well. Lacewing larva are cannibalistic, resulting in some conspecific mortality, especially when prey levels are low. Adult green lacewings do not feed on insects and rather feed on pollen and nectar, which are absent in cannabis production systems. As with parasitoids, this can result in decreased performance from lacewings in cannabis compared to other crops.

Brown lacewing (*Micromus variegatus*) can provide superior control as the adult life stage will feed on aphids and other insects, unlike green lacewing adults. Adults of both species are nocturnal, flying to forage for aphid colonies at night. Adults are also attracted to lights, and can be drawn away from crops by lighting sources such as streetlights or security lights. Lacewing eggs are white ovals that become green and are suspended at the end of a slender stalk. The lifecycle from egg to adult takes approximately 25 days at 25°C, and adults live for another 40 days.

Minute pirate bug (*Orius insidiosus*) can feed on small cannabis aphid and—while not sufficient on its own—likely offers some level of additional control.

Dicyphus hesperus will feed on aphids and are well adapted predators for plants with dense and sticky trichomes. However, other than being observed in field-grown cannabis and hemp, little is known about *D. hesperus*' effectiveness as BCAs in cannabis. For more information on *D. hesperus*, see Section 9.2.9. Both these BCAs can be maintained on banker plants with *D. hesperus* on mullein (*Verbascum thapsus*) and *O. insidiosus* on ornamental peppers (*Capsicum annuum* cv. 'Purple Flash') and sweet alyssum (*Lobularia maritima*). See Section 9.2.7 for more information on *O. insidious*.

Ladybird beetles (Coccinellidae) are valuable generalist predators. Certain species of ladybird beetles such as *Adalia bipunctata* are more sustainable, as they are mass-reared and provide superior control of aphids in controlled environments. Wild-harvested ladybird beetles such as *Hippodamia convergens* and *Harmonia axyridis* have been found to be less effective, as they are harvested from natural populations during their overwintering phase when they dormant. Furthermore, the sustainability of wild-harvested ladybird beetles has become an issue. Supply of these wild-harvested ladybird beetles is inconsistent, and some suppliers have stopped selling them. While ladybird beetle adults will feed on aphids, the larval stage consumes many more aphids. Larval development takes about three weeks before they pupate for 3–5 days. Daylengths below 14 hours at 23°C will induce diapause in *A. bipunctata* and other ladybird beetle species, which can reduce their effectiveness in the flowering stage.

Hover flies (*Eupeodes americanus*) are excellent aphid predators. The adults of many hover fly species are often mistaken as wasps due to their similar appearance. The adult are strong fliers and will lay white oblong eggs near aphid colonies. *Eupeodes americanus* larva are voracious aphid predators and grow up to 11 mm long. The larvae are translucent beige with a gray and brown mottling. The larval stage lasts for approximately two weeks before pupating in soil and emerging as an adult fly approximately one week later. *Eupeodes americanus* are not currently available commercially in North America, but are available in Europe. They are abundant outdoors in North America, and will naturally enter facilities without vent screening. Adults require pollen and nectar before they will begin to produce eggs. Additional pollen or nectar bearing plants should be provided if *E. americanus* or other hover flies are released.

Soil predators such as the rove beetle *Dalotia coriaria* and the predatory mite *Stratiolaelaps scimitus* were commonly recommended as BCAs to control rice root aphids, but there is no evidence that they provide any control. These BCAs are frequently used for the control of fungus gnat and thrips. Entomopathogenic nematodes have not been found to be effective at controlling rice root aphid.

9.2.3.2 Cultural Control

Cultural control of aphids relies predominantly on exclusion and sanitation. See Section 9.1.1 for general exclusion and sanitation techniques. Containers and substrates should not be reused to avoid contamination of new plants with root aphids. Yellow sticky cards can catch winged aphids, but provide negligible control overall. In fact, yellow sticky cards likely catch more flying BCAs like *Aphidius* than aphids. Wrapping the lower stem of a plant with sticky tape is an effective monitoring tool for root aphids and provides some control by preventing the aphids from climbing up the stem to disperse or move down to the rootzone to form a new colony.

De-leafing can reduce aphid populations and can provide significant control when aphid populations are high. In severe infestations, aphids can be aspirated using a vacuum. Neither of these two methods will eliminate aphids, but they can be used to rapidly reduce the population so that biological control has a chance.

A crop-free period is likely the most effective way to manage cannabis aphid long-term. Most aphids can only survive for a few days without food, so a week-long period without cannabis or other host plants will kill any remaining aphid. This can be done to individual production areas one at a time if they are sufficiently isolated and care is taken to prevent aphids from moving from infested areas into clean areas.

Ant populations in the cannabis production environment should be controlled to eliminate any protection from natural enemies by attending ants.

Excessive fertilization can increase the susceptibility to aphids and makes the plants more favorable for their development.

To date, there are no known cannabis cultivars with resistance to aphids. However, cannabis aphid infestations appear to occur more commonly on certain cannabis cultivars. It is unknown if this is due to an attraction to or preference for certain cultivars, or if cannabis aphid performs better on certain cultivars.

9.2.3.3 Chemical Control

Of the available chemical control options, horticultural oil products and insecticidal soap products provide adequate control of the above aphids if sufficient contact is made. These products work by suffocating (oils) or desiccating (soaps) the aphids. Aphids killed by these products often appear flattened and dried with a darker red or brown color. Foliar applications of *Beauvaria bassiana* provide limited control of aphids but prolonged high humidity is required (> 70% for 5–6 hours).

Soil drenches of *B. bassiana* and *Metarhizium anisopliae* products have provided effective control for root aphids and some *B. bassiana* products received an emergency registration for this use in Canada (June 2020). *Isaria fumosorosea* and *Lecanicillium lecanii* are other entomopathogenic fungi that have some efficacy in controlling rice root aphids (Cranshaw and Wainwright-Evans 2020).

9.2.4 MITES

Mites are arachnids rather than insects like the other pests mentioned in this chapter. While insects have three pairs of legs, adult mites have a fourth pair. The larval stages only have three pairs of legs. The fourth pair of legs develops during the nymphal stage. Mites also have a different lifecycle, with an active larval and resting nymphal life stage which is similar to the pupal stage in holometabolic insects. Unlike insects, mites have just two body segments: the gnathosoma and the idiosoma. The former is essentially just the mite's mouthparts. There is no segmentation on their body, as the gnathosoma and idiosoma are fused together. Mites have an adapted piercing/sucking mouthpart to ingest cell content rather than plant fluids from the phloem or xylem. Finally, mites are entirely wingless.

Mites can be one of the most damaging pests in indoor-grown cannabis, due to their ability to reach large populations before they are noticed. Because of their small size, mites also are often moved between facilities via contaminated cuttings or other shared plant materials. People are also a common vector of mites into and within production facilities.

To scout for mites, look for stippling on leaves or deformed foliage, and check the underside for mite colonies. Without a hand lens, spider mite colonies often look like grains of sand. A 10–20× hand lens can be used to confirm two-spotted spider mites. Broad mites and hemp russet mites require more magnification, and a microscope capable of 40–60× magnification is required. Randomly inspecting foliage under a microscope is the best way to find hemp russet mites before obvious damage symptoms appear. However, the damage caused by these mites, detailed in their respective subsections following, is often how they are found.

9.2.4.1 Two-Spotted Spider Mite

Two-spotted spider mite (TSSM) (*Tetranychus urticae*) is one of the most cosmopolitan and polyphagous pests, recorded in over 124 countries on more than 1,100 host plants. It is the most common type of mite found in greenhouses and other controlled environment systems. Adults are small and easy to miss during plant inspection, measuring only 0.5 mm in length. Immature mites are clear with a slight yellow color, but can also be green to green-brown in color. Over time, adults develop two dark spots on their dorsal side, which are formed as a result of stored gut content that is visible through their translucent body (Figure 9.11). Short-day lengths (< 10 hours of light) and

FIGURE 9.11 Two-spotted Spider mite (*Tetranychus urticae*).

Source: Copyright Dr. Guido Bohne

FIGURE 9.12 Damage caused by: A) two-spotted spider mite (*Tetranychus urticae*); and B) onion thrips (*Thrips tabaci*).

Sources: Copyright: A) Kadie Britt, University of California, Riverside; B) Whitney Cranshaw, Colorado State University, Bugwood.org

cool conditions (< 18°C) can initiate the development of a diapausing form whereby females are orange in color and no longer reproduce. These females can be confused with the predatory mite *Phytoseiulus persimilis*. However, *P. persimilis* lacks the two spots on either side of its body which can be seen with a hand lens. Two-spotted spider mite eggs are very small and spherical in shape, and are clear with a slight yellow or white tint. While TSSMs do not have wings, they are able to disperse by producing silk strands to hang from plants and be carried by air currents. Mites are commonly found on the underside of leaves, but when populations numbers are high, they can be seen on the upper side of leaves and even the flower buds.

Spider mites feed by puncturing plant cells with their stylet mouthparts and ingesting cellular contents. This feeding damage causes a stippling effect visible from both the upper and lower surfaces of the foliage. The stippling damage of TSSM is smaller and more consistent in shape than the damage caused by thrips (Figure 9.12). Spider mites produce webbing that covers the foliage they are on. Few natural enemies can predate on spider mites once they are protected by their webbing. Without adequate scouting, webbing-covered foliage is often the first sign that TSSM is established in the crop. Two-spotted spider mite can be a serious problem in controlled environment production, as they thrive in the sheltered, hot and dry conditions (30°C and < 40% RH) that can occur in cannabis production facilities. Two-spotted spider mites can develop from egg to adult in seven days at 30°C. Small infestations of TSSM do not cause substantial yield losses. However, populations can increase rapidly, and high populations can cause stunted plant growth and poor plant health. Considerable losses of harvestable flower can arise due to the contamination by mites and their webbing.

There are records of carmine spider mite (*Tetranychus cinnabarinus*) as a pest on cannabis. However, recent findings suggest that *T. cinnabarinus* and TSSM are the same species (Auger et al. 2013). Reports of *T. cinnabrinus* on cannabis are likely the orange/red overwintering form of TSSM.

9.2.4.2 Hemp Russet Mite

Like cannabis aphid, hemp russet mite (HRM) (*Aculops cannabicola*) has been understudied due to its specialization on cannabis. Eriophyoid mites, like HRM, are the most highly host-adapted group of plant feeding mites, having developed very close associations with a narrow range of host

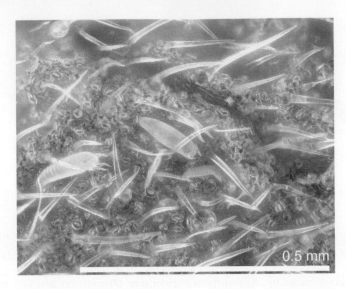

FIGURE 9.13   Hemp russet mite (Aculops cannabicola) on hemp between foliar trichomes.

Source: Copyright Matt Bertone

plants (Krantz and Lindquist 1979). Hemp russet mite is not known to feed or reproduce on plants other than *C. sativa*. Eriophyoids only feed on the fluids of cells on the surface of the plant and do not remove all the fluid or any chloroplast from those cells. They are better adapted as parasites, keeping their host alive, than most other mites.

Hemp russet mite is smaller than TSSM, measuring 0.2 mm in length. A 10–20× hand lens, at minimum, is required to see HRM. Hemp russet mites are beige and have a tapered worm-like shape (Figure 9.13). As with all Eriophyoid mites, HRM is unique in having only two pairs of legs rather than four. Both pairs of legs project out from near the mouthparts. The life history of HRM is not well understood, but based on current information and the life history of the related tomato russet mite (*Aculops lycopersici*), it is likely that the lifecycle takes about 14 days at 25°C, and adults live for another three weeks. Due to their small size, hemp russet mites are not particularly mobile on their own, but they are easily spread by air currents in controlled environments. Humans can also spread HRM, but they are not a significant source of dispersal. It was previously thought that HRM could feed and travel on cannabis seeds, but this has been disproven. Infested plant material remains the primary method for the introduction of HRM into cannabis facilities.

Hemp russet mite feeds on the upper surface of leaflets and on petioles. At low populations, the damage is difficult to see. As populations increase, feeding on the upper surface of the leaf causes a slight upwards curling along the edge. However, this is not a consistent symptom, and may possibly be cultivar dependent. Mites feeding on the petiole render it brittle and prone to breaking off. Further increases in population result in russeting that may dull the color of the leaves. Severe infestation can lead to necrosis and premature defoliating. High populations of hemp russet mite will also feed on inflorescence stunting their growth and on the stalks of trichomes significantly reducing the production of cannabinoids (McPartland and Hillig 2003).

9.2.4.3   Broad Mite

Broad mites (BM) (*Polyphagotarsonemus latus*) are smaller than HRM, measuring 0.1–0.2 mm in length as adults and can be hard to see even with a 10–20× hand lens. At 24°C, the lifecycle takes five days from egg to adult. Adults live for another seven days. Adults are translucent white or yellow in color, and females have a white stripe on their backs (Figure 9.14). While adult BM have four pairs of legs, the hind pair of legs on female mites are reduced to thread-like appendages that

FIGURE 9.14 Various life stages of broad mite (Polyphagotarsonemus latus): A) Adult broad mite with egg; B) eggs; and C) broad mite larva.

Source: Copyright Justin Renkema, Agriculture and Agri-Food Canada

are not used for walking. Conversely, adult males have enlarged hind legs used to grasp and carry female nymphs and adults for mating. Broad mite eggs are often laid on the underside of leaves and are clear with white bumps.

Broad mites are a tropical pest, but arrive in temperate areas on cuttings or other plant materials and can only survive the winter months inside warm controlled environments. They prefer a warm climate with high humidity (> 70%), and they are often found in sheltered areas of the plant such as growing points and flower buds. Damage is most often in these areas. Broad mite feed on epithelial cells of the plant, injecting a toxin as they feed. Feeding on the underside of partially emerged foliage initially causes some downward curling of the leaf. Less commonly, they feed on the upper surface, causing upwards curling of the foliage. When they feed on younger, developing foliage, the result is greater twisting and distortion, and the emerging foliage can appear stunted and hardened. Corking or scabbing can also appear. Feeding on flower buds can cause them not to open or even to abort. Broad mites can be difficult to diagnose, as the damage can resemble injury caused by herbicide, environmental factors, phytotoxicity and certain viruses. As few as ten BM per plant have been found to cause economic damage on chili pepper plants, but these thresholds have not been investigated in cannabis. Like other mites, BM is mostly likely to enter a production facility on infested cuttings. Broad mites do not disperse far on their own, rarely leaving the plant on which they were born. They are dispersed by wind or air currents, and are also known to be transported by other insects, such as whiteflies (Fan and Petitt 1998).

9.2.5 Mite Control

9.2.5.1 Biological Control

Phytoseiulus persimilis is an excellent TSSM predator and can effectively predate on this pest, even in the presence of webbing. However, it will not survive on alternate food sources and therefore cannot be used preventatively. It is recommended to release this BCA directly into TSSM colonies. Another limitation for *P. persimilis* is that they are less effective at temperatures above 30°C and in dry conditions. In greenhouse cannabis production it is not uncommon to find *P. persimilis* effectively controlling TSSM lower in the canopy, while TSSM is essentially untouched higher in the canopy where it is too warm and dry for *P. persimilis*.

Other **predatory mites** such *Neoseiulus fallacis, Amblyseius californicus, Amblyseius andersoni* and *Galendromus occidentalis* will feed on TSSM and alternate food sources, including thrips and plant pollen and can be sustained in the absence of TSSM. These mites are also better suited for warmer temperatures, with *N. fallacis* and *N. californicus* tolerating temperatures up to 32°C, while *A. andersoni* and *G. occidentalis* can tolerate temperatures up to 40°C and 43°C, respectively. These predatory mites perform better in humid environments (> 50% RH), but *N. californicus* can tolerate drier conditions down to 40% RH and is an excellent option for preventative TSSM control in controlled environment cannabis production. *Galendromus occidentalis* performs well on trichome dense plants, but the resinous nature of cannabis trichomes may still affect its performance. Temperatures below 26°C may induce diapause in *G. occidentalis*. None of these mites will control BM or HRM. *Amblyseius swirskii* is another predatory mite that, while commonly used for control of thrips, will feed on TSSM.

Stethorus punctillum is a beetle, and an excellent predator of TSSM. Adults are strong fliers and can find small TSSM colonies in the crop. Adults are small (1.5 mm) and black, while the larvae are 0.5 mm long and gray in color. Both the larvae and adults are predacious. *Stethorus punctillum* completes its lifecycle in approximately 14 days at 25°C. Adults can live for up to two years. It can tolerate a wide range of environmental conditions from 16–35°C and relative humidity from 30–80%. *Stethorus punctillum* is compatible with predatory mites, but if spider mite populations become too low, it will feed on predatory mites. Even with its good searching ability, *S. punctillum* should be considered as a hot spot treatment used in combination with a preventative predatory mite program. Unlike *P. persimilis*, *S. punctillum* will struggle to control TSSM once there is significant webbing.

Feltiella acarisuga is a predatory midge with a similar appearance to *Aphidoletes aphidimyza*. It requires high humidity (> 70%) and temperatures from 20–27°C. Adults are good searchers and will deposit eggs directly in TSSM colonies. Adults are short lived, surviving only 3–4 days. Larva grow to 2.0 mm in length and have a creamy yellow or orange color. Larva feed for up to a week before pupating on the underside of leaves. *Feltiella acarisuga* has been found feeding on HRM, but does not provide adequate control.

Generalist predators such as green and brown lacewings, *Orius insidiosus* and *Dicyphus hesperus* will feed on two-spotted spider mites. For more information on lacewings, see Section 9.2.3. For more information on *O. insidiosus*, and *D. Hesperus*, see Section 9.2.7.

Biological control for mites other than TSSM is not very effective. In fact, for both BM and HRM, it's better to remove and destroy infested plants immediately rather than try to manage the infestation with BCAs. To remove infested plants, it is best to carefully place them in a bag first to prevent them from infesting other plants. *Amblyseius andersoni* and *O. insidiosus* will feed on HRM, but will not provide adequate control for even moderate infestations. Predatory mites such as the *A. swirskii* and *N. cucumeris* have shown some effectiveness at controlling BM, but the control is not reliable. If BM are found when these predatory mites are already in the crop, it is strongly recommended that infested plants are removed and destroyed rather than releasing additional predatory mites.

9.2.5.2   Cultural Control

Exclusion and sanitation are incredibly important for both HRM and BM, and can greatly reduce the likelihood of TSSM infestations. The most likely way these pests enter a controlled environment is on infested plant material. All incoming plant material should be carefully inspected, quarantined and inspected again. Should either HRM or BM be found, the infested plants, as well as neighboring plants, should be carefully bagged, removed and destroyed, even if there are no signs of damage. The area around the infested plants should also be disinfected. If a severe outbreak occurs in the production area, or either BM or HRM becomes established, a prolonged host-free period of 2–4 weeks will be the only way to eradicate the pest. Ensure that no weeds, such as chickweed (*Stellaria media*), creeping woodsorrel (*Oxalis corniculata*) or henbit (*Lamium amplexicaule*) are present in or around the production facility that could harbor populations of TSSM.

Hot water baths at 46°C for 15 minutes can kill the adults and eggs of TSSM and BM, and possibly HRM adults, if a surfactant such as soap is used. Information on water baths to kill HRM eggs is not yet available. Temperatures of 46°C for 15 minutes appear to be the limit to avoid damaging the plant material, but this should be confirmed on a small number of plants for each cultivar before implementing. Each cultivar should be tested separately, as they may have differing tolerance to hot water.

When scouting, it is important to avoid areas with known mite infestations until the end. Always scout or complete any task in clean areas first. Humans are much more likely to disperse TSSM than either BM or HRM.

There are no cultivars with known resistance to any of these mites.

9.2.5.3 Chemical Control

Foliar sprays of horticultural oils can provide some control against these mites, especially for TSSM. However, they will not provide sufficient control to rescue a crop from a mite outbreak. Oil sprays need good contact to kill the mites, and so multiple applications may be necessary. It may be hard to reach HRM and BM, due to their small size and proclivity to be in protected spaces. Foliar application of sulfur has shown good effectiveness at controlling HRM. However, these sprays can severely impact most BCAs in the crop. Be sure to test each cultivar for phytotoxicity before applying oils or sulfur. Horticultural oils used as a dip have shown promise in reducing mite numbers on incoming plant material. Insecticidal soaps do not provide control of these mites.

When conditions are favorable for spider mites, foliar sprays can result in TSSM outbreaks due to the sprays killing the beneficial insects. Biological control and cultural control of mites are effective in cannabis, and strongly recommended.

9.2.6 THRIPS

Like aphids and mites, thrips are a very common pests in controlled environments. They have a slender and elongated body (1–3 mm long). Thrips feed by piercing plant cells on the surface of the foliage and ingesting cell fluids. This created damage similar to that caused by two-spotted spider mites (TSSM) (Figure 9.12). However, thrips feed more on the upper surface of the leaf and the feeding creates slightly larger and more irregularly shaped stippling than TSSM. Furthermore, thrips leave small black flecks of frass (excrement) on or near their feeding sites. Adult thrips have fringed wings but are poor fliers, and larvae are wingless. Thrips disperse rapidly through controlled environments, leaping from leaf to leaf or by short flights from plant to plant. Thrips will also disperse long distance by air currents. They often enter controlled environments through vents and doorways, but also enter on incoming plant material.

9.2.6.1 Western Flower Thrips and Onion Thrips

Two species of thrips are commonly found in cannabis production facilities: onion thrips (OT) (*Thrips tabaci*) and western flower thrips (WFT) (*Frankliniella occidentalis*). Most thrips species have similar lifecycles. Eggs are deposited into plant tissue and hatch a couple days later. Thrips larvae have three instars—the first two instars are active and feed, while the last instar, effectively a prepupal stage, is immobile and does not feed. Onion thrips and WFT drop to the soil to pupate, but they can also pupate on the plant in leaf axils or other crevices. A complete lifecycle can happen in as short as ten days at 30°C but can take up to 20 days at 20°C.

Onion thrips and WFT have very similar appearances (Figure 9.15). Both species are pale yellow to gray in color. Onion thrips adults (1.3 mm long) are slightly shorter than WFT (1.7 mm long) and have a somewhat paler appearance. However, size and color alone should not be used to differentiate these two species. Larvae of the two species are yellow in color and smaller than the adults, but they cannot be differentiated at this stage. A taxonomic key developed by Summerfield and Jandricic (2021) can be used to identify the adults of many common greenhouse thrips species including both OT and WFT. A simple way to differentiate OT and WFT is to look at adult thrips under a microscope with at least 45× magnification. Onion thrips have three gray ocelli, while WFT have three red ocelli (Figure 9.16). Ocelli appear as dots on the top of their head between their eyes. They are

FIGURE 9.15   Adult western flower thrips (*Frankliniella occidentalis*) (left) and adult onion thrips (*Thrips tabaci*) (right).

Source: Copyright Ashley Summerfield, University of Guelph

FIGURE 9.16   Head and pronotum of: A) western flower thrips (*Frankliniella occidentalis*); and B) onion thrips (*Thrips tabaci*). Note the red ocelli and presence of long coarse hairs on the top of the pronotum on western flower thrip; onion thrips have gray ocelli and lack the long coarse hairs on the top of the pronotum.

Source: Adapted from Summerfield and Jandricic (2018)

a light-sensing organ similar to a second type of eye. Onion thrips also lacks the long coarse hairs on the top of their pronotum (first body segment immediately after their head).

It is important to note that while thrips are commonly found in cannabis production facilities, it is generally accepted that they rarely cause economically significant yield losses when feeding on healthy plants. It is possible that in the presence of other pests, thrips feeding could result in enough stress to reduce yields and they should be managed. However, McCune et al. (2021) recently found that plants initially infested a greater number of OT had a lower fresh weight yield than plants that started with fewer thrips. Very high thrips populations can also cause enough foliar damage to reduce photosynthesis and plant growth, and cause foliage to die.

To scout for thrips, shake foliage and flowers over a white piece of paper. Thrips will fall off the plant and can be easily seen against a white background. Adult and larval thrips can often be seen on the upper and lower surfaces of cannabis leaves. Thrips should be inspected under a microscope to confirm the species identification.

9.2.7 Thrips Control

9.2.7.1 Biological Control

Biological control of thrips in cannabis can be difficult. Most biological control programs for thrips are developed around western flower thrips. Yet in other crops, especially floriculture, these programs are not as effective for the control of onion thrips. Preventative control of thrips is important, as regaining control after an outbreak can be difficult. The management options discussed in what follows can provide enough control to keep thrips populations below levels at which yield losses can occur from thrips feeding alone. Eradication of either OT or WFT is not likely to occur with these options.

> **Predatory mites** (*Amblyseius swirskii* and *Neoseiulus cucumeris*) are effective predators against early instar larva. They will also provide some non-consumptive control by disturbing thrips larvae, reducing their feeding activity and decreasing their survival. Both species of mites are about 0.5 mm long, pear shaped and beige in color. Females prefer to lay eggs on trichomes. *Amblyseius swirskii* will feed on more WFT and OT than *N. cucumeris* and is more effective at higher temperatures (> 30°C). For these reasons, growers often use *A. swirskii* in the summer when temperatures and thrips pressure are greater, and switch to *N. cucumeris* in the cooler months to save money (as *N. cucumeris* is often less expensive). Both species of mite prefer humid environments, around 70% relative humidity. *Amblyseius swirskii* also has a faster lifecycle than *N. cucumeris*, going from egg to adult in five days compared to nine days at 25°C. Adults of both species can live for upwards of 30 days, laying eggs after just two days. Application method is also important. Both *A. swirskii* and *N. cucumeris* are commonly available in controlled release sachets that provide continuous supply of mites for up to six weeks. This is useful in flowering plants, as you do not contaminate the flower by broadcasting the mites with their carrier material. The sachets are simply hung in the canopy, and the mites leave to search for food and oviposit in the crop. In rooting/cloning, and even in the vegetative stage when plants are very close together, broadcasting loose *A. swirskii* and *N. cucumeris* is an option. *Amblyseius andersoni*, often used for spider mite prevention, will feed on thrips in the absence of spider mites. Similarly, *Amblyseius swirskii* will feed on spider mites and white flies, and actually performs better on a mixed prey diet. Both *A. swirskii* and *N. cucumeris* can be sustained on pollen when provided as an alternate food source, but their longevity and fecundity are reduced on a pure pollen diet compared to a live prey diet. Controlled release sachets of these mites can be used in inoculative programs and should be placed in the canopy where they are sheltered from direct sunlight and sprays. They should be placed away from the surface of the growing media, where they can absorb water. The sachets will

protect the mites inside against most pesticide sprays, but the contents can still become wet and begin to ferment. Controlled release sachets should be checked periodically by tearing open a few sachets from the crop and using a hand lens or microscope to determine if the predatory mites and the feeder mites inside are still active.

Minute pirate bug (*Orius insidiosus*) is an effective generalist that has a preference for thrips. It will feed on all mobile life stages of thrips. They also kill more thrips than they eat. *Orius insidiosus* nymphs are orange and gradually appear more like the adults. Development takes approximately ten days at 20°C. Nymphs may struggle on flowering cannabis plants with dense trichomes. Both nymphs and adults are predacious, and the adult live for 3–4 weeks. *Orius* will diapause when day lengths are less than 14 hours, which could reduce their effectiveness in the flowering stage. *Orius* populations can be maintained on banker plants with ornamental peppers (*Capsicum annuum* cv. 'Purple Flash') and sweet alyssum (*Lobularia maritima*). These banker plants can be further augmented by providing *Orius* with a supplemental food such as *Ephestia kuehniella* eggs.

Entomopathogenic nematodes such as *Steinernema feltiae* provide control against soil-dwelling stages (prepupal and pupal stages) of thrips. Nematodes should be applied to the rootzone and not the foliage. Be sure to read and follow the storage and application instructions when using nematodes. They are sensitive to growing media moisture levels, and can be easily killed if the media gets too dry or too wet. Nematodes are easily washed through rockwool based media, reducing their effectiveness.

When used as part of a fungus gnat control program, **soil predators** such as the mite *Stratiolaelaps scimitus* (formerly *Hypoaspis miles*), *Gaeolaelaps gillespiei* and the rove beetle *Dalotia coriaria* (formerly *Atheta coriaria*) will both provide additional control against the soil-dwelling life of thrips.

9.2.7.2 Cultural Control

Exclusion remains the most effective control for thrips in controlled environments. Thrips proof screening (< 215 μm mesh size) should be installed on all vents to prevent their entry into the facility. This size screening will also prevent most pests other than mites from entering the facility through the vents. Screening vents will reduce airflow, so plan accordingly. Maintaining a positive pressure inside the facility will also help prevent thrips from flying in. Both WFT and OT have very broad host ranges, so weeds growing near vents or other entry points should be managed. In facilities without vent screens, thrips populations in the facility can increase very suddenly due to activities from nearby fields such as mowing or spraying and harvesting of nearby crops that create large flight events. Incoming plant material should be quarantined and inspected for the presence of thrips. Hot water baths with temperatures of 41°C for 15 minutes can kill adult and larva western flower thrips, as well as their eggs.

Yellow sticky cards or tape can be effective for mass trapping WFT, but are not effective for OT. As OT is much more common than WFT in cannabis, this makes sticky cards a poor monitoring tool for thrips. Yellow sticky cards are more attractive to thrips than blue cards, but they also catch more beneficial insects such as *Aphidius* and *Eretmocerus*.

Plant health and nutrition is an important factor in thrips susceptibility. Drought-stressed plants are more prone to thrips damage. Similarly, excess nitrogen in foliar tissue can result in larger thrips populations due to increased fecundity.

9.2.7.3 Chemical Control

Western flower thrips has quickly developed resistance to many insecticides, while onion thrips has been slower to develop resistance. If you chose to spray chemical insecticides to control these pests, carefully identify the thrips species in your facility, and determine if you have a mixed population. Chemical insecticides may reduce the population of OT, but not that of WFT, while also killing most BCAs, leading to a WFT outbreak.

Foliar applications of horticultural oils (mineral and vegetable oils) can provide some control of the mobile life stages (larva and adult). Thorough coverage is necessary, as thrips are often in tight and protected areas. Oils will not control eggs or pupa, so repeated applications will be necessary.

The microbial biopesticide *Beauveria bassiana* is very effective at controlling adult thrips in controlled environments. However, it will not control larva, eggs or pupa. *Beauveria bassiana* will affect various BCAs including *Aphidius spp.*, *Aphidoletes aphidimyza*, *Chrysoperla rufilabris* and *Orius insidiosus*. *Metarhizium anisopliae* is also effective as either a foliar spray for adults or as a soil drench against pupa.

Dip treatments with horticultural oils, insecticidal soaps and microbial biopesticides on incoming cuttings have been effective at reducing thrips. However, label restrictions may prevent this treatment method.

Effective control of thrips requires the integration of many control methods including exclusion and mass trapping (cultural/physical control), BCAs such as *O. insidiosus*, predatory mites and entomopathogenic nematodes (biological control) and microbial biopesticides such as *Beaveuria bassiana* (chemical/biological control). Each of these controls can be used to target a different life stage of thrips.

9.2.8 WHITEFLY

Whiteflies can be a serious pest of cannabis produced in controlled environments. Two species of whiteflies have been recorded as pests on cannabis: greenhouse whitefly (GHW) (*Trialeurodes vaporariorum*) and sweetpotato whitefly (SPW) (*Bemisia tabaci*). Some authors have reported a third whitefly species, the silverleaf whitefly (*Bemisia argentifolii*, also known as *Bemisia tabaci* Biotype B). It is now generally accepted that *B. tabaci* and *B. argentifolii* are the same species.

Adult whiteflies rest on the underside of foliage and look like small white moths (Figure 9.17), but they are neither moths nor flies. They are more closely related to aphids. Adults are 1.5–2 mm long with white wings and a yellow body. Adults lay eggs on the underside of leaves; GHW eggs hatch in four days, while SPW eggs hatch in seven days. Eggs are small (0.2 mm), pear shaped and

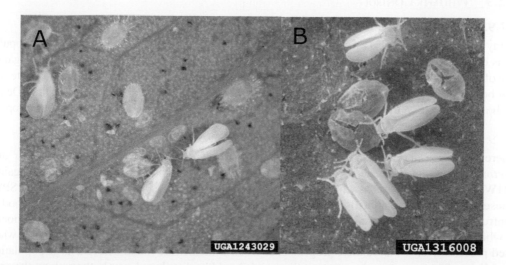

FIGURE 9.17 A) Adult and nymph greenhouse whitefly (GHW) (*Trialeurodes vaporariorum*); and B) adult Sweetpotato whitefly (SPW) (*Bemisia tabaci*) with empty pupal casings. Note the distinctly tent-like shape the wings are resting on SPW and the absence of hair-like filaments on the pupal casings compared to GHW.

Sources: Copyright: A) Whitney Cranshaw, Colorado State University, Bugwood.org; B) Scott Bauer, USDA Agricultural Research Service, Bugwood.org

are initially white. Greenhouse whitefly eggs will turn gray over time, while SPW eggs will become brown. Nymphs are flattened and a translucent white or yellow, with two yellow spots. Emerging nymphs move a short distance, then remain in one spot to feed until they pupate. The pupae do not feed and develop red eyes shortly before adult emergence. Sweetpotato whitefly develops slightly faster than GHW at 30°C, completing its lifecycle in as few as 16 days compared to 22 days. However, at cooler temperatures (20°C), GHW will develop 5–8 days faster than SPW. Adults live for up to 50 days at 16°C, but it can be as short as ten days in warmer temperatures (30°C).

Differentiating the two species can be difficult. As adults, SPW is slightly smaller than GHW. Sweetpotato whitefly will also hold its wings tent-like above its body, white GHW holds its wings flatter and parallel with the surface that it is resting on (Figure 9.17). The pupa can be differentiated, as GHW pupa are more a white or cream color and have long waxy hairs around the edge of their body. Conversely, SPW pupa are more yellow in color and lack the hairs around their body.

Whiteflies feed like aphids, inserting their stylet-like mouthparts into the plant to ingest phloem sap. This can reduce plant vigor and cause signs of nutrient deficiency or wilting, and large populations can reduce plant growth. This feeding mechanism means that whiteflies, like aphids, produce honeydew to expel excess sugars. Sooty mold can grow on whitefly honeydew, contaminating the flower and reducing transpiration and photosynthetic capacity in severe infestations. Sweetpotato whitefly is a tropical pest that does not overwinter in temperate areas. It is unlikely to be a common pest in controlled environments in areas where it does not overwinter, as the primary method of introduction would be on infested plant material. However, if established in a production facility, SPW can become a significant pest. Whitefly are also known to carry broad mites on their legs, acting as a mode of dispersal between plants for a pest that rarely moves far on its own.

To scout for whiteflies, look at the underside of leaves randomly, or where there are is honeydew or sooty mold. The different life stages of whiteflies are often stratified on the plant, with adults moving up to the youngest foliage, leaving nymphs and pupa on older leaves lower in the canopy. Shaking the plant will cause adults to leave the plant. SPW are more active than GHW, but adults of both species will only fly a short distance before landing again after being disturbed.

9.2.9   WHITEFLY CONTROL

9.2.9.1   Biological Control

Many of the parasitoids and predators described in the following subsections can reduce the populations of both species of whitefly. However, they likely have a preference or are more effective against one of the two species, so identifying the whitefly species is important. Biological control of whiteflies can be successful, but it may take a long time to find a program that works for your facility. Dense trichomes can reduce the efficiency of these parasitoids, as has been found in some vegetable crops.

9.2.9.1.1   Parasitoids

There are three main parasitoids of whiteflies: *Encarsia formosa*, *Eretmocerus eremicus* and *Eretmocerus mundus*. *Encarsia formosa* will parasitize SPW, but it is much more effective against GHW. *Eretmocerus eremicus* will parasitize both species, and *E. mundus* will only parasitize SPW. These parasitoids have difficulty walking through large colonies of whiteflies and are not good at controlling established whitefly populations or hotspots. Consider some of the predators mentioned in what follows for inundative releases against outbreaks. These parasitoids are most effective when used preventatively in inoculative programs, or when whiteflies are found before a large population establishes. These parasitoids are weak fliers and should be distributed evenly throughout the crop to help their searching efficiency.

Encarsia formosa is a small wasp, about 0.6 mm long. They have a black head and the rest of their body is yellow. Encarsia populations are almost entirely female. *Encarsia formosa* prefers to lay its eggs in third and fourth instar nymphs, but it can parasitize nymphs of any age. Adult wasps will also feed directly on early instar nymphs, a behavior known as host feeding. Their development varies significantly based on the host species and the host plant. *Encarsia formosa* takes longer to develop on SPW than on GHW. How long *E. formosa* development takes on cannabis is unknown. On most vegetables, larva develop in about 20 days at 25°C. Over time, parasitized GHW pupae turn black, while SPW turn brown. Adults are active at temperatures from 20–30°C and a relative humidity around 50–70%. Outside of that temperature range, activity and longevity drop significantly. Adults live for ten days at 30°C but can survive up to 30 days at 20°C. *Encarsia formosa* are sold as parasitized whitefly nymphs. They can be purchased loose or glued to cards. When using cards, hang them in the lower leaves of the canopy, as adults fly upwards immediately after emerging. Low light levels will reduce *E. formosa* activity, so if supplemental lighting is not used, avoid this BCA during low light times of the year.

Eretmocerus eremicus looks like *E. formosa* except that it is entirely yellow. Parasitized whitefly nymphs turn a less obvious yellow (GHW) or brownish-yellow (SPW). Adults are most effective at higher temperatures than *E. formosa* (25–29°C) and stay active at temperatures above 30°C, unlike *E. formosa. Eretmocerus eremicus* lays its egg underneath a second or third instar nymph, rather than inside it. A few days after the egg hatches, the larva will chew its way inside the nymph and wait until it pupates. *Eretmocerus eremicus* larva will then feed on the whitefly pupa. The entire lifecycle takes 17–20 days. Adult *E. eremicus* must host feed to start producing eggs. Adults live for 1–2 weeks, but lay most of their eggs in the first six days.

Eretmocerus mundus has a similar lifecycle as *E. eremicus*, but only parasitizes SPW. It is better at controlling SPW than *E. eremicus*. However, *E. mundus* may no longer be available in North America.

When examining whitefly colonies for parasitism, the color of the pupa will help identify parasitized pupae. It is also possible to check empty whitefly pupal cases for their emergence holes. Adult whiteflies emerge from a "T"-shaped emergence hole, and parasitoids emerge from a circular hole.

9.2.9.1.2 Predators

Various generalist predators will feed on whitefly. Combining predators with parasitoids is common for whitefly control in other crops. The following predators will all feed on both GHW and SPW.

Delphastus catalinae (formerly *Delphastus pusillus*) is a small black beetle roughly 1.4 mm long. It is an excellent whitefly predator. Both adult and larva feed on all life stages of whitefly. They even prefer to feed on unparasitized whiteflies, making them an excellent addition when *E. formosa* or *E. eremicus* have been released. Adults live for 6–9 weeks. Larvae are 4–5 mm long and cream in color. *Delphastus catalinae* is often used as a hot spot treatment, due to its voracious feeding habits and because it needs to consume prey before producing eggs. It does not diapause, but can be difficult to maintain in high numbers through the winter when temperatures are often cooler (about 20°C).

Predatory mites (*Amblyseius swirskii* and *Amblydromalus limonicus*) will feed on whitefly eggs and first instar whitefly nymphs. *Amblyseius swirskii* is commonly used as part of a thrips control program, but will feed on whitefly eggs. For more information on *A. swirskii*, see Section 9.2.7.1. *Amblydromalus limonicus* is another generalist predatory mite that is gaining attention as a potential BCA for whitefly control. It feeds on both thrips and whitefly, including eggs and all ages of whitefly nymphs. *Amblydromalus limonicus* is more effective at controlling whitefly than *A. swirksii* and provides nearly equal control of thrips, but it is likely not cost effective as a control for just thrips.

Dicyphus Hesperus is a generalist that will feed on most pests found in cannabis, but it is mostly used as a predator of whiteflies in vegetable crops. It is well adapted for plants with dense trichomes due to its larger size (6 mm) and long legs. Adults are long and slender. They are initially green, but become black with beige translucent wings. *Dicyphus hesperus* has a long lifecycle, taking five weeks from egg to adult at 25°C. However, both the adult and nymph stages will feed on insects with a preference for whitefly eggs and nymphs. Adults live for up to nine weeks. The adults are good searchers, flying to forage for prey. Mass-produced *D. hesperus* do not diapause. *Dicyphus hesperus* can be maintained on mullein (*Verbascum thapsus*) as a banker plant. It is an omnivore and needs to consume water from plants to survive. In other crops such as gerbera, it can cause cosmetic damage to the flower when their populations are very high (> 100/plant) or in the absence of prey. This is not likely to be a problem in cannabis.

Green Lacewings (*Chrysoperla carnea* or *Chrysoperla rufilabris*) larva will feed on all life stages of whitefly. For more information on green lacewings, see Section 9.2.3.1.

9.2.9.2   Cultural Control

Whiteflies can easily be excluded from controlled environments with vent screening and proper biosecurity. Incoming plant material should be inspected for all stages of whiteflies, including eggs.

Both species of whiteflies are trapped on yellow sticky cards or tape, but these traps also catch many BCAs including *Aphidius* and *Eretmocerus* wasps. As whitefly tend to congregate on leaves, vacuuming or de-leafing can be an option for a rapid reduction in the number of adults present. Vacuuming will only remove adults, and infested leaves can be missed while de-leafing, as well as disturbing adults, causing them to disperse, so neither of these methods alone will manage whitefly—but when combined could offer significant control in hotspots.

There are no known cultivars that are resistant to whitefly. Greater trichome densities may be related to a decreased susceptibility to whitefly, as has been found in some vegetable crops.

9.2.9.3   Chemical Control

Whitefly, especially SPW, develop resistance to chemical insecticides very quickly. Horticultural oil and insecticidal soaps can reduce whitefly populations. These products do not provide great control against adults, as they can fly away before the products contact them. These products will kill whitefly nymphs and pupa.

Various microbial biopesticides can infect whitefly, including *Beauvaria bassiana*, *Isaria fumosorosea*, *Lecanicillium lecanii* and *Metarhizium anisopliae*. These products work primarily against the nymphs and adult life stages. *Beauvaria bassiana* and *Lecanicillium lecanii* provide the best control of these products.

9.2.10   FUNGUS GNATS AND SHORE FLIES

Fungus gnats (*Bradysia* spp.) are small slender flies that resemble mosquitos and can be confused for some BCAs like *Aphidius* wasps, *Feltiella acarisuga* or *Aphidoletes aphidimyza*. Fungus gnat adults are 3–4 mm long and black (Figure 9.18). Adults are weak fliers but are very active, especially when disturbed. They can often be seen on growing media, where they lay their eggs. Adult fungus gnats feed on water or nectar and do not cause any direct damage themselves; although, in cannabis, adults are prone to getting stuck in trichomes, contaminating the flower. Adults can transmit root diseases, going from plant to plant and walking across the media (Gillespie and Menzies 1993). The eggs hatch in as few as two days and 2–4 mm long, translucent white larvae with black heads feed for 1–2 weeks before pupating for five days. The entire lifecycle takes approximately 21 days at 25°C. Larvae feed mostly on decaying organic matter, but also feed on fine roots and can tunnel into the tender parts of the crown and stem of the plant, which can affect plant vigor and health.

Fungus gnats are often confused for shore flies (*Scatella stagnalis*), as they have a similar but more robust appearance. They can be identified by the five white dots on their wings. Shore fly larvae lack the black head of fungus gnat larvae. The main difference is that shore fly larvae do not

FIGURE 9.18 Adult dark-winged fungus gnat (*Bradysia* sp.).

Source: Copyright Whitney Cranshaw, Colorado State University, Bugwood.org

feed on healthy plant material. The adults can transmit root diseases like fungus gnats. Shore fly development is also quicker than fungus gnats, taking just 14 days to complete the lifecycle.

9.2.11 Fungus Gnat and Shore Fly Control

9.2.11.1 Biological Control

Soil-dwelling predators used to target thrips pupa are even more effective against fungus gnats and shore flies. *Dalotia coriaria* is a small elongated (3–4 mm) beetle that spends most of its time on or in the growing media. Both the adults and larvae are predacious. They feed on fungus gnat/shore fly larva. They also may feed on small root aphids, but they do not provide effective control of this pest. Adults live for up to 21 days. They curve their abdomen upwards when disturbed, appearing like a scorpion. They are most effective at temperatures from 20–25°C. The predatory mites *Stratiolaelaps scimitus* (formerly *Hypoaspis miles*) and *Gaeolaelaps gillespiei* are also effective as soil-dwelling predators that feed on fungus gnat and shore fly larvae and eggs. Both are small (< 1 mm) beige mites that complete their lifecycle in 18 days at 20°C. Both mites are most effective at temperatures from 15–25°C. *Gaeolaelaps gillespiei* is more active on the surface of the media and is more effective at controlling fungus gnats than S. *scimitus*. *Gaeolaelaps gillespiei* is most effective in coco fiber and rockwool-based media.

Entomopathogenic nematodes such as *Steinernema feltiae* applied to the media provide excellent control against soil-dwelling stages of both pests. For more information on nematodes, see Section 9.2.7.1.

9.2.11.2 Cultural Control

Both fungus gnats and shore flies can be easily managed by allowing the surface of the growing media to dry out, as the larva of both species are prone to desiccation.

Do not allow puddles or algae growth on the ground or on growing surfaces such as benches and troughs, as these serve as breeding areas for both pests. Peat moss or coco fiber–based growing media are more prone to fungus gnat and shore fly infestations, due to the organic matter.

Fungus gnats can be mass trapped using yellow sticky cards or tape, and shore flies are more effectively trapped on blue sticky cards or tape.

9.2.11.3 Chemical Control

Fungus gnats and shore flies rarely require chemical control. Horticultural oils and insecticidal soaps are not effective as making good contact with the adults or larvae can be difficult. *Bacillus thuringiensis* subsp. *Israelensis* is a microbial biopesticide that can be used to control the larvae in the growing media.

9.2.12 Other Pests

With more than 300 species of insects recorded on cannabis (McPartland et al. 2000; Cranshaw et al. 2019), there are undoubtedly more species of pests that will be encountered in controlled environment cannabis production than are covered in this chapter. Fortunately, with proper cultural controls (e.g., screening vents, biosecurity—discussed earlier), most of these insects can be easily excluded from production facilities. In what follows, we will briefly discuss some additional pests that may be found in cannabis facilities either because they are common greenhouse pests (e.g., leafhopper, lygus bugs) or because they can cause serious damage to cannabis plants if they manage to get into the facility (e.g., caterpillars).

9.2.12.1 Caterpillars

Many species of caterpillar are known to feed on cannabis. Adult moths of these species are large (up to 45 mm wide), so even the most basic vent screening can exclude them.

Corn earworm (*Helicoverpa zea*) is a Lepidopteran (moth) insect and one of the most important pests of outdoor cannabis especially in the United States. It is a significant pest of corn, but is also attracted to cannabis after corn has stopped silking and cannabis is flowering. Larvae, or caterpillars, are the damaging life stage, chewing through developing flowers and leaving behind large fecal pellets. Corn earworm caterpillars are variable in color (Figure 9.19) and feed

FIGURE 9.19 Different instars of corn earworm (*Helicoverpa zea*) caterpillars, displaying the various coloring and sizes as the caterpillars develop.

Source: Copyright Whitney Cranshaw, Colorado State University, Bugwood.org

FIGURE 9.20   Cabbage looper (*Trichoplusia ni*) caterpillar.

Source: Copyright Whitney Cranshaw, Colorado State University, Bugwood.org

for approximately 23 days at 25°C before pupating. Chemical control options include formulated insecticides containing *Helicoverpa armigera* Nucleopolyhedrovirus BV-0003, which is specific to corn earworm, or *Bacillus thuringiensis* subsp. *kurstaki* can control most Lepidopteran pests. In less severe infestations, physical controls such as manually removing caterpillars is also an option.

Other moths such as the cabbage looper (*Trichoplusia ni*) are also common in greenhouses. Adult moths are a gray-brown, while caterpillars are bright green (Figure 9.20). Caterpillars complete their development in up to two weeks, and while they initially do not eat much, later instars can consumer a significant amount of plant material, up to three times their body weight daily. Chemical control options include formulated products containing *Autographa californica* Nucleopolyhedrovirus FV11, which is specific to cabbage looper, or *Bacillus thuringiensis* subsp. *kurstaki*, which can control most Lepidopteran pests. *Trichogramma*, a small parasitoid wasp, can provide some control against cabbage looper caterpillars. Again, like with corn earworm, lighter infestations can be managed by physically removing the pest.

Pheromone-baited traps for both of these moths are available commercially and can be used to monitor for these pests. If these pests frequently enter the facility, then pheromone traps can help detect them. However, screening vents can very effectively exclude these Lepidopteran pests.

9.2.12.2 Leafhoppers

Many different species of leafhoppers (Cicadellidae) can enter greenhouses, and some polyphagous species are likely able to feed and reproduce on cannabis. Many species have been recorded on outdoor cannabis. Leafhoppers are highly variable in color, but have a distinct wedge shape (Figure 9.21) and hop off foliage and take flight when disturbed. Nymphs are flightless, but move quickly and can hop off the foliage. Adults and nymphs ingest plant fluids like aphids, and similarly produce honeydew. Leafhoppers are difficult to control, as they can escape insecticide sprays. There is no commercially available biological control for leafhoppers. They can be mass trapped using sticky tape, but this only targets the adults. Exclusion and crop-free periods are the best way to manage this pest. Keeping the area around the facility free of weeds can help reduce the chance leafhoppers enter the facility.

FIGURE 9.21 Adult potato leafhopper (*Empoasca fabae*) (image of a nymph in the insert). Potato leafhopper is one of many species of leafhopper recorded on cannabis, and not necessarily the only species you will find in controlled environment production facilities. Notice the distinct wedge shape of the body.

Source: Copyright Steve L. Brown, University of Georgia, Bugwood.org

FIGURE 9.22 Adult tarnished plant bug (*Lygus lineolaris*) (image of a nymph in the insert).

Source: Copyright Scott Bauer, USDA Agricultural Research Service, Bugwood.org

9.2.12.3 Lygus Bugs

Lygus bugs, such as the tarnished plant bug (*Lygus lineolaris*), are a common pest of cannabis grown outside. Their feeding is much more destructive than other piercing/sucking insects. However, they are unlikely to cause significant yield losses. Nymphs are green and turn a beige-brown as adults (Figure 9.22). Adults will fly away or drop to the ground when disturbed. Nymphs are flightless but will run to the underside of leaves to hide when disturbed. Like leafhoppers, no biological controls are available and chemical control is not effective. Physical controls such as exclusion by using screens on vents works well. Cultural controls, such as maintaining a weed-free perimeter around the controlled environment facility, and crop-free periods are also effective ways to manage this pest.

BIBLIOGRAPHY

Auger, P., A. Migeon, E. A. Ueckermann, L. Tiedt, M. Navajas, and M. Navajas. 2013. Evidence for synonymy between *Tetranychus urticae* and *Tetranychus cinnabarinus* (Acari, Prostigmata, Tetranychidae): Review and new data. *Acarologia* 53(4): 383–415. https://www1.montpellier.inra.fr/CBGP/acarologia/export_pdf.

Blackman, R. L., and V. F. Eastop. 2008. *Aphids on the world's herbaceous plants and shrubs, 2 volume set.* West Sussex: John Wiley & Sons.

Buitenhuis, R. 2017. *Grower guide: Quality assurance of biocontrol products.* www.vinelandresearch.com/wp-content/uploads/2020/02/Grower-Guide.pdf.

CABI. 2015. *Bemisia tabaci (MEAM1) (silverleaf whitefly): Invasive species compendium.* www.cabi.org/isc/datasheet/8925 (accessed November 20, 2021).

Carmichael, E. J. 2020. A survey of aphid species and their associated natural enemies in Fraser Valley hop fields and an exploration of potential alternative Summer hosts of the Damson-Hop Aphid, Phorodon humuli (Hemiptera: Aphididae). Master's thesis. Virginia Polytechnic Institute and State University. https://vtechworks.lib.vt.edu/bitstream/handle/10919/98493/Carmichael_2019_Hop_Field_Survey_and_Transfer_Experiments.pdf.

Cranshaw, W., S. Halbert, C. Favret, K. Britt, and G. Miller. 2018. *Phorodon cannabis* Passerini (Hemiptera: Aphididae), a newly recognized pest in north america found on industrial hemp. *Insecta Mundi* 0662: 1–12. https://journals.flvc.org/mundi/article/view/0662/102363.

Cranshaw, W., M. Schreiner, K. Britt, T. Kuhar, J. McPartland, and J. Grant. 2019. Developing insect pest management systems for hemp in the United States: A work in progress. *Journal of Integrated Pest Management* 10(1): 1–11. https://academic.oup.com/jipm/article-pdf/10/1/26/33022980/pmz023.pdf.

Cranshaw, W., and S. Wainwright-Evans. 2020. *Cannabis sativa* as a host of rice root aphid (Hemiptera: Aphididae) in North America. *Journal of Integrated Pest Management* 11(1). https://academic.oup.com/jipm/article-pdf/11/1/15/33514867/pmaa008.pdf.

Fan, Y., and F. L. Petitt. 1998. Dispersal of the broad mite, *Polyphagotarsonemus latus* (Acari: Tarsonemidae) on *Bemisia argentifolii* (Homoptera: Aleyrodidae). *Experimental & Applied Acarology* 22(7): 411–415. https://link.springer.com/content/pdf/10.1023/A:1006045911286.pdf.

Fasulo, T. R. 2019. Broad mite, *Polyphagotarsonemus latus* (Banks) (Arachnida: Acari: Tarsonemidae). *EDIS.* https://edis.ifas.ufl.edu/pdf/IN/IN34000.pdf.

Ferguson, G., G. Murphy, and L. Shipp. 2014a. *Whiteflies in greenhouse crops – biology, damage and management.* www.omafra.gov.on.ca/english/crops/facts/14-031.htm (accessed November 20, 2021).

Ferguson, G., G. Murphy, and L. Shipp. 2014b. *Fungus gnats and shore flies in greenhouse crops.* www.omafra.gov.on.ca/english/crops/facts/14-003.htm (accessed November 20, 2021).

Food and Agriculture Organization of the United Nations. 2022. *How to practice integrated pest management.* www.fao.org/agriculture/crops/thematic-sitemap/theme/spi/scpi-home/managing-ecosystems/integrated-pest-management/ipm-how/en/ (accessed November 20, 2021).

Gates, C., M. Brownbridge, R. Buitenhuis, G. Ferguson, G. Murphy, and S. Jandricic. 2015. Biocontrol-based IPM: Applied knowledge for greenhouse pest control. *Greenhouse IPM.* http://greenhouseipm.org/ (accessed November 20, 2021).

Giladi, Y., L. Hadad, N. Luria, W. Cranshaw, O. Lachman, A. Dumbrovsky, and Y. Giladi. 2019. First report of beet curly top virus infecting *Cannabis sativa* in Western Colorado. *Plant Disease* 104(3): 999. https://apsjournals.apsnet.org/doi/suppl/10.1094/PDIS.2020.104.issue-3/suppl_file/cover.pdf.

Gillespie, D. R., and J. Menzies. 1993. Fungus gnats vector *Fusarium oxysporum* f.sp. radicislycopersici1. *Annals of Applied Biology* 123(3): 539–544. https://onlinelibrary.wiley.com/doi/epdf/10.1111/j.1744-7348.1993.tb04926.x.

Gotoh, T., Y. Kitashima, and T. Sato. 2013. Effect of hot-water treatment on the two-spotted spider mite, *Tetranychus urticae*, and its predator, *Neoseiulus californicus* (Acari: Tetranychidae, Phytoseiidae). *International Journal of Acarology* 39(7): 533–537. www.tandfonline.com/doi/abs/10.1080/01647954.2013.857720.

Hadad, L., N. Luria, E. Smith, N. Sela, O. Lachman, and A. Dombrovsky. 2019. Lettuce chlorosis virus disease: A new threat to cannabis production. *Viruses* 11: 802, August. www.mdpi.com/1999-4915/11/9/802/pdf.

Harris, M. A., R. D. Oetting, and W. A. Gardner. 1995. Use of entomopathogenic nematodes and a new monitoring technique for control of fungus gnats, *Bradysia coprophila* (Diptera: Sciaridae), in floriculture. *Biological Control* 5(3): 412–418. https://www-sciencedirect-com.subzero.lib.uoguelph.ca/science/article/pii/S1049964485710493/pdf.

Hoddle, M., R. Van Driesche, and J. Sanderson. 1998. Biology and use of the whitefly parasitoid *Encarsia formosa*. *Annual Review of Entomology* 43: 645–649, February. www.annualreviews.org/doi/pdf/10.1146/annurev.ento.43.1.645.

Kogan, M. 1998. Integrated pest management: Historical perspectives and contemporary developments. *Annual Review of Entomology* 43(1): 243–270. www.annualreviews.org/doi/pdf/10.1146/annurev.ento.43.1.243.

Krantz, G. W., and E. E. Lindquist. 1979. Evolution of phytophagous mites (Acari). *Annual Review of Entomology* 24(1): 121–158. www.annualreviews.org/doi/pdf/10.1146/annurev.en.24.010179.001005.

McCune, F., C. Murphy, J. Eaves, and V. Fournier. 2021. Onion thrips, *Thrips tabaci* (Thysanoptera: Thripidae), reduces yields in indoor-grown cannabis. *Phytoprotection* 101(1): 14–20. www.erudit.org/fr/revues/phyto/2021-v101-n1-phyto05944/1076365ar.pdf.

McPartland, J., R. C. Clarke, and D. P. Watson. 2000. *Hemp diseases and pests: Management and biological control: An advanced treatise.* Wallingsford: CABI.

McPartland, J., and K. W. Hillig. 2003. The hemp russet mite. *Journal of Industrial Hemp* 8(2): 107–112. www.tandfonline.com/doi/abs/10.1300/J237v08n02_10.

Qiu, Y. T., J. Van Lenteren, Y. Drost, and C. Posthuma-Doodeman. 2004. Life-history parameters of *Encarsia formosa*, *Eretmocerus eremicus* and *E. mundus*, aphelinid parasitoids of *Bemisia Argentifolii* (Hemiptera: Aleyrodidae). *European Journal of Entomology* 101(1): 83–94. www.eje.cz/pdfs/eje/2004/01/17.pdf.

Sani, I., S. I. Ismail, S. Abdullah, J. Jalinas, S. Jamian, and N. Saad. 2020. A review of the biology and control of whitefly, *bemisia tabaci* (Hemiptera: Aleyrodidae), with special reference to biological control using entomopathogenic fungi. *Insects* 11(9): 619. www.mdpi.com/2075-4450/11/9/619/pdf.

Schmidt, H. E., and E. Karl. 1970. Ein Beitrag Zur Analyse Der Virosen Des Hanfes (*Cannabis sativa* L.) Unter Berucksichtigung Der Hanfblattlaus (*Phorodon cannabis* Pass.) Als Virusvektor. Als Virusvektor. *Zentralblatt Fur Bakteriologie, Parasitenkunde, Infektionskrankheiten Und Hygiene* 125: 16–22.

Schreiner, M., and W. Cranshaw. 2018a. *Pest management of hemp in enclosed production: Hemp russet mite.* https://webdoc.agsci.colostate.edu/hempinsects/PDFs/HempRussetMite_New_6-28-18.pdf (accessed November 20, 2021).

Schreiner, M., and W. Cranshaw. 2018b. *Pest management of hemp in enclosed production: Rice root aphid.* https://webdoc.agsci.colostate.edu/hempinsects/PDFs/RiceRootAphidextrainfo.pdf (accessed November 20, 2021).

Schreiner, M., and W. Cranshaw. 2018c. *Pest management of hemp in enclosed production: Cannabis aphid.* https://webdoc.agsci.colostate.edu/hempinsects/PDFs/Cannabis Aphid Management_.pdf (accessed November 20, 2021).

Schreiner, M., and W. Cranshaw. 2018d. *Pest management of hemp in enclosed production: Thrips.* https://webdoc.agsci.colostate.edu/hempinsects/PDFs/Thrips extra info.pdf (accessed November 20, 2021).

Summerfield, A., and S. Jandricic. 2021. *Simple key to important thrips pests of Canadian greenhouses* (2nd ed). GreenhouseIPM.org/pests/thripskey (accessed February 20, 2022).

van Lenteren, J. C. 2003. *Quality control and production of biological control agents: Theory and testing procedures.* Wallingford: CABI.

Wimmer, D., D. Hoffmann, and P. Schausberger. 2008. Prey suitability of western flower thrips, *Frankliniella occidentalis*, and onion thrips, *Thrips tabaci*, for the predatory mite *Amblyseius swirskii*. *Biocontrol Science and Technology* 18(6): 533–542. www.tandfonline.com/doi/abs/10.1080/09583150802029784.

10 Harvest and Post-Harvest

Deron Caplan, Philipp Matzneller and Juan David Gutierrez

CONTENTS

DOI: 10.1201/9781003150442-10

10.1   INTRODUCTION

Producing high-quality cannabis goes well beyond cultivation. The harvest and post-harvest processes are crucial to preserve and improve crop quality. Ineffective post-harvest management can easily ruin a crop, lower its value or make it unsalable.

Section 10.2 examines how to assess crop maturity and time the harvest. Following sections describe the theory and practice surrounding post-harvest procedures, including trimming, drying, curing, and storage, based on scientific literature, personal experience and consultation with experienced cannabis cultivators. In the final section, some methods to calculate and report crop yield are presented.

10.2   HARVEST

10.2.1   Harvest Timing and Maturity

As a cannabis plant nears the end of its lifecycle, vegetative growth slows while its inflorescences swell to their maximum size. The stigmas of the inflorescences turn orange or brown, and the resinous trichome glands change in color from clear to opaque, to yellow or amber, eventually becoming brown (Mahlberg and Kim 2004). If left unharvested, flowering plants will continue through a degenerative process termed "senescence" until they eventually die. This process varies by cultivar and growing environment, so selecting a harvest window can be challenging based on visual observation alone. This section describes some methods that can be used to make harvesting decisions and outlines factors that can influence crop maturation rates.

There are no universally accepted methods to characterize crop maturity or readiness for cannabis; however, some practices are useful regardless of how maturity is assessed. For a commercial grower, crop-level maturity is far more important than that of individual plants. Crop maturity can be assessed by inspecting multiple inflorescence sites on a representative number of plants. It is best to inspect inflorescences from different locations of a plant since, generally, those at the top of the plant mature faster than those at the bottom.

Harvesting subsets of plants within a crop based on relative maturity (i.e., selective harvesting) ensures uniform quality and maturity of the final product. It also allows for lower inflorescences to continue maturing, as they typically mature slowest. Conversely, selective harvesting imposes operational challenges such as long harvest events and variable starting points for post-harvest processing. Selective harvesting is usually better suited for small operations whereby quality is more important than operational efficiency.

10.2.2   Harvesting to Mitigate Spoilage

As plants reach the end of the flowering stage, they become more susceptible to fungal pathogens. Large inflorescences are particularly vulnerable to infection; their size creates high-humidity microclimates within the inflorescence, which encourages fungal growth.

One of the most common diseases during late flowering is gray mold or "bud rot," caused by *Botrytis cinerea* and some other genera of fungi. The onset of bud rot can be slow and difficult to identify, but infection rates can quickly increase, so diligent scouting is required during late flowering. The growing environment should also be carefully controlled and kept consistent; specifically, the air movement around the large inflorescences should be sufficient to prevent water from condensing. Moist conditions in and around the inflorescences expedite the spread of fungal pathogens. In our experience, bud rot damage tends to be most severe in locations where the air movement is less than 0.5 m/s. Maintaining an air movement of 0.5–1.0 m/s above the canopy using circulation fans is recommended throughout the flowering stage.

To avoid the spread of bud rot, it helps to remove infected inflorescences as soon as symptoms are observed. An early harvest can also prevent crop damage and product loss when bud rot symptoms are severe.

10.2.3 Trichome Head Color

The most common method for assessing inflorescence maturity relies on the color or opacity of the inflorescence's glandular trichome heads.

There are several types of trichomes on cannabis plants, but those which contribute most to resin production in the inflorescences are the capitate-stalked type. These trichomes are at their highest concentrations on the surfaces of calyxes and bracts around cannabis inflorescences (Mahlberg and Kim 2004). The formation and size of capitate-stalked trichomes increase as the plant matures, especially during flowering, though the rate of trichome and resin production depends both on genotype and environmental conditions, and varies by location on the plant.

It is commonly believed that cannabinoid content is at its maximum when capitate-stocked trichomes appear translucent (milky). Further, when they appear amber, the inflorescence is over-ripe, indicating a loss of volatile terpenes (smell and flavor) and THC's and CBD's breakdown into CBN (loss of potency). Some growers use trichome color to determine when plants are ready to be harvested based on these factors. It can be difficult to accurately gauge maturity when using this practice, because not all trichomes mature at the same time. Though this practice is widespread, the relationship between trichome color and cannabinoid or terpene composition is still poorly understood.

10.3 FACTORS THAT INFLUENCE INFLORESCENCE MATURITY

10.3.1 Cultivar Differences

Clonal offspring from the same cultivar typically ripen at around the same time if grown under similar conditions. As with other cultivated species, there is a genotypic component that determines the maturation rate. Plants from different genotypes (different cultivars) can reach maturity at drastically different times. Most modern cultivars were bred for controlled environment production and mature in 45–70 days after induction of flowering. Cultivars that take longer than 70 days to flower may be of lower economic value due to the added resources needed to extend the flowering period and a typically lower annualized yield compared to faster-maturing cultivars.

10.3.2 Environment

It is generally accepted that the growing environment plays a role in the maturation rates of a cannabis crop. Higher temperatures, specifically average 24-hour temperature, may increase maturation rates. Similar to other crops, higher light intensity during the flowering stage may also increase maturation rates (Adams et al. 2001). It is speculated that reducing the light intensity a few days before harvest will lead to higher terpene content. The lower light intensity and lower radiant heat associated with it could prevent heat-induced terpene volatilization in the inflorescences.

10.3.3 Controlled Stress

Controlled stress can—in some cases—increase secondary metabolite production in essential oil–producing plants. A prominent example is in herbs and spices cultivated in semi-arid regions such as the Mediterranean. Intermittent drought and high solar radiation in these areas have been attributed to aromatic herbs and spices with abundant essential oil (Kleinwächter and Selmar 2015). Controlled drought applied during late flowering may also increase THC and CBD yield in cannabis (Caplan et al. 2019). Cannabis growers commonly use drought and other stressors to influence various aspects of their crops. For most controlled stress treatments applied to cannabis, the efficacy and mechanisms of action are still poorly understood. Further, their effects depend on the cultivar and growing environment-specific conditions. More academic research and grower trials are needed to understand the utility of controlled stressors to optimize inflorescence yield and quality in a

TABLE 10.1

Summary of Controlled Plant Stressors Used in Controlled Environment Cannabis Production at the End of the Flowering Stage

Treatment	Scientific Evidence	Target or Proven Effect
UVB radiation during flowering	Yes	Yield and cannabinoid concentration of indoor-grown cannabis are not improved with long-term exposure to ultraviolet-B radiation (Rodriguez-Morrison et al. 2021)
Controlled drought stress	Yes	Increase total THC and CBD concentration and yield (Caplan et al. 2019)
Low atmospheric CO_2 levels	No	Slow inflorescence growth and concentrate cannabinoids and terpenes
Pre-harvest growing media leaching	No	Reduce chlorophyll content in inflorescences to increase smoke quality
Low relative humidity during last weeks of flowering	No	Increase trichome production and density; lower risk of microbial spoilage
Wounding	No	Increase cannabinoid and terpene production
Cold (growing media or air temperature)	No	Increase cannabinoid and terpene production; promote blue and purple pigmentation of inflorescences

consistent, safe and cost-effective way. Table 10.1 summarizes some pre-harvest stressors that are commonly used by controlled environment cannabis growers.

10.3.4 Pre-Harvest Growing Media Leaching (Flushing)

Arguably, the most widespread pre-harvest stressor used by cannabis growers is growing media leaching, commonly known as flushing. Typically, fertilizer application is stopped during the last week or two of flowering and only fresh water is used for irrigation. The result is a drop in growing media electrical conductivity (EC) and therefore fertilizer content. Theoretically, the maturing plant becomes reliant on internal energy reserves to continue its growth. The reduction of mineral nutrition at the end of the crop is thought to produce inflorescences with a smoother smoke and white-colored ash after combustion.

A comparable practice is used in the curing step of tobacco processing. Curing involves the enzymatic breakdown of chlorophyll to preserve quality, flavor and aroma (Sumner and Moore 2009). If tobacco plants are grown with an oversupply of nitrogen, curing is negatively affected, leading to a harsh or bitter taste when smoked. Similar reasoning can be applied to the practice of flushing cannabis; lower mineral nutrition before harvest could result in a smoother smoking experience. More research is needed to validate the impacts of flushing and its effects on cannabis smoke quality.

10.4 THE HARVESTING PROCESS

Once the grower determines that the crop is mature enough to deliver a desirable end-product, the harvesting process begins. Plants are typically cut just above the growing media, or individual branches are cut one at a time. Branches are usually cut with pruners then placed on drying racks or transported in bins or bags to the processing area. To prevent contamination, inflorescences that touch the growing media, bench or floor can be discarded. Inflorescences are sometimes loosely attached to the stems and may fall off during harvest, so gentle handling is essential. When hang drying, it is good practice to keep the inflorescences on the stem, leaving enough stem material to hang or clip each branch to a drying rack.

When transporting freshly harvested plant material, avoid sealing it in airtight containers or packing too densely. After harvest, the plants continue to transpire, leading to moist conditions in transport containers and an increased risk of spoilage. Pack plants in open containers or loosely in closed containers and process them soon after harvest. When using hard-bottomed transport containers, the plants can be arranged standing upright with their stems' base at the bottom of the container; this avoids compressing the inflorescences during transport.

10.5 POST-HARVEST

The post-harvest process includes all the procedures conducted after the crop is harvested until packaging, including trimming, drying, curing and storage. These practices are critical to producing a high-quality end-product that meets consumer expectations and resists spoilage.

10.5.1 INTENDED END-USE OF THE CROP

The harvest methodology often depends on the intended end-use of the crop. If the crop is to be sold as whole dry inflorescence, a producer may take extra precautions to maintain the inflorescence's integrity during the harvesting process. The extra care often lowers harvest efficiency and increases labor costs, but results in a more intact final product. Conversely, if the crop is to be processed as biomass for extraction, gentle handling may be less important and more efficient technologies or practices can be employed.

10.5.1.1 Dry Smokable Inflorescences

Consumers often pay close attention to the visual appearance of cannabis inflorescences in addition to their aroma, quality of smoke and physiological effects. To that end, producers can tailor their harvesting, trimming, drying and curing methodologies to preserve the integrity of their dry smokable inflorescence. For instance, if the market demands a hand-trimmed product, a grower can alter their processing methodology to meet that demand, or—if high total terpene content is important—the grower may elect to decrease their harvest-to-sale duration, since terpene content decreases quickly after harvest (Milay et al. 2020).

10.5.1.2 Biomass for Extraction or Derivative Products

A crop destined for extraction can be processed using cost-effective and automated methods or trimming, and curing can be omitted altogether. If the biomass is to be extracted for full-spectrum products that include all secondary metabolites such as minor cannabinoids, terpenes and flavonoids, then the aim should be to preserve these compounds during processing. Conversely, if the goal is to extract cannabinoid isolates from the biomass, it may be feasible to speed-dry at higher temperatures, which would otherwise cause evaporation of terpenes in the final product.

10.6 TRIMMING

The process of removing branches and leaves from cannabis inflorescences is called trimming. The goal is to preserve desirable plant parts, usually based on their composition of cannabinoids and terpenes. Most of the tetrahydrocannabinol (THC) and cannabidiol (CBD) are found in inflorescences, but the cannabinoid content in the inflorescence leaves is also considerable; about half that of the inflorescences. In the fan leaves, cannabinoid concentration is only about 10% of that in the inflorescences (Nirit et al. 2019). As such, fan leaves are often considered a byproduct and are discarded.

Trimming is a ubiquitous practice among cannabis cultivators, but to the best of our knowledge, no scientific studies are published evaluating cannabis trimming techniques. For now, growers rely on practices based on anecdotal evidence that were developed from experience or years of trial and error.

The primary objective of trimming is to create a product that meets consumer expectations. Typically, for consumers of smokable dried cannabis, this means that inflorescence leaves are completely removed from the inflorescences and the trimming process does not alter the natural shape of the inflorescence (Figure 10.1). If too much plant material is removed, the inflorescences can develop a round shape, diminishing their aesthetic appeal. Behavioral studies on fruit and vegetable sales have shown that consumers expressly avoid unattractive produce nearly all the time (Grewal et al. 2019), and only 36% of online shoppers in the United States would consider buying imperfect fruit or vegetables (Henderson 2017). It stands to reason that aesthetic appeal also plays an essential role in the perceived quality and resultant sales of cannabis.

In tobacco, the green color of cured leaves indicates a high chlorophyll content and is associated with a harsh, bitter taste (Sumner and Moore 2009). In cannabis, there is more chlorophyll in the leaves than in the inflorescences (Nirit et al. 2019). Therefore, trimming inflorescence leaves reduces the overall chlorophyll content and theoretically increases the smoke quality of inflorescences.

For the most part, terpenes account for the aroma and flavor of cannabis inflorescences—and for some consumers, an intense aroma indicates higher quality. Inflorescence leaves contain approximately 50% of the terpene content of the inflorescences, and inflorescence leaves account for between 20% and 40% of the total yield (excluding stems and fan leaves). Therefore, by including inflorescence leaves in the final product, the total terpene concentration is diluted. The same concept holds for cannabinoids. If the potential cannabinoid concentration of a trimmed inflorescence is 20%, but the inflorescence leaves are not removed, the total cannabinoid content is reduced to around 17%.

10.6.1 Trimming Methodology

A thoughtful trimming methodology is critical, considering its impacts on consumer preference, smoke quality and the cannabinoid/terpene content of the final product. There are two primary methods to trim cannabis: wet trimming and dry trimming. In the former, leaves are removed before drying, while in the latter, the plant is dried before leaves are removed.

One consideration when selecting a methodology is the risk of spreading bacteria and fungi that were present while the crop was growing. Interestingly, more mold has been observed in wet-trimmed inflorescences compared to an untrimmed control (Punja et al. 2019). The increased mold may be attributed to wounding during the trimming process, which is known to increase colonization by mold in various fruits (Vilanova et al. 2014; Kavanagh and Wood 1967).

Fungal pathogens are particularly problematic; they can infect plants during the crop cycle, but are also responsible for product losses after harvest. Powdery mildew is a common pathogen on cannabis leaves, while *Penicillium*, *Botrytis* and *Fusarium* are often present to varying extents in inflorescences. Inflorescence leaves have been found to have higher *Cladosporium* and *Penicillium* counts than inflorescences (Punja et al. 2019). In locations with limited control over temperature, relative humidity and air movement, eliminating potentially fungus-infected leaves before drying could reduce the risk of mold development.

10.6.1.1 Wet Trimming

The first step in processing freshly harvested cannabis plants is to transfer whole or partial plants out of the growing area to a separate and clean space where trimming can begin. The fan leaves are removed from the plant manually by pressing the lower part of the petiole downwards towards the stem until it breaks. Trimming scissors can also be used, but in our experience, removing fan leaves by hand accelerates the process with the same effect.

Next, the inflorescence leaves are trimmed. This can be done manually with scissors or by using an automated trimming machine (Figure 10.2). In both cases, trimming is easier and faster when the inflorescence leaves are still turgid. After being cut from the plant, leaves start to lose water and begin wilting. Plants visibly wilt after losing 1–40% of their water content (Juneau and Tarasoff 2012); therefore, cannabis should be wet trimmed within a maximum of a couple of hours after

FIGURE 10.1 Stages of cannabis trimming. Top left: Untrimmed inflorescence with inflorescence leaves. Top right: Trimmed inflorescence with some inflorescence leaves. Bottom left: Trimmed inflorescence with no inflorescence leaves and natural shape. Bottom right: Over-trimmed inflorescence, which lost its natural shape.

harvest. Some commercial growers store the harvested plants at low temperatures (4–10°C), which slows the loss of turgidity if plants cannot be processed quickly.

Trichome preservation is critical during the trimming process. Trichomes separate more easily from the plant when the water content is low during trimming (dry trimming). Since trichomes are more prone to separating from dry plant material during processing, wet trimming likely has the advantage of better preserving trichomes in the final product. Handling of the plant material can also lead to separation and loss of trichomes. Inflorescences should be touched as little as possible, and only using clean gloves or tools, to both reduce trichome loss and prevent microbial contamination.

10.6.1.2 Dry Trimming

In dry trimming, inflorescence leaves are removed after drying, and fan leaves are removed either at harvest or after drying. As the size of cannabis production facilities increases, operational efficiency becomes more important for the business to succeed. Dry trimming allows for a faster harvest because plants do not need to be processed before hanging them in the dry room. In contrast, hang drying requires more space than rack drying, which is normally used in wet trimming.

On a fresh weight basis, cannabis inflorescences are composed of 46% flowers, 24% inflorescence leaves, 16% fan leaves and 14% stems (Nirit et al. 2019). These can be used as rough guidelines, since the proportions will vary based on factors like pruning and de-leafing during cultivation. Following these proportions, 30% more moisture must be removed from the plans when they are dried with the fan leaves still attached. This requires a greater dehumidification capacity or a longer drying time. Fan leaves also typically have a relatively high concentration of mold spores, which can germinate under favorable conditions (Punja et al. 2019). Removing dry fan leaves is also more time-consuming than removing them from fresh plants. The additional dehumidification, mold risk, and labor requirements make a compelling argument for removing fan leaves before drying.

Cleanliness also needs to be considered during trimming, both to prevent spoilage and ensure proper operation of tools. Glandular trichomes extrude a sticky resin as the plants mature. Although the resin contains cannabinoids and terpenes, it poses a challenge while trimming. The resin sticks to the surfaces of tools, equipment and hands or gloves, which must be regularly cleaned and disinfected to ensure proper operation and avoid cross-contamination. During drying, resin loses water and becomes less sticky. As a result, dry trimming usually requires reduced cleaning of surfaces that come in contact with plant material.

10.6.1.3 Hand Trimming

Until only recently, all cannabis was trimmed by hand, and many growers and consumers still prefer hand-trimmed cannabis. Both wet and dry inflorescences can be trimmed by hand, typically using scissors with a narrow, pointed blade. The following are general instruction for hand trimming.

1. Using gloved hands, hold a branch containing an untrimmed inflorescence at its base.
2. Handle the inflorescences as little as possible, so that trichomes do not break off into your hands.
3. Using scissors, remove the inflorescence leaves as close to their base as possible, being careful not to remove bracts.
4. Separate the inflorescence leaves from the trimmed inflorescences in separate containers.
5. Cut the inflorescences from the stem. This is commonly termed de-stemming or bucking.

In our experience, a trimmer can process from 300–1,000 grams of dry inflorescences in an eight-hour shift. The principal factors that influence trimming speed are experience, desired consistency of the final product and the cultivar's physical characteristics. Some cultivars have more inflorescence leaves and thus take longer to trim.

Overall, hand trimming is labor-intensive and—for commercial operations—can contribute to more than 50% of the total post-processing costs. However, despite the added expense, hand trimming is still common practice for many commercial growers. Although machine trimming technology is advancing rapidly, narrowing the quality gap compared to hand trimming, some consumers are still willing to pay a premium for hand-trimmed product. Some processors use a hybrid process, in which inflorescences are machine-trimmed and finished by hand, so the product appears hand trimmed.

Hand trimming can be physically demanding. It involves repetitive movements and sitting for long periods. It is important to provide ergonomic workstations and that trimmers take regular breaks to avoid musculoskeletal disorders. Keeping scissors clean and sharp makes hand trimming easier. Sticky residue builds up on the scissors over time, which increases the force required for cutting (Descatha et al. 2009).

10.6.1.4 De-Stemming (Bucking)

To prepare inflorescences for machine trimming, the stems are typically first cut away. This can be done by hand or using devices called buckers or de-stemmers. Buckers work by pulling a branch through the hole of a metal plate using powered rollers, which shear off the inflorescences from the stem. The inflorescences fall into a container for further processing.

10.6.1.5 Machine Trimming

The market for legal cannabis has expanded rapidly over the past five years, and the scale of commercial cannabis producers has vastly increased. Some producers, especially in Canada and the United States, have build facilities with over 10 hectares of cultivation area and potential annual yields of more than 100,000 kg. Several technological advances have emerged during the last decade to efficiently process such high volumes; one of the most important is the trimming machine.

The simplest machine is the bowl trimmer, which is operated by manually cranking a handle. The action turns both a blade and rubber, silicone or leather arms inside the container, trimming the inflorescence leaves. The simplest electric variety is the automatic hand trimmer, a plug-in or battery-powered device with oscillating blades. Some automatic hand trimmers have vacuum attachments to collect the trimmed leaves. Bowl and hand trimmers are mainly used by home growers or small commercial growers.

Commercial-scale trimming machines can be categorized into trimmers with horizontal blades or grates, similar to a motorized bowl trimmer and tumbling trimmers (Figure 10.2).

In recent years, the latter has become the most widely used in commercial settings. In a tumbling trimmer, inflorescences are fed into a grated stainless steel or wire-mesh drum, which is continuously tumbling. In 3–10 minutes, the product moves from one end to the other of the tumbler while the inflorescences are trimmed. When trimming wet inflorescences, blades are positioned outside of

FIGURE 10.2 Commercial-scale tumbling trimmer (left) and bowl-type trimmer (right).

the tumbler, which trim inflorescence leaves through the drum's openings. With dry inflorescences, blades are not always required; the tumbling motion is sometimes sufficient to remove inflorescence leaves. The size and capacity of tumbling trimmers have steadily increased; industrial-scale trimmers can process more than 250 kg of wet inflorescences per hour. For small and medium-scale facilities, models with a 15–100 kg per hour capacity are available.

10.7 DRYING

Immediately after harvest, cannabis inflorescences have a moisture content of around 70–80%, and if they are not preserved in some way, they spoil quickly. The purposes of drying cannabis inflorescences are to create a smokable product, prevent immediate spoilage and preserve quality: the structural integrity, aroma and chemical composition.

Trichome preservation is just as critical during the drying process as it is while trimming. Compared to other types of trichomes, the capitate-stalked type contain the most resin; but they are fragile organs that are prone to snapping at their base. When dry, trichomes are especially brittle; proper care must be taken so that as few of them as possible are separated while drying.

Drying also affects the trichomes that remain attached to the inflorescence; specifically, their terpene content and composition. Terpenes are responsible for the characteristic smells and flavors of cannabis, and play a role in consumer experience and preference. Some terpenes are easily volatilized (evaporated) under hot and dry conditions; likewise, heat and low humidity facilitate faster drying, which reduces the risk of spoilage. Balancing these two factors makes drying cannabis a difficult process to master.

This section describes the major considerations when drying cannabis, with a focus on mitigating spoilage and maximizing inflorescence quality, followed by a description of various methods of drying.

10.7.1 MITIGATING SPOILAGE DURING DRYING

When inflorescences are infected with molds or other pathogens, they begin to degrade, affecting their physical structure, smell and chemical composition. The degradation can make the final product unappealing to consumers and reduce its value. The key to preventing spoilage is lowering the water content of fresh cannabis. Bacteria and fungi require available moisture to grow and reproduce, with each microorganism having different moisture requirements. Water content must be reduced below critical values specific to each organism to prevent their growth and thus prevent spoilage. The concept of water activity is used to assess what moisture levels are safe and is described in detail in the next subsection.

Some cannabis-infecting pathogens may be harmful to the consumer. Cannabis inflorescences can harbor bacterial pathogens such as *Salmonella* and *E. coli*, as well as fungal pathogens like some species of *Aspergillus* (Kagen et al. 1983), all of which pose risks to human health (Holmes et al. 2015). Proper post-harvest and sanitation procedures can help keep these pathogens at low enough levels to mitigate the risk.

While the plant is still alive, cannabis has mechanisms to protect itself against pathogens. Some cannabinoids and terpenes have anti-fungal and anti-bacterial properties. For instance, all five major cannabinoids—Δ9-tetrahydrocannabinol (THC), cannabidiol (CBD), cannabigerol (CBG), cannabichromene (CBC) and cannabinol (CBN)—have antibiotic activity against a variety of methicillin-resistant *Staphylococcus aureus* (Appendino et al. 2008), which can cause skin infections and food poisoning in humans. There is a greater risk of pathogen infection once the plants have been harvested. Processing is the production step when most human handling occurs, and human pathogens are easily transferred from workers to the product. We recommend that those involved in processing wear clean gloves, face masks, hair and/or beard coverings, and clean clothing that completely covers the arms.

While most infection occurs after harvest, there are some key practices to mitigate pathogen spread in growing plants. Controlled environment producers can filter incoming and recirculated air, and keep surfaces and equipment within the growing areas clean; this is more difficult for outdoor producers. A clean growing environment helps reduce airborne fungal and bacterial spores and their ability to inoculate cannabis inflorescences. While difficult to eliminate, lower quantities of inoculant during cultivation and harvest reduce the spread and severity of infections. A thorough sanitation program that includes cleaning and disinfecting tools, surfaces and equipment, in addition to proper air filtration, are some of the best tools for indoor growers in their battle against mold.

10.7.2 WATER ACTIVITY AND MICROBIAL CONTAMINATION

Significant effort has gone into developing guidelines for the safe consumption of pharmaceutical and agricultural products. The fact that cannabis is widely consumed by inhalation differentiates it from other products, so using safety standards for non-inhaled products may not be appropriate. Various approaches have been taken by regulators across jurisdictions with regulated cannabis production. For example, the state of Washington based its safety guidelines on the American Herbal Pharmacopoeia's Cannabis Monograph; Colorado developed testing requirements from existing literature on cannabis microbiology, and Canada adopted microbial contaminant limits found in the European Pharmacopoeia.

The water content in a perishable product, like cannabis, affects its rate of spoilage. The bacteria and fungi responsible for spoilage require available moisture to grow and reproduce, with each microorganism having different moisture requirements. Water activity (aW) is a measure of available water that can support microbial growth; aW ranges from 0 1. Most microorganisms cannot grow below values of 0.9 and microbial growth typically stops below 0.6 (Grant et al. 2004). Products with high moisture content are more likely to spoil, but if they also contain high levels of salt or sugar, microbial growth is inhibited because the salt or sugar make some of the water functionally unavailable for the microbes to use (lower aW). Regulations in Colorado, Washington state and California dictate that dried cannabis inflorescences must have a water activity below 0.65 (Table 10.2).

Water activity can be measured in several ways. When selecting an instrument, its cost, accuracy and measurement time should all be considered. Water activity meters are available for less than 1,000 USD, while some cost over 10,000 USD. For home growers, estimating water activity from inflorescence moisture content is a more cost-effective approach. The moisture content of cannabis samples can be measured using a moisture analyzer or using the "loss on drying" (LOD) method described in Section 10.7.5. The relationship between moisture content and water activity is described using moisture sorption isotherms. The shape of the isotherm is unique to each product. In a generalized isotherm model developed for cannabis, moisture content of 14% (± 1%) corresponds to a water activity of 0.65 (Figure 10.3).

10.7.3 DRYING TO PRESERVE QUALITY

Another objective of the drying process is to alter moisture content to a desired and uniform level. For smokable inflorescence, the target moisture content is typically 8–13% (w/w) to ensure uniform combustion and a pleasant smoking experience. Drying too quickly results in a fragile, crumbly inflorescence, while drying too slowly increases the risk of spoilage.

To our knowledge, there is no published information on the relationship between moisture content and smoking experience; however, there are serious implications to the producers based on what final moisture content they select, even within the 8–13% range. A lower moisture content inflates cannabinoid and terpene content since both are measured on a weight-for-weight (w/w) basis, and water contributes to the total weight. THC content is currently one of the major factors driving consumer buying decisions in legal markets (Ontario Cannabis Store 2020), so growers

TABLE 10.2

Cannabis Finished-Product Tolerances for Microbial Contamination, Water Activity and Moisture Content in the United States and Canada

	Colorado (Marijuana Enforcement Division 2021)	Washington (Washington State Legislature 2019)	California (Bureau of Cannabis Control 2020)	Canada (Council of Europe 2017)
	Limit			
Water activity (aW)	< 0.65	< 0.65	< 0.65	
Moisture content		< 15%		
Escherichia coli	Absent (1g)	Absent (1g)	Absent (0.5g)	Absent (1g)
Salmonella spp.	Absent (1g)	Absent (1g)	Absent (0.5g)	Absent (25g)
Total yeast and mold	< 10^4 CFU (1 g)			< 10^4 CFU (1 g)
Total aerobic microbial count				< 10^5 CFU (1 g)
Aspergillus			Absent (0.5 g)	
Bile-tolerant gram-negative bacteria		< 10^4 CFU (1 g)		< 10^4 CFU (1 g)
Candida albicans				
Aflatoxins (B1, B2, G1, and G2)		< 20 ppb	< 20 ppb	< 4 ppb
Ochratoxin A		< 20 ppb	< 20 ppb	
Aflatoxins B1				< 2 ppb

Notes: CFU = colony-forming unit. (Grams) signify prescribed samples size

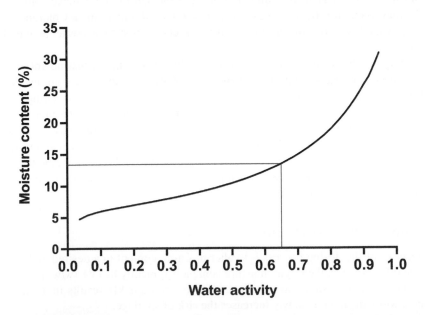

FIGURE 10.3 Cannabis moisture sorption isotherm. Four independent isotherms were generated from unique samples of dried cannabis inflorescences and used as inputs to generate a generalized isotherm model for cannabis. The curve shown represents average values; moisture content at water activity 0.65 typically corresponds to a moisture content within two percentage points of 14%.

Source: Data were generated by AquaLab and CannaSafe Analytics using an AquaLab Vapor Sorption Analyzer (Holmes et al. 2015)

TABLE 10.3

Example of the Effect of Inflorescence Moisture Content on Key Production Indexes

Moisture Content (%)	Total THC Content (%)	Yield (g)	Revenue[1] (USD)
13	20.0	10,000	$50,000
8	21.1	9,457	$47,283
0	23.0	8,700	$43,500

Notes: [1]Assumed sale price of $5 per gram of dry inflorescence

may be inclined to over-dry their products to inflate THC content. If this practice is taken to the extreme, the consumers may experience a dry, crumbly, difficult product that is difficult to roll into a joint or is harsh to smoke. Conversely, cannabis with higher moisture content is heavier, and since it is sold by weight, higher final moisture content can create higher revenues for the same amount of harvested product. Table 10.3 provides an example of the impacts of moisture content on yield, THC content and revenue.

Preserving cannabinoids and terpenes is another key outcome of the drying process. Cannabinoid content typically changes minimally during the drying process. Turner et al. (1973) found that cannabis inflorescence stored at 37°C had only marginal cannabinoid degradation after 24 hours, and notable degradation was only seen at 50°C. These temperatures far exceed those used in traditional air-drying techniques. Conversely, terpenes are easily volatilized in moderately warm and dry conditions. Over a range of temperatures from −80 to 25°C, terpene content in dry inflorescences can decrease over 50% after four months, while total THC content (THCA × 0.877 + THC) decreases only marginally over the same period (Milay et al. 2020).

Terpenes are a diverse group of biochemical compounds that are categorized based on their structural composition. The two most prevalent groups in cannabis are the monoterpenes (e.g., myrcene and α-pinene) and the sesquiterpenes (e.g., β-caryophyllene and α-humulene). Among the terpenes, monoterpenes are typically more prone to volatilization at lower temperatures; therefore, monoterpene loss should theoretically be greater than that of other terpenes during drying and storage. This, in fact, was recently disproven as researchers found that monoterpene and sesquiterpenes losses were similar between temperature treatments over a one-year storage period (Milay et al. 2020). As a processor, measuring and tracking total terpene content during the drying process may be an effective tool to optimize your drying strategy. In theory, if you are preserving total terpene content, the overall chemical composition of your product should also be preserved.

Other environmental factors that should be considered during drying are atmospheric oxygen concentration and light exposure, which are covered in Section 10.9. The effect of these factors on inflorescence degradation holds true during drying, curing and storage. In short, darkness should be maintained while drying, as well as sufficient air exchange. An excess of 20 air exchanges per hour in a drying room provides adequate ventilation from our experience.

10.7.4 DRYING METHODOLOGY

There are various methods for drying cannabis, but most rely on the same basic principles, following a process called "slow drying." Fresh product is dried in a sealed or semi-sealed room that is well-ventilated and climate controlled. For a smokable product, temperatures are typically maintained at 18–21°C and relative humidity at 45–60%. Depending on the inflorescences' size, it takes around 5–7 days to reach the desired final moisture content of 8–13% (w/w) using traditional methods.

Exact temperatures and relative humidity setpoints for mitigating spoilage and preserving volatiles during drying have yet to be established, and, in our experience, they vary greatly by producer.

Drying speed can be controlled by adjusting the drying room's air temperature and relative humidity. To increase drying speed, raise the temperature; higher temperature air has more drying potential at constant relative humidity. Conversely, higher temperatures will result in more volatile loss.

Producers of cannabis destined for extraction may have different final moisture content targets. For example, suppose the end-product is THC or CBD distillate, in which terpenes are mostly excluded. In that case, there is no concern for terpene loss associated with higher-temperature drying methods. In these circumstances, growers may elect to dry to < 8% moisture content to increase shelf life and reduce weight during shipping.

10.7.5   MEASURING MOISTURE CONTENT

Inflorescence moisture content can be measured using the "loss on drying" (LOD) principle by heating samples to evaporate moisture while taking continuous weight measurements. The lowest-cost method utilizes a benchtop scale and a convention oven to dry the sample. First, grind and weigh the sample, then dry it at 105°C for at least two hours. Allow the sample to cool for about one hour, then weigh again. Repeat this process until there is a negligible weight loss between drying periods. The initial weight of the sample, minus its stable dried weight, divided by the initial weight, multiplied by 100 gives the moisture content of the sample as a percentage (Equation 10.1). Once the approximate drying time has been established for a particular size and type of sample, some of the earlier weighing events can be omitted in subsequent tests. This method is typically reproducible to ± 7% of the true moisture content for plant tissue samples (Kalra 1998).

Equation 10.1 Calculating Moisture Content Using the Loss on Drying Method

$$\text{Moisture content (\%)} = \frac{\text{Initial weight} - \text{dry weight}}{\text{Initial weight}} \times 100 \qquad (10.1)$$

A benchtop moisture analyzer can also be used to make quick and reliable moisture content measurements of small samples. These devices work using the "loss on drying" (LOD) principle by heating samples to evaporate moisture while taking continuous weight measurements using an automated scale. When the weight has stabilized, the sample is considered dry, and the loss of weight can be used to calculate the sample's initial moisture content. Moisture analyzers can be sourced from well-reputed laboratory suppliers, like Mettler Toledo (www.mt.com), Leco (www.leco.com) and Sartorius (www.sartorius.com).

Using the LOD principle, the weight loss during drying is equated to the moisture content, not its water content. The moisture of a material includes water and other volatile compounds, like fats and oils, that can be released from the sample when heated. For dry cannabis inflorescences, the terpenes (sometimes > 3% [w/w]) can be volatilized during the heating process and would therefore not be included in the final weight, leading to an overestimated moisture content. To mitigate this effect, most moisture analyzers have built-in or programmable heat settings for more accurate measurements. For instance, setting the drying temperature around 80°C has been shown to provide a close approximation of true water content (Gast et al. 2014). Temperatures lower than 80°C may not effectively remove moisture from the sample, while higher temperatures result in volatile loss.

10.7.6   HANG DRYING

Hang drying has a long and successful history for both small-scale and large-scale producers. The process begins directly after harvest and involves hanging whole or partially broken-down plants upside down (Figure 10.4) and drying slowly under ambient environmental conditions. Keeping the stem and some leaves attached facilitates a slower drying process. Theoretically, moisture evaporates

FIGURE 10.4 Fresh cannabis hanging to dry.

from the leaves and inflorescences while simultaneously, more moisture is pulled from the stem. For a smokable product, temperatures are typically maintained at 18–21°C and relative humidity at 45–60%. Depending on the size of the inflorescences, it takes around 5–7 days to reach the desired final moisture content of 8–13% (w/w). Once the stem holding the inflorescences snaps cleanly or the inflorescence moisture content has reached a target value, de-stemming and trimming can begin.

10.7.7 Screen Drying

Screen drying, also known as rack or tray drying, is mostly used in conjunction with wet trimming. Once the inflorescences have been wet trimmed, they are laid out on screens or grated trays (Figure 10.5). The plant material is typically mixed several times during the process to ensure

FIGURE 10.5   Fresh, trimmed cannabis drying on a grated tray.

uniform drying. Since there is less plant material attached to the inflorescences, drying occurs slightly faster than when hang drying. Usually, 4–6 days is sufficient to reach a final moisture content of 8–13% (w/w).

10.7.8   CONVEYOR DRYING

Conveyor drying is essentially an automated version of screen drying. A grated conveyor belt moves and mixes fresh material at a constant speed, while controlling environmental parameters such as temperature and air movement. Automating the turning and mixing steps reduces labor requirements compared to screen drying; however, the continuous movement may result in over-handling, resulting in trichome loss. Large-scale producers typically use these devices to speed-dry cannabis destined for extraction, but with proper optimization, conveyor drying could be effective for high-quality smokable inflorescences.

10.7.9   FREEZE-DRYING

Freeze-drying, also known as lyophilization, removes moisture under low pressure through sublimation, i.e., directly from a solid state to a gaseous state. Low pressures are maintained in a sealed container using a vacuum pump, and freezing temperatures are maintained using liquid nitrogen or a compressor-based refrigeration system. This technology is widely used to preserve perishable

TABLE 10.4

Comparison of Commercially Available Drying Methods Used to Process Smokable Cannabis

Drying Method	Drying Time for Smokable Inflorescences	Relative Terpene Loss	Relative Capital Investment
Hang Drying	5–7 days	High	$
Screen Drying	4–6 days	High	$
Conveyor Drying	4–6 days	High	$$
Freeze-Drying	10–15 hours	Low	$$$$
Microwave Vacuum Drying	< 45 minutes	Medium	$$$

foods and pharmaceuticals, but has only recently been applied commercially in cannabis processing. Freeze-drying is among the best drying methods to preserve both the chemistry and physical integrity of perishable goods. Still, it is also among the most expensive, about 10 times more costly than traditional drying methods (Challa et al. 2020). Freeze-drying cannabis preserves its cannabinoids and terpenes and mitigates shrinkage during drying. Some cannabis-specific freeze-drying systems are available, including Cyro Cure (www.cryocure.com).

10.7.10 MICROWAVE VACUUM DRYING

Microwave vacuum drying removes moisture using microwave radiation, low temperatures and low pressure. The low temperatures prevent the loss of volatiles, while the microwave radiation evaporates moisture throughout the inflorescence. The result is a very fast drying process (sometimes less than 45 minutes) and good preservation of the original chemistry. EnWave Technologies (www. enwave.net) offers an industrial-scale drying system using microwave vacuum technology. EnWave claims less than 5% loss of THC and less than 35% loss of terpene content for cannabis.

10.8 CURING

Curing is likely the least understood step in cannabis processing; nonetheless, it is nearly universally practiced by producers of smokable inflorescences. Curing is thought to improve the smoothness and taste of cannabis smoke, preserve existing secondary metabolites and homogenize the moisture within the inflorescence. To our knowledge, there are no peer-reviewed publications on cannabis curing; the information in this section is based purely on processor anecdotes, personal experience and parallels from other crops, like tobacco. In traditional curing methods, trimmed, dried cannabis is packed loosely in airtight or semi-airtight containers, which are opened to allow air exchange one or more times per day. Temperatures are usually maintained constant, from 15–20°C.

After drying, the moisture content between inflorescences can vary, based on their size or location in the drying room. Smaller inflorescences may dry faster than larger ones, and drying racks situated closer to circulation fans or heaters may end up drier. From experience, we have found that bulking inflorescences into a sealed or semi-sealed container results in a homogenization of moisture within the batch. Further, the surfaces of inflorescences tend to dry faster than their insides; curing may also homogenize the moisture content within individual inflorescences.

Consumer and grower anecdotes suggest that curing increases the smoothness and flavor of smoked cannabis. This could be attributed to the enzymatic breakdown of chlorophyll and the hydrolysis of starches into sugars during curing. In the curing of tobacco leaves, chlorophyll and starch breakdown are attributed to improved smoking quality. Cannabis and tobacco smoke share similar physical and chemical properties (Graves et al. 2020); therefore, it could be that these same chemical processes are responsible for improving smoking quality in cannabis.

While the objectives of curing cannabis and tobacco are similar—improving flavor, aroma and quality of smoke—there are some stark differences in their curing processes. There are several ways to cure tobacco, including flue-curing, air curing and fire curing. Each has its advantages and produces a final product with a unique flavor or smoke quality. Consistent among these methods is that curing occurs before drying. For example, in flue-curing, tobacco leaves are harvested green and fresh from the plants and are cured for 20–60 hours. The available moisture in the fresh material maximizes the conversion of starches to sugars and chlorophyll's enzymatic breakdown. Once the tobacco leaves have become sufficiently yellow, signifying a reduction of chlorophyll, they are then completely dried. This killing step halts the chemical reactions within the leaves. Afterwards, moisture is re-added, usually by raising the ambient relative humidity until the leaf moisture content reaches about 15% (Sumner and Moore 2009). Chlorophyll and starch breakdown in cannabis could also be occurring mostly during the "slow drying" step, rather than during the subsequent curing step, or potentially occurring during both steps. These unknowns underline the need for more research on the physical and chemical reactions occurring during the drying and curing of cannabis.

10.9   LONG-TERM STORAGE

Both dry cannabis inflorescences and extracted oils are perishable, but proper storage can extend their shelf life. Generally, light and oxygen should be avoided, cool to ambient temperatures should be maintained, and relative humidity should be sufficient that inflorescences do not desiccate. Of course, there are some exceptions to these rules; for instance, dry air and freezing temperatures can preserve material destined for extraction.

During storage, the chemistry within cannabis inflorescences slowly changes. Cannabinoids and terpenes decompose through oxidization, dehydrogenation, isomerization and polymerization (Turek and Stintzing 2013), resulting in new compounds. The altered chemistry of these biologically active compounds means that the consumers may experience different effects, depending on product age. Physical changes are also apparent over time, including changes in color and texture.

In fresh inflorescences, cannabinoids exist primarily as carboxylic acids such as THCA and CBDA. These acids slowly de-carboxylate during storage to become neutral cannabinoids, such as THC and CBD. While cannabinoid loss is slow, terpene loss during storage is severe; even under proper storage conditions, total terpene content in dry inflorescence can decrease over 50% after four months, while total THC content (THCA × 0.877 + THC) only marginally decreases over the same period (Milay et al. 2020). Since terpenes largely account for the taste and aroma of cannabis, preserving them in storage is critical to maintaining product quality.

Understanding how the chemical composition changes in stored inflorescences allows growers to monitor product degradation. Cannabinol (CBN) and cannabinolic acid (CBNA) tend to accumulate with product age (Trofin et al. 2012), and total terpene content tends to decline (Milay et al. 2020). Serial sampling of these chemicals is informative when deciding an acceptable storage period or shelf life. Degradation cannot be stopped entirely; ultimately, product quality is at its highest immediately after curing. If storage is required, critical storage parameters to consider are light, temperature and oxygen.

10.9.1   LIGHT

Both UV and visible light cause the oxidative breakdown of dried cannabis and cannabis extracts. In a study on long-term cannabis storage, inflorescences stored for one year under indirect light had a 55% loss of THC content. In contrast, similar cannabis inflorescences kept in the dark only had a 13% loss (Fairbairn et al. 1976). Notably, exposure to light has been shown not to increase CBN content (Fairbairn et al. 1976); therefore, CBN content changes would be a poor indicator of light-related degradation. Furthermore, terpenes (especially monoterpenes) degrade faster in the presence of light (Misharina et al. 2003).

10.9.2 TEMPERATURE

Temperature also has a considerable influence on cannabis stability during storage. Most chemical reactions accelerate as temperature increases, including those involved in cannabinoid and terpene degradation. Conversely, low temperatures increase oxygen solubility in liquids, which may accelerate oxidative deterioration during storage. In a recent study, cannabis stored at 4°C retained more terpenes and cannabinoids after four months compared to storage at 25°C, −30°C and −80°C. The study did not identify the effects of temperature on physical parameters such as density and color; however, we have no reason to think that storage at 4°C should have a negative effect on these factors.

10.9.3 OXYGEN

The presence of atmospheric oxygen during storage is another contributor to cannabis degradation. Monoterpenes, for example, will undergo oxidation reactions, resulting in new chemical compounds. Oxygen also encourages the growth of aerobic microorganisms that can cause spoilage. Compared to light and temperature, however, oxygen is less destructive to cannabis inflorescences; this could be because the resin is stored in glands, reducing the surface area exposed to the air (Challa et al. 2020). In extracted products, the resin is removed from the glands and may be more vulnerable to oxidative damage. Generally, the key to avoiding oxygen-induced degradation is either reducing exposure to the air (i.e., reducing vessel head space or tight packing) or lowering the atmospheric oxygen in the vessel. This principle holds in the storage of essential oils; when stored in containers with more headspace (i.e., half-filled containers), they tend to degrade faster than if there is little headspace (El-Nikeety et al. 2000).

Replacing the air inside a storage vessel with more inert gases like nitrogen (N_2) or carbon dioxide (CO_2) can help mitigate oxidate degradation. This is called modified atmosphere packaging (MAP) and is common in the storage of food and pharmaceuticals. Growers considering MAP storage should keep in mind that light and temperature play a larger role in degradation and ensure these factors are addressed first. Further, consider the packaged product's expected shelf life; if oxidative damage is minimal over six months, for example, then non-MAP packaging could provide acceptable conditions.

10.10 YIELD: CALCULATING AND REPORTING

Yield is the weight of a salable product produced for a particular crop. Depending on its end-use, cannabis yield can be calculated in one of the following two ways.

1. Trimmed inflorescence yield: If the crop is to be sold as dry inflorescences, then the yield includes trimmed inflorescences that are dried to the desired moisture content for sale.
2. Untrimmed inflorescence yield: If the crop is to be sold as a feedstock for extraction, then yield may either be expressed the same as for dry inflorescences or may include the weight of inflorescence leaves (either trimmed from or still attached to the inflorescence), again dried to the desired moisture content for sale.

Maintaining accurate yield data is critical to optimizing productivity. Product quality (i.e., cannabinoid/terpene content or inflorescence density) may be exceptional, but to achieve profitability, a grower must produce at a sufficient scale. At present, there is no industry-standard method for reporting yield. Growers sometimes report yield per plant; however, this metric ignores the planting density, giving an incomplete representation of an operation's productivity. Understanding the yield per unit canopy area, either for a single crop or on an annualized basis, is of far greater importance than the yield from a given plant (Table 10.5). The canopy area is calculated as the area—either in square feet or square meters—where flowering plants are grown.

TABLE 10.5
Examples of the Effect on the Number of Crops per Year on Annualized Yield per Unit Area

Yield per Unit Area	500 grams/m²	46 grams/ft²
Number of Crops per Year	Annualized Yield	
4	2,000 grams/m²/year	186 grams/ft²/year
5	2,500 grams/m²/year	232 grams/ft²/year
6	3,000 grams/m²/year	279 grams/ft²/year

Equation 10.2 Yield per Unit Flowering Area

$$\text{Yield per unit area} = \frac{\text{total yield of one crop in grams}}{\text{flowering footprint}} \tag{10.2}$$

Equation 10.3 Annualized Yield per Unit Area

$$\text{Annualized yield per unit area} = (\text{yield per unit area}) \times (\text{number of crops per year within that area}) \tag{10.3}$$

REFERENCES

Adams, S. R., K. E. Cockshull, and C. R. J. Cave. 2001. Effect of temperature on the growth and development of tomato fruits. *Annals of Botany* 88(5): 869–877. https://academic.oup.com/aob/article/88/5/869/2587168.

Appendino, G., S. Gibbons, A. Giana, et al. 2008. Antibacterial cannabinoids from *Cannabis sativa*: A structure-activity study. *Journal of Natural Products* 71(8): 1427–1430.

Bureau of Cannabis Control. 2020. *California code of regulations title 16 division 42.* https://bcc.ca.gov/law_regs/readopt_text_final.pdf.

Caplan, D., M. Dixon, and Y. Zheng. 2019. Increasing inflorescence dry weight and cannabinoid content in medical cannabis using controlled drought stress. *HortScience* 54(5): 964–969.

Challa, S. K. R., N. N. Misra, and A. Martynenko. 2020. Drying of cannabis – state of the practices and future needs. *Drying Technology* 1–10. https://doi.org/10.1080/07373937.2020.1752230.

Council of Europe. 2017. *European pharmacopoeia 5.1.8-C,5.1.4,2.8.18.* https://www.edqm.eu/sites/default/files/medias/fichiers/European_Pharmacopoeia/The_European_Pharmacopoeia/European_Pharmacopoeia_10th_Edition/Index/european_pharmacopeia_10.0_english_index.pdf.

Descatha, A., Y. Roquelaure, J. F. Chastang, B. Evanoff, D. Cyr, and A. Leclerc. 2009. Description of outcomes of upper-extremity musculoskeletal disorders in workers highly exposed to repetitive work. *Journal of Hand Surgery* 34(5): 890–895, May.

El-Nikeety, M. M. A., A. T. M. El-Akel, M. M. I. A. El-Hady, and A. Z. M. Badei. 2000. Changes in physical properties and chemical constituents of parsley herb volatile oil during storage. *Egyptian Journal of Food Science* 26–28: 35–49.

Fairbairn, J. W., J. A. Liebmann, and M. G. Rowan. 1976. The stability of cannabis and its preparations on storage. *Journal of Pharmacy and Pharmacology* 28(1): 1–7, January. http://doi.wiley.com/10.1111/j.2042-7158.1976.tb04014.x.

Gast, J., A. Darling, L. Allen, L. Corporation, and S. Joseph. 2014. *Determination of moisture/loss on drying in cannabis by various analytical techniques.* https://knowledge.leco.com/images/e-seminar_archive/posters/Analytical-Organic/MOISTURE_LOD_CANNABIS_TGM_RC612_JG_PCON20_LPO-019.pdf.

Grant, W. D., M. J. Danson, D. J. Scott, et al. 2004. Life at low water activity. *Philosophical Transactions of the Royal Society B: Biological Sciences* 359: 1249–1267. https://royalsocietypublishing.org/doi/10.1098/rstb.2004.1502

Graves, B. M., T. J. Johnson, R. T. Nishida, et al. 2020. Comprehensive characterization of mainstream marijuana and tobacco smoke. *Scientific Reports* 10(1): 1–12.

Grewal, L., J. Hmurovic, C. Lamberton, and R. W. Reczek. 2019. The self-perception connection: Why consumers devalue unattractive produce. *Journal of Marketing* 83(1): 89–107, January.

Henderson, K. 2017. 73% of shoppers open to buying wonky fruit and veg, study finds. *The European Supermarket Magazine*.

Holmes, M., J. M. Vyas, W. Steinbach, and J. Mcpartland. 2015. Microbiological safety testing of cannabis. *Cannabis Safety Institute* 1–54, May.

Juneau, K. J., and C. S. Tarasoff. 2012. Leaf area and water content changes after permanent and temporary storage. Ed. Ben Bond-Lamberty. *PLoS One* 7(8): e42604, August.

Kagen, S. L., V. P. Kurup, P. G. Sohnle, and J. N. Fink. 1983. Marijuana smoking and fungal sensitization. *The Journal of Allergy and Clinical Immunology* 71(4): 389–393.

Kalra, Y. P. 1998. *Handbook of reference methods for plant analysis.* Boca Raton: CRC Press. https://onlinelibrary.wiley.com/doi/abs/10.2135/cropsci1998.0011183X003800060050x.

Kavanagh, J. A., and R. K. S. Wood. 1967. The role of wounds in the infection of oranges by *Penicillium digitatum* Sacc. *Annals of Applied Biology* 60(3): 375–383, December.

Kleinwächter, M., and D. Selmar. 2015. New insights explain that drought stress enhances the quality of spice and medicinal plants: Potential applications. *Agronomy for Sustainable Development* 35(1): 121–131, January 21. http://pcp.oxfordjournals.org/cgi/doi/10.1093/pcp/pct054.

Mahlberg, P. G., and E. S. Kim. 2004. Accumulation of cannabinoids in glandular trichomes of Cannabis (*Cannabaceae*). *Journal of Industrial Hemp* 9(1).

Marijuana Enforcement Division. 2021. *Code of Colorado regulations – Colorado marijuana rules.* https://www.sos.state.co.us/CCR/DisplayRule.do?action=ruleinfo&ruleId=3314&deptID=19&agencyID=185&deptName=Department%20of%20Revenue&agencyName=Marijuana%20Enforcement%20Division&seriesNum=1%20CCR%20212-3.

Milay, L., P. Berman, A. Shapira, O. Guberman, and D. Meiri. 2020. Metabolic profiling of cannabis secondary metabolites for evaluation of optimal postharvest storage conditions. *Frontiers in Plant Science* 11: 1–15, October.

Misharina, T. A., A. N. Polshkov, E. L. Ruchkina, and I. B. Medvedeva. 2003. Changes in the composition of the essential oil in stored marjoram. *Prikladnaia Biokhimiia i Mikrobiologiia* 39(3): 353–358.

Nirit, B., J. Gorelick, and S. Koch. 2019. Interplay between chemistry and morphology in medical cannabis (Cannabis sativa L.). *Industrial Crops and Products* 129: 185–194, March 1.

Ontario Cannabis Store. 2020. *A year in review (2019–2020): Ontario's first full year of legal cannabis operations.* https://ocs.ca/pages/insights-publication.

Punja, Z. K., D. Collyer, C. Scott, S. Lung, J. Holmes, and D. Sutton. 2019. Pathogens and molds affecting production and quality of cannabis sativa L. *Frontiers in Plant Science* 10: 1120, October.

Rodriguez-Morrison, V., D. Llewellyn, and Y. Zheng. 2021. *Cannabis inflorescence yield and cannabinoid concentration are not improved with long-term exposure to short-wavelength ultraviolet-B radiationno,* June. https://www.preprints.org/manuscript/202106.0317/v1.

Sumner, P. E., and J. M. Moore. 2009. Harvesting and curing flue-cured tobacco. *The University of Georgia Cooperative Extension* 1–16.

Trofin, I. G., G. Dabija, D. I. Váireanu, and L. Filipescu. 2012. The influence of long-term storage conditions on the stability of cannabinoids derived from cannabis resin. *Revista de Chimie* 63(4): 422–427.

Turek, C., and F. C. Stintzing. 2013. Stability of essential oils: A review. *Comprehensive Reviews in Food Science and Food Safety* 12(1): 40–53.

Turner, C. E., K. W. Hadley, P. S. Fetterman, N. J. Doorenbos, M. W. Quimby, and C. Waller. 1973. Constituents of cannabis sativa L. IV: Stability of cannabinoids in stored plant material. *Journal of Pharmaceutical Sciences* 62(10): 1601–1605, October. https://linkinghub.elsevier.com/retrieve/pii/S0022354915413036.

Vilanova, L., I. Viñas, R. Torres, J. Usall, G. Buron-Moles, and N. Teixidó. 2014. Increasing maturity reduces wound response and lignification processes against *Penicillium* expansum (pathogen) and *Penicillium digitatum* (non-host pathogen) infection in apples. *Postharvest Biology and Technology* 88: 54–60, February.

Washington State Legislature. 2019. *WAC 314-55-102, WAC 246-70-050 – Quality assurance testing.* https://apps.leg.wa.gov/wac/default.aspx?cite=314-55-102%7B%5C%7Dpdf=true.

Index

Note: Page numbers in *italics* indicate figures; page numbers in **bold** indicate tables.

Printed and bound by CPI Group (UK) Ltd, Croydon, CR0 4YY

24/10/2024

01778292-0008